Department of **Library Services**	**Normal Loan** Kimberlin Library
De Montfort University	www.**library**.dmu.ac.uk

Please return this item on or before the date indicated below (if you have used Self Service please write in the date yourself as a reminder – the due date is printed on your receipt).

Items are issued subject to reservation recall. See Library Guide for details. Fines will be charged for the late return or late renewal of items.

24 hour renewals can be made by telephoning **(0116) 257 7043**, or you can renew online via the Library Catalogue under 'My Account'.

General library information: (0116) 257 7042 (service hours)

ᴄ 3 FEB 2009	
1 2 FEB 2009	
0 5 MAR 2009	
⊢ 2 DEC 2009	
2 2 MAR 2011	
- 7 NOV 2011	
- 3 JAN 2012	

Form 14 PC28

RAPID PROTOTYPING

RAPID PROTOTYPING
Laser-based and Other Technologies

Patri K. Venuvinod and Weiyin Ma
Department of Manufacturing Engineering and Engineering Management
City University of Hong Kong

Kluwer Academic Publishers
Boston/Dordrecht/London

Distributors for North, Central and South America:
Kluwer Academic Publishers
101 Philip Drive
Assinippi Park
Norwell, Massachusetts 02061 USA
Telephone (781) 871-6600
Fax (781) 871-6528
E-Mail <kluwer@wkap.com>

Distributors for all other countries:
Kluwer Academic Publishers Group
Post Office Box 322
3300 AH Dordrecht, THE NETHERLANDS
Telephone 31 78 6576 000
Fax 31 78 6576 474
E-Mail <orderdept@wkap.nl>

 Electronic Services <http://www.wkap.nl>

Library of Congress Cataloging-in-Publication

Title: Rapid Prototyping
       Laser-based and Other Technologies
Author (s): Patri K. Venuvinod and Weiyin Ma
ISBN: 1-4020-7577-4

Copyright © 2004 by Kluwer Academic Publishers

Table of Contents

Preface

Since the dawn of civilization, mankind has been engaged in the conception and manufacture of discrete products to serve the functional needs of local customers and the tools (technology) needed by other craftsmen. In fact, much of the progress in civilization can be attributed to progress in discrete product manufacture.

The functionality of a discrete object depends on two entities: form, and material composition. For instance, the aesthetic appearance of a sculpture depends upon its form whereas its durability depends upon the material composition. An ideal manufacturing process is one that is able to automatically generate any form (freeform) in any material. However, unfortunately, most traditional manufacturing processes are severely constrained on all these counts.

There are three basic ways of creating form: conservative, subtractive, and additive. In the first approach, we take a material and apply the needed forces to deform it to the required shape, without either adding or removing material, i.e., we conserve material. Many industrial processes such as forging, casting, sheet metal forming and extrusion emulate this approach. A problem with many of these approaches is that they focus on form generation without explicitly providing any means for controlling material composition. In fact, even form is not created directly. They merely duplicate the external form embedded in external tooling such as dies and molds and the internal form embedded in cores, etc. Till recently, we have had to resort to the 'subtractive' approach to create the form of the tooling. The production of such tooling can be quite expensive and time consuming, thus making unit costs highly sensitive to production volume.

The subtractive approach involves taking a block of material and chipping away unwanted segments. This is the way Michelangelo had created his brilliant sculptures. In modern industry, CNC machines work on the subtractive principle. An advantage of CNC is that it can utilize information embedded in a CAD model of the part. Further, form generation depends on the relative motion between the subtractive tool (e.g., an end mill) and the blank. In other words, it is not necessary to have tooling embedded with the required form, so small-volume production becomes possible. As a result, much of the tooling industry today depends upon CNC machining. However, when applied to the direct manufacture of products, CNC machining is not economical for high production volumes. Further, only those form features accessible by the subtractive tools can be created. This means that reentrant corners and internal form cannot be created. Another problem is that, although the machine is computer controlled, the physical side of machining requires manual attention. Lastly, like the conservative approach, the subtractive approach merely focuses on the generation of form without providing a means of controlling material composition.

The 'additive' approach starts with nothing and builds an object incrementally by adding material. The material added each time can be the same or different. Thus, one is able to address the problems of form generation and material composition at once through the same process. The smaller the volume of each material increment, the greater are the form accuracy achievable and the degree of control on material composition. In the ultimate, the principle is capable of even achieving the dream of "from bits to atoms". The immediate advantage, however, is that, in principle, any solid 3D freeform can be generated without the aid of external tooling with embedded form, so most of the problems associated with the conservative and subtractive methods are totally sidestepped.

Unfortunately, till recently, the additive principle could not be implemented in industry owing to the lack of suitable materials and supporting technologies. However, by the 1980s, progress in photopolymers, laser technologies, CAD modeling, etc. had matured sufficiently to enable layer-wise additive creation of 3D physical objects through selective polymerization of a photosensitive resin. The first commercially available equipment based on this principle was StereoLithography Apparatus-1 (SLA-1) released by 3D Systems, Inc. in 1987. This started a new revolution in manufacturing. The revolution is still in progress as evident from the growing number of commercially available SFF technologies: selective laser sintering (SLS), fused deposition modeling (FDM), layered object manufacturing (LOM), 3D Printing (3DP), etc.

From a different viewpoint, the arrival of the additive SFF technologies was timely. By the 1990s, the world of manufacturing was experiencing an unprecedented transformation owing to rapidly changing customer attitudes and globalization of manufacture. Increasingly affluent consumers had started demanding greater product variety. This meant that production runs had to become smaller and the pace of product innovation had to pick up by an order of magnitude. In addition to productivity and quality, time-to-market became an important competitive weapon. Companies had to conceive and deliver products as rapidly as possible.

But, product development is an iterative process requiring evaluation of intermediate designs through virtual (computerized) and physical prototypes. Fortunately, progress in CAD modeling and virtual reality (VR) enabled overcome several bottlenecks in the creation of virtual prototypes, whereas the new SFF technologies were found to be helpful in overcome bottlenecks in preparing physical prototypes. Naturally, these technologies started being referred to as rapid prototyping (RP) technologies.

While a variety of RP technologies have been developed in the last fifteen or so years, they share many common features. Almost all of them are capable of creating an object in a layered manner directly from information c ontained i n a C AD model a nd w ithout the n eed f or e xternal tooling. Many of them use lasers as the energy source to solidify selected voxels within a layer. Together, these technologies are capable of producing objects from a variety of materials including polymers and, metallic/ceramic powders. The objects can be of a single material, multiple-material, or, even functionally graded. Initially, RP was used to facilitate visualization of design concepts via prototypes produced from wax and other polymers. However, through the development of new materials and processes, RP is now being applied to produce functional prototypes and tooling components (rapid tooling—RT). RP is also being applied to direct manufacture of components in small to medium sized batches (rapid manufacture—RM). Thus, RP has now become a billion dollar industry.

The authors of this book believe that RP has now matured to a level that a course on RP can be included within an undergraduate industrial/manufacturing/mechanical engineering curriculum. In particular, a course on RP can act as an integrative or 'capstone' course. The study of RP requires basic concepts drawn from several fields such as computational geometry, polymer chemistry, control, heat t ransfer, m etallurgy, ceramics, optics, fluid mechanics, and manufacturing technology to be integrated. RP machines are truly mechatronic in nature, i.e., they synergistically combine mechanical, electrical/electronic and embedded computer technologies. RP can be viewed as high technology ('hi-tec') since it uses the latest developments in lasers, CNC control, computers, etc. The study of RP also provides insights into the emerging fields of MEMS and nano-technology. By studying RP, a student becomes aware of the latest trends in several engineering fields. The authors have already introduced such a course at the Department of Manufacturing Engineering and Engineering Management at the City University of Hong Kong. It appears that similar courses are being introduced in several other universities around the world.

A look at the books published so far on RP (surprisingly, there are already over 20 of them) reveals that few of these are immediately suitable for use by undergraduate engineering students. Firstly, owing to the rapid and continuing progress in the rapid prototyping and tooling (RP/T) technologies, several of these books are already partly outdated. Further some books are biased towards the particular RP techniques that the author(s) had helped develop. This book intends to fill these gaps by providing a balanced and updated understanding of the wide range of RP/T technologies available today. At the same time, we hope that the book would be useful to R&D personnel and professional engineers working directly or indirectly in areas related to RP.

The book consists of ten chapters. Chapter 1 provides an introduction to the general field of RP/T by clarifying why it is important to adopt 'rapid' technologies in contemporary industry. Next, a broad overview of the various RP/T technologies is given while emphasizing the common threads that runs through all of them. The historical development of RP/T technologies is also outlined so as to strengthen the view of RP/T as an emerging field.

Much of the success of RP process technologies is derived from successes in the development of new materials. Hence, a deep understanding of RP processes and their capabilities in terms of product accuracy and integrity requires an equally deep understanding of the physical, chemical, and mechanical characteristics of the materials involved. The materials involved can be broadly grouped into polymers, metallic and ceramic powders, and composites. This, in turn, requires the student to have a basic understanding of atomic structure and bonding. Almost all the available books on RP/T assume that the reader is already equipped with knowledge of such materials basics. However, often, this is not true. Chapter 2 is therefore designed to fill this gap.

Lasers represent yet another technology underpinning the development of RP/T. Processes such as SL, SLS, and LOM would not have been possible if there were no lasers. Further, the capabilities of laser-based SFF processes are intimately tied to our ability to control the laser beam characteristics and manipulate the laser beam. Again, many RP books assume that the reader possesses sufficient understanding of lasers to fully appreciate these issues. This book deviates from this assumption by devoting a full chapter (Chapter

3) to laser basics. The chapter utilizes the principles of atomic structure outlined in Chapter 2 to explain the principle of lasers. The discussion then moves on to certain characteristics of lasers that are of particularly useful in understanding most laser-based RP processes. Finally, the general characteristics of some important lasers used in RP are described.

The starting point of an RP exercise is a CAD model of the object. Owing to its importance to engineering in general, courses on CAD are usually included in most engineering curricula. Hence, a separate chapter on CAD is not included in this book. However, many times, we do not have a CAD model available. In the automobile industry, e.g., we often design the external aesthetic shape of a car by manipulating clay or wooden models, the so-called mockups. We then have to turn the physical mockup in digital form in order to survive with today's NC technologies, including rapid prototyping. At other times, we might have a piece of patient's bone from which we need to quickly produce a prosthetic. Such problems are solved in modern industry through a technology called reverse engineering (RE). This technology is of particular importance to industries engaged in original equipment manufacturing (OEM), as in China and other developing economies. Interestingly, few books on RP include a detailed discussion of RE. This book deviates from this tradition by including a full chapter (Chapter 4) on reverse engineering techniques and preparing a CAD model on the basis of RE data.

While materials, lasers and precision mechatronics provide the essential substance and infrastructure for rapid prototyping, a CAD model, however, has to be processed in order to make the RP equipment function. Chapter 5 is devoted to discussions of various topics in data processing for rapid prototyping. One of the important topics addressed is on CAD/RP interfacing. With today's RP industry, the de-facto standard is the STL interface that has been adopted by almost all RP equipment manufacturers as a standard input. With the STL interface, an object is defined by tessellating an original CAD model as an approximate triangular faceted surface model. Due to the simplicity of the geometric algorithms in processing a tessellated model, many data processing steps can be made automatic. Next, the determination of an optimal build orientation and support structure design are addressed. Model slicing is then presented followed by discussions on direct and adaptive slicing. An algorithm for selective hatching/solidification among different layers is also introduced in this chapter to further improve the building speed. The chapter is closed with discussions on tool path generation.

Chapters 6 to 8 discuss the various processing technologies available for RP in as much detail as possible within the confines of the available space. Since SL (stereolithography) presently holds the major share in the RP industry, a full chapter (chapter 6) is devoted to it. The discussion on SL also acts as an introduction to subsequent discussions on other RP processes by clarifying issues related to laser utilization, implications of layered fabrication with regard to product quality, etc. Chapter 7 provides a fairly detailed description of selective laser sintering (SLS) processes and issues concerning their modeling. A full chapter is devoted to this process in view of potential of SLS in the fabrication of functional parts and hard tooling. Chapter 8 discusses a host of other processes including laser cladding, FDM and LOM.

Chapter 9 is entirely devoted to the important area of rapid tooling (RT). Both direct and indirect tool fabrication techniques are discussed.

Finally, Chapter 10 discusses the major areas of application of RP techniques: Among the areas discussed are the fabrication of heterogeneous and functionally graded objects, assemblies and MEMS. Further areas discussed include architectural and medical applications.

Acknowledgments

The authors wish to gratefully acknowledge the Department of Manufacturing Engineering and Engineering Management (MEEM) of City University of Hong Kong (CityU) for providing various resources to the authors' RP-related teaching and research activities.

The authors also acknowledge the financial support provided by the Research Grants Council of Hong Kong Special Administrative Region (HKSAR) through CERG research grants CityU1041/96E and CityU1131/03E, and the City University of Hong Kong through Strategic Research Grants 7000589, 7000861, 7001074, 7001241 and 7001296.

The authors also wish to acknowledge the funding support received from Industry Department of the Government of HKSAR for setting up the Rapid Prototyping Technology Center (RPTC) jointly set up by the MEEM department of City University of Hong Kong and the Hong Kong Productivity Council (HKPC). In connection with RPTC, the authors wish to thank Mr. H.Y. Wong, Dr. Edmund H.M. Cheung, Dr. I.K. Hui, Dr. K.S. Chin, Dr. Ralph W.L. Ip, Dr. Meng Hua, Dr. H.W. Law and Dr. Ricky W.H. Yeung from CityU. Thanks are also extended to Dr. T.L. Ng, Dr. S.W. Lui, Mr. L.M. Li, Dr. L.K. Yeung, Mr. Lawrence Cheung, Mr. Thomas Chow, and Mr. Norman Chan from HKPC, and Ms. Annie Choi and Mr. K.S. Chiang from the Industry Department for collaborating on various R&D projects.

The authors also wish to gratefully acknowledge input from their coauthors of related publications. Among others, particular mention is made of Prof. Jean-Pierre Kruth of Katholieke Universiteit Leuven, and Dr. S. Liu, Dr. P. He, Dr. N. Zhao, Mr. W.C. But and Mr. K.Y. Chu from City University of Hong Kong. Special thanks also go to Dr. K.M. Yu (Hong

Kong Polytechnic University) for providing valuable information on some related topics, and Mr. C.C. Leung and Mr. Y.S. Tai of the Integrated Design and Prototyping Laboratory (MEEM, CityU) and the CAD and Software Applications Laboratory (MEEM, CityU), respectively, for their daily support and maintenance of related facilities.

The authors also wish to thank Mr. Deepak Patri for the book cover design, Mr. Michael Hackett (Kluwer Academic Publishers) for his patient support throughout the writing of the book and Dr. D.R. Vij (Kurukshetra University, India) for communications relating to the publication.

We would also like to express our special gratitude to our families for their support and understanding in the preparation of this publication.

Patri K. Venuvinod and Weiyin Ma

Chapter 1

INTRODUCTION

1.1 THE IMPORTANCE OF BEING RAPID

Global experiences over the last two hundred years point to three broad trends. Firstly, almost every large and developed economy in the world has achieved material progress through three sequential but overlapping movements: consolidation and modernization of agriculture, growth of domestic manufacturing followed by its gradual integration into global manufacturing, and growth of the service sector. Secondly, there has been relentless and ever-increasing penetration of technology into every aspect of industry. Thirdly, owing to the emergence of affluent societies, the world market is becoming more and more customer-oriented (Venuvinod 2000).

The maturation of manufacturing enterprises has generally followed three successive but overlapping phases of competitive emphasis - see Figure 1-1. The first phase is characterized by competition through productivity (P). In the second phase, the competitive strategy shifts to achieving higher quality (Q), i.e., ensuring greater consumer satisfaction, while maintaining productivity. The competitive focus in the third phase is on gaining further market share through innovation, i.e., through "new ways of delivering customer value (O'Hare 1988). Clearly, introducing new products at ever increasing rates is crucial for remaining successful in today's competitive global economy. At the same time, decreasing product development cycle times and increasing product complexity require new approaches to product realization. In response to these challenges, academia and industry have come up with a spectrum of technologies that are helpful in developing new products and broadening the number of product alternatives. Examples of these technologies are feature-based design, design for manufacturability analysis, simulation, computational prototyping, and virtual and physical prototyping. It is now generally agreed that "getting physical fast" is critical in exploring novel design concepts. The sooner designers experiment with new products, the faster they gain inspiration for further design changes. Thus, time is of the essence in all the three phases (P, Q, or I).

To clarify the above observations, let us consider a deliberately simplified product realization scenario. Assume that we are a design and manufacturing company. Our business is based on a single product conceived, designed, tooled and manufactured entirely in-house. Our first goal is to maximize the profit derivable from this venture. A model for estimating the profit is shown in Figure 1-2. It is evident that, other elements remaining the same, our profit will increase if we are able to reduce our manufacturing cost. This is the P-strategy.

According to Figure 1-2, two important components of the manufacturing cost are the labor and overhead costs that increase in proportion to the manufacturing time, t_m. In other words, *rapid manufacturing* (RM) is a highly desirable intermediate goal.

What can we do to reach this goal? The answer is in technology. We need to utilize a superior manufacturing technology capable of producing the same product in a shorter time. However, in doing so, we must make sure that the costs of the new technology and the associated material and tooling are not so high as to offset the cost decrease we have achieved through a reduction in t_m.

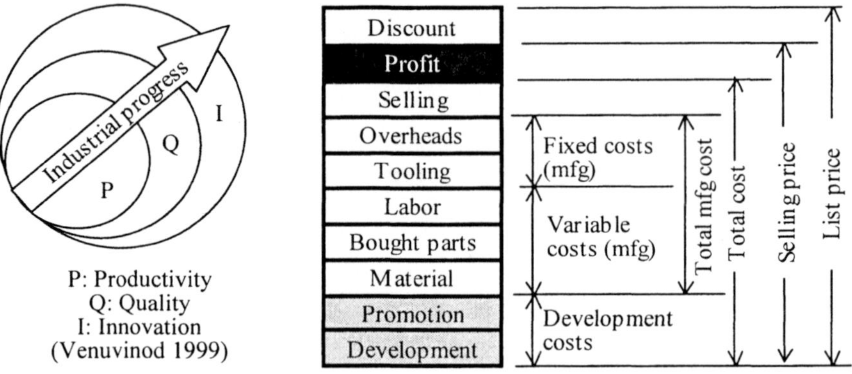

P: Productivity
Q: Quality
I: Innovation
(Venuvinod 1999)

Figure 1-1: Changing competitive strategies.

Figure 1-2: Product realization cost structure.

The notion of 'rapidity' enters the analysis of productivity in yet another way. The tooling required for producing our product can be viewed as a product on its own, so the tooling cost can be estimated in a manner analogous to Figure 1-2. This implies that *rapid tooling* (RT) is a highly desirable complementary goal. Of course, we must be careful that the tooling technology we adopt does not in itself need further tooling, i.e., our RT technology should be able to directly produce the tooling we need.

The manufacturing technology that we choose will depend on the type of production and production volume (Figure 1-3). The upper curve in the figure refers to the traditional manufacturing technologies while the lower two curves relates to computer-based manufacturing technologies. Since, the latter use computers, they are more flexible and are effective regardless of the geometric complexity of the parts being produced. They also involve significantly less tooling, so they are particularly effective in prototyping and low-to-medium batch production. Many of the technologies to be discussed later in this book present an emerging alternative to NC/CNC technologies.

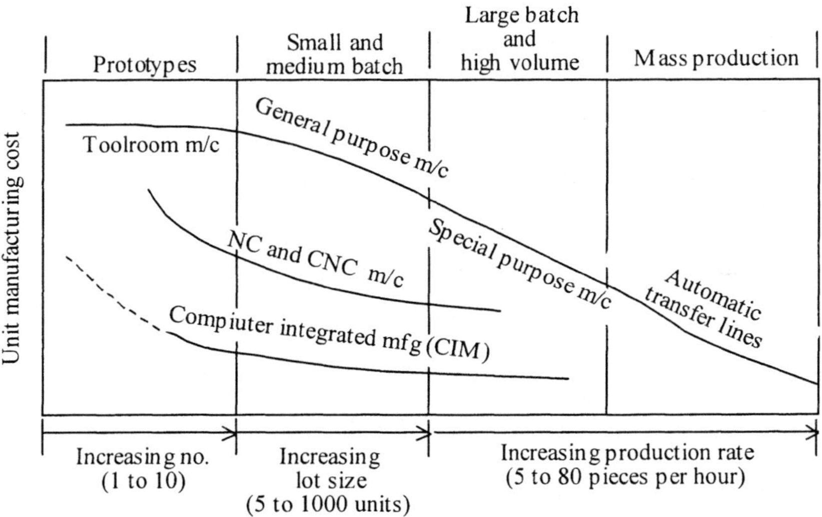

Figure 1-3: Impact of technology on unit manufacturing cost.

Having done our best to improve our manufacturing productivity, we can now look at other ways of increasing our profits. It follows from Figure 1-1 that, other conditions remaining the same, our profits will increase if we are able to command a higher selling price for our product. Our success will depend upon how well we are able to satisfy, delight or, even, surprise our customers. Customer satisfaction and delight depend on quality. However, customers finding new features and capabilities in our products will be pleasantly surprised. In other words, it is useful to explore how we might start competing on the basis of quality (Q) and innovation (I). This requires us to be sensitive to the attitudes and preference s of today's customers.

Customers these days generally want to be treated as individuals. They expect low volume, high quality, and customized products. This means we need to find ways of minimizing time-to-market (time elapsed from the time we conceive our product to the time we start delivering it to our customers). Otherwise, we might lose our market share to a competitor who has come up with a similar or better product in a shorter time. Clearly, rapid product too is a highly desirable goal.

Since product development is a creative process, it is also an iterative process (Figure 1 -4). A s each d esign c oncept i s d eveloped, it n eeds to be evaluated preferably with the aid of a prototype, so opportunities for improvement identified are identified. This cycle needs to be repeated until a satisfactory f inal design i s r eached. The m ain a dvantage o f p rototyping i s that it gives the various interested parties (e.g., engineering, sales and marketing, manufacturing, part suppliers, and subcontractors) a better sense for the product Prototypes can be of many types. Some merely exhibit the appearance. Some demonstrate functional attributes. Yet others can point our manufacturing issues.

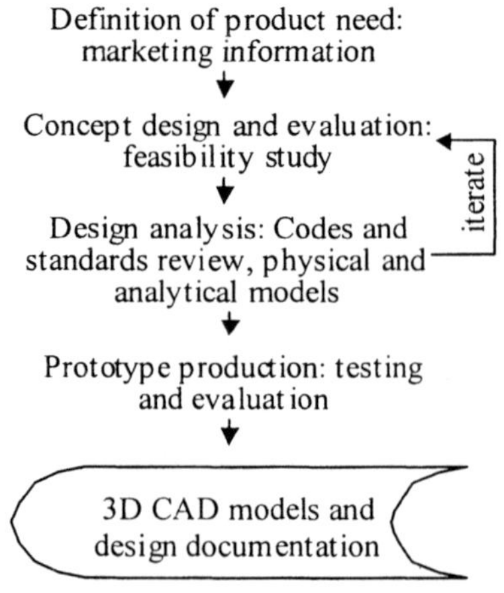

Figure 1-4: Modern Product development.

In traditional prototyping, parts drawings are passed on to a number of different shops to produce the diverse components of the product. This may involve machining some parts, fabricating others out of metal or plastic, giving some a surface treatment in different finishes, and so on. If the product contains components that cannot be made in-house (e.g., circuit boards), they are handed over to a subcontracting company. Once all the parts are available, they get assembled into the complete product either in-house or by yet another company. The problem with this approach is that it is too time consuming to be able to meet today's short-time-to-market demands. Fortunately, several rapid prototyping technologies (RP) have become commercially available over the last fifteen or so years. In particular, two classes of technologies are of particular interest: Rapid Prototyping (RP), and Rapid Tooling (RT). These form the main theme of this book.

So far we have assumed that our company is engaged in original design and manufacture (ODM), i.e., our company's operations straddle both development/design and manufacturing of products. But, what if we are just an original equipment manufacturing company (OEM), i.e., we obtain our product designs from some one else and all we do is manufacture them after adjusting the designs to suit the particular manufacturing facilities available with us? Or we could simply be a service bureau that merely fabricates prototypes for others. In either case, we will have to obtain our design documentation from our clients. Fortunately, this can be done quite rapidly these days via the Internet. We can also use the Internet to form a member of a virtual product development team in collaboration with our client company (Burns and Howison 2001 and Shyamsundar and Gadh 2002). Internet-based collaborative design enables designers located around the world to arrive at commonly agreed virtual prototypes (VP). In principle, the design activity can go on 24 hours a day, 7 days a week, and 365 days an year. A key technological facilitator of this is three-dimensional computer-aided design (3D CAD) augmented for the purposes of Internet-based collaborative design. Several software-implementation tools and standards supporting such augmentation have appeared in recent years (Diehl 2001): Continuous Data Streaming, VRML (Virtual Reality Modeling language), Quicktime VR (Virtual Reality), MetaStream, MPEG (Moving Picture Experts Group) standards, Pure Java 1.1 Applets, Java3D, X3D, and CORBA (Common Object Request Broker Arbitrator). Several application software using such tools are commercially available now, e.g., 'Envision 3D' from Adaptive Media, 'Pivotal Studio' from centric Software, 'One Space' from CoCreate, and 'ConceptWorks' from RealityWorks.

So far we have assumed that we are able to use computerized 3D models of our products, so we can proceed with RP/T. But what if they are not available? Such instances are many. We might be asked to reproduce a heritage artifact, a natural object such as an ancient human scull dug up by a paleontologist, or a bone from the leg of an injured patient. Or an engineering company, our client, had developed a product some fifteen years ago on a 2D drafting system and we are asked to reproduce it now. In all such instances our starting point is merely an object-in-hand from which we will have to create a 3D CAD model rapidly. Fortunately several technologies under the generic name of Reverse Engineering (RE) are now available to solve this problem. A commonly used RE technique is to measure the locations of a large number of points on the surface of the given object with the help of a 3D laser scanner. This step is called 3D digitization. The point cloud data so obtained is then processed in software to construct the CAD solid model needed for implementing the subsequent rapid prototyping and tooling operations. We will study RE techniques in Chapter 4.

Combining RP/T with RE, we have a computer-based technology analogous to 'Xeroxing' that, according to some, is not always good. For instance, a former employee of Xerox Corporation, Merle L. Meacham, has said (rather unkindly) that "[i]n a few minutes a computer can make a mistake so great that it would take many men many months to equal it", so Xerox can be viewed as a "trademark for a photocopying device that can make rapid reproductions of human error, perfectly." But product development, by nature, is a trial-and-error process. It is only through recognizing our errors quickly that we can arrive at a superior solution. Viewed thus, Meacham's comments become cautionary statements rather than constituting a negation of the worth of RE-RP/T. The caution is that our computational procedures should not themselves introduce errors.

1.2 THE NATURE OF RP/T

The term 'rapid prototyping' immediately suggests "speedy fabrication of sample parts for demonstration, evaluation, or testing (Cambridge Scientific Abstracts)." While this statement does succinctly capture *what* RP does, it does not throw any light on *how* the so-called speedy fabrication is achieved. How can we define RP as a process? The answer is not simple since RP is still an emerging field, so its terminology has not yet stabilized. In fact there is no commonly accepted title to this field. Among the titles used so far are Direct CAD Manufacturing, Automated Fabrication, Solid Freeform Fabrication (SFF), Fast Freeform Fabrication, Layered Manufacturing (LM), Material Increase Manufacturing, Instar Manufacturing, 3D Printing, Desktop Manufacturing, and 'Fabbing'.

As illustrated in Figure 1-5, an RP implementation starts with a 3D geometric model of the object to be prototyped. Usually, the design team prepares the model. Alternatively the model may be obtained by subjecting an available object to RE, CT, or MRI.

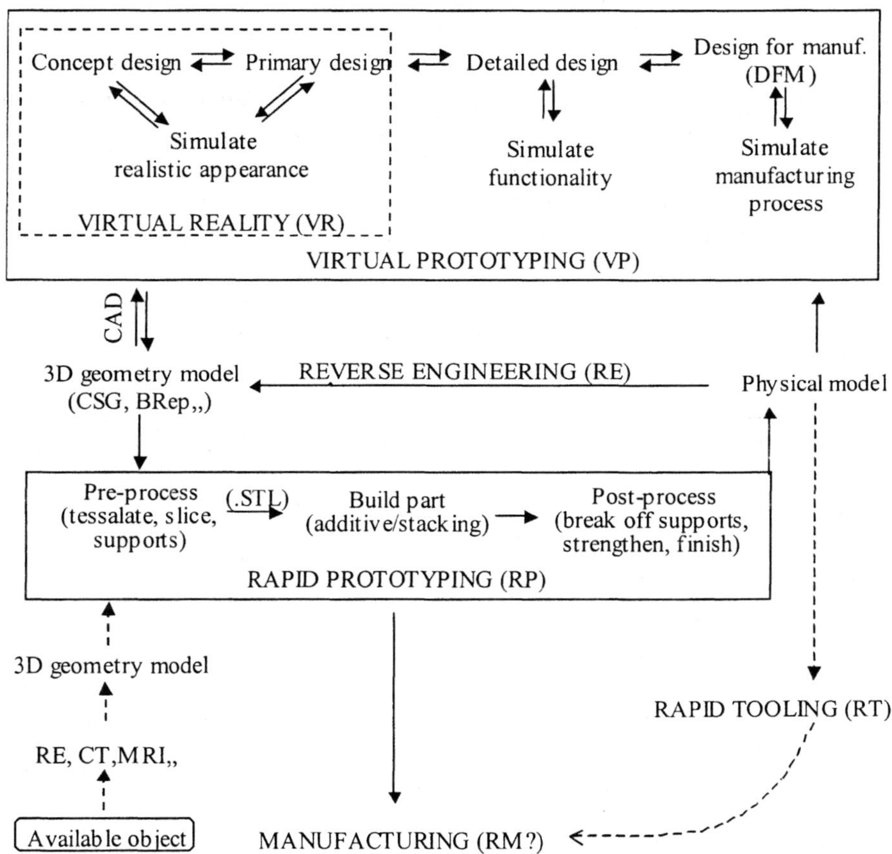

Figure 1-5: Integration of 'rapid technologies'.

Two basic types of 3D modeling approaches are available: B-rep, and CSG. With B-rep (boundary representation), the boundaries of the part are explicitly represented through a graph structure in which nodes represent the faces, edges, and vertices of the object whereas arcs between nodes represent adjacency relationships. Face nodes contain the parameters of surface equations, edge nodes the parameters of curves, and vertex nodes the 3D coordinate values. If the model needs to be subjected to subsequent geometric feature recognition (GFR), it is useful to prepare a higher level graph called multi-adjacency attributed graph (MAAG) where both the nodes and arcs contain certain high-level attributes derived from low-level object data (Venuvinod and Yuen 1994 and Yuen and Venuvinod 1999). B-rep is usually built according to an orientation convention based on the outward pointing normals of faces, so one can recognize the interior and exterior of the object.

With CSG (constructive solid geometry) the geometry of the object is modeled as a binary tree. The leaf nodes of the tree are scaled and oriented instances of primitive shapes such as the parallelepiped, the sphere, the cylinder, the cone, the torus, and certain linearly swept 2D profiles. Primitives can be subjected to certain set operations. The operations are 'regularized', so non-solid artifacts such as dangling edges and repeated faces are avoided. For the same object, CSG yields a much more compact representation than B-rep. However, unlike B-rep, a CSG representation is implicit, so it becomes necessary to evaluate the model for deriving explicit surface information such as needed for RP and finite element analyses.

The surfaces on many industrial objects (e.g., turbine blades) can be complexly curved, so the true CAD models can become too large and unwieldy for down stream RP analyses. Hence, most RP systems require the object's surface to be approximated by a mesh of non-overlapping triangular facets. This process is called tessellation. The resulting faceted geometry is usually transmitted to the RP machine in a standard file format called .STL developed originally by 3D Systems, Inc.

Many CAD vendors have adopted the .STL format since it is sufficient for most visualization applications. However, the tessellated model could be too coarse for accurate patterns and some functional parts. In principle, the accuracy of the model can be improved by adopting a finer mesh. But, this will make the .STL file too large. An alternative is to adopt a higher order geometric description such as IGES (Initial Graphic Exchange Specification) 4.0 or STEP (Standard for Exchange of Product Data Model).

Now we are approaching the essence of RP viewed as a process. How does RP output a physical model of the object of interest from the digital information contained in the .STL file (or a higher order file)?

W.D. Rase, a geographer, says "the different ways to produce prototypes can be subsumed under three groups which have their equivalents in the arts:"

- *Michelangelo,* Buonarotti (1475-1564, Italy): "A block of material is treated with a tool until the final form is reached. A block of marble is transformed into a sculpture."
- *Rodin:,* Auguste (1840-1917, France): "The model is built by accumulating small amounts of material. A sculpture or a scaled model is aggregated from plaster or other plastic material."

- *Chillida*, Eduardo (1924-2002, Spain): "The material is formed by strong forces, like a blacksmith transforms a rod of iron into a sickle (Rase 1997)."

The only difference between the shape generation approaches used by sculptors and in modern industry is in the way energy is applied to shape the raw material. Sculptors mainly use their hands while modern manufacturing technology uses machines.

One can arrive at a similar classification by applying shaping science (Kochan *et al.* 1999). Michalengelo's style is 'subtractive' since it arrives at the desired shape by removing material from a blank. In modern industry, we resort to CNC machining, electric discharge machining (EDM), laser cutting, etc. to obtain a similar effect. A disadvantage of this approach is that only shapes or forms with features accessible by the subtractive tools can be realized. For instance, reentrant corners and internal openings cannot be created via CNC milling. In other words, a subtractive process cannot be a freeform process.

Likewise Chillida's style is 'forced' since it exploits the deforming ability of solid or liquid material. Some authors refer to this approach also as 'net-shaping'. Examples of the forced approach are foundry, forging, sheet metal forming and plastic injection molding. A disadvantage of these methods is that they need molds, dies or other tools.

Till recently, almost every solid fabrication implementation was either subtractive or forced. However, the emergence of RP in the late 1980s introduced a new solid fabrication paradigm that either adds or stacks material to obtain the desired shape (Rodin's style). This 'additive/stacking' approach views the solid "as a spatial body which is stacked by points and surfaces according to the discretion of its geometric information (Kochan *et al.* 1999)." The main advantage of this approach is that it is capable of creating freeform solids without using external tools.

It should be noted, however, that building up structures in layers is not a new idea; in fact, it goes back to the days of the Egyptian pyramids that were likely built block-by-block, layer-by-layer. Stacking up layers of individually s haped m aterial layers a lso h as a l ong tradition in a range o f manufacturing applications such as tape casting and shape melting. However, the traditional methods were not automated. Here we are concerned with automated prototype construction.

(According to shaping science, there is a fourth approach called 'growth shaping'. For instance, trees and animal skeletons grow in a certain manner to attain a particular shape. However, there have been few applications of this approach in modern industry).

As we will see later, there are many ways of implementing the additive/stacking process. In fact, describing these myriad ways is a major purpose of this book. For the moment though, let us examine how the process is implemented in one RP process called stereolithography (SL). We pick this particular process simply because, amongst the existing RP technologies, it was the first one to be commercialized. Further, even today, it is the most used. Figure 1-6 illustrates the working principle of a stereolithography apparatus (SLA).

SL uses a liquid photopolymer held in a vat as the raw material. The part is built by selectively polymerizing the liquid resin in a layered manner. A servo-controlled vertically movable table (elevator) holds a base-plate (platform) on which the prototype would be built. Upon receiving the .STL file, a candidate orientation of the part is chosen manually. Next a software module slices the part into a number of horizontal layers of sub-millimeter thickness (see Figure 1-5). In each layer, the regions covered by the part are identified. Since the part will be selectively solidified within the liquid polymer volume, overhanging and cantilever features of the part will not be supported, so structures in the form of thin vertical columns need to be designed to support such regions. The .STL files for the object and supports are then merged and the total build time calculated in software. The calculation i s r epeated f or d ifferent o bject o rientations a nd t he o rientation yielding the lowest build-time is selected.   Initially the moveable table is placed at a position just below the surface of the vat. A knife-edge levels the liquid surface by sweeping over it. A scanner system controls a UV laser beam in the X and Y directions (horizontal plane), o it can be moved over the surface of the liquid photopolymer to trace the geometry of the cross-section of the object. This causes the liquid to harden only the desired regions of the layer to be hardened. The exact pattern that the laser traces is a combination of the object information contained in the CAD system, and information from the rapid prototyping application software that optimizes the faithfulness of the fabricated object. Upon completion of selective polymerization of the particular layer, the elevator moves down to dip the object by one layer thickness, so the next layer can be processed. The entire process is repeated until the object along with the supports is fully built. Once completely built, the object is removed and cleaned. Quite often, the 'green' parts produced by the primary additive/stacking operation are not sufficiently strong to be handled easily or be functional. Hence, the primary shaping process(es) is usually followed by secondary post-processing operations capable of strengthening parts to their full potential. Usually the object is given a final cure by bathing it in intense light in a box resembling an oven called a Post-Curing Apparatus (PCA). After final cure, the supports are cut off from the object and surfaces are sanded or otherwise finished.

Now we are in a position to list the characteristics of RP from a process viewpoint. Firstly, it is digital in nature since the information needed for fabrication comes from a digitally encoded CAD, CT, or MRI model of the object. Secondly, the object is built by means of an additive/stacking process. The stacking may be along surfaces or points. The former feature has led to the alternate name of layered manufacturing (LM). Often these layers are plane and horizontal. The layers may be subdivided into tiny three-dimensional volumes called 'voxels' that are analogous to pixels in 2D, so part building can proceed voxel by voxel. In principle, different layers or voxels within them can be filled with different materials.

In short, from the process-viewpoint, RP can be defined as a "special class of machine technology that quickly produce models and prototype parts from 3D data using an additive approach to form the physical models" (Wohlers 2000).

The benefits derivable from the additive/stacking approach are many. Firstly it enables objects with unlimited geometric complexity to be fabricated (solid freeform fabrication —SFF). Very complex parts even with hollow structures can be built without unduly increasing the build time. Secondly, owing to the possibility of filling different layers/voxels with different materials, RP can also be used to build multi-material and functionally graded objects (Fessler *et al.* 1997, Gasdaska *et al.* 1998, Jackson *et al.* 1999, Rodrigues *et al.* 2000 and Shisovsky 2001a). If the process resolution is sufficiently fine, it is possible to fabricate even articulated assemblies (Zhu 2002). Thirdly, the processes are easily automated owing to the methodical filling of voxel after voxel. In contrast, the full automation of milling and other subtractive processes has been more difficult to accomplish. Fourthly, this is achieved without needing additional tooling. The advent of NC machine tools previously had eliminated the need for much tooling. However, complex parts still require expensive 5-axis machining. This penalty does not exist with layered manufacturing (Bjørke 1991and Bjørke 1992). Finally, since most RP processes are additive in nature, no waste material is produced as in subtractive processes (e.g., as in machining).

How expensive is RP? The answer depends upon the part material, the desired part accuracy, the RP technology used, and how well the equipment has been tuned for the particular job. Hopkinson and Dickens recently compared the costs of producing thermoplastic components weighing in the range 3.6-760g by injection molding and SL (Hopkinson and Dickens2001). Figure 1-7 illustrates their results for 3.6g components. Note that while the unit cost associated with injection molding falls steeply with increasing production volume, that associated with SL is insensitive to production quantity. This is because the cost of SL has no upfront tooling costs, i.e., its cost is almost entirely made up of machine and material costs. The figure also highlights the importance of tuning the RP process for the particular job—the break-even quantity could be increased from 7500 to 27000 through tuning. Clearly, with further reduction in machine and material costs, RP can start competing in high volume production too. For the moment though, the cost analysis combined with opinions generated in an Internet conference have suggested that the most suitable candidates for rapid manufacturing are small parts with high geometric complexity to be made in relatively small volumes (Hopkinson and Deckard 2001).

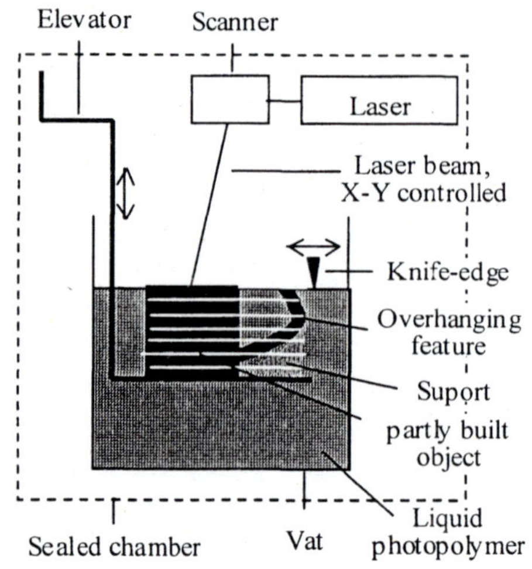

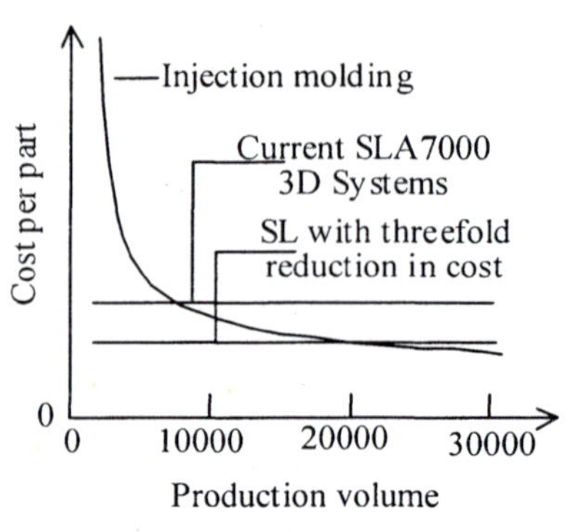

Figure 1-6. Stereolithography Apparatus (SLA).

Figure 1-7. RP economics (Hopkinson and Dickens 2001).

A problem with all layered manufacturing processes though is the poor surface finish arising from the stairstepping effect. The effect is unavoidable owing to stacking of layers of finite thickness. A smaller layer thickness would result in lower surface roughness. However, that strategy would increase build time proportionately. Stairstepping is more pronounced over object surfaces closer to the horizontal plane. This insight has led to adaptive slicing; thinner layers are used in problematic object regions.

Some RP processes are hybrid in nature, i.e., they eclectically combine two different shaping approaches. For instance, there are many RP processes that resort to an additive/stacking approach for building the part and then subject selected surfaces to CNC machining (a subtractive shaping approach) so as to ensure the required accuracy and surface finish (through removal of stairsteps).

In RT, hybridization goes a step further by combining an additive/stacking step with a range of conventional fabrication processes including and beyond machining. So, in the context of RT, one does not talk about an RT process but an RT process route. The tooling components producible via RT include cavity and core components of injection molds, sheet metal dies, and electrodes for EDM. A major difference between RT and R P i s t hat t he c omponents p roduced v ia t he f ormer n eed t o b e m uch more accurate and durable than those from the latter. Tooling capable of surviving tens of shots is called soft tooling. If the number of shots is a few thousands, it is called hard tooling. Tooling applicable to intermediate volumes is called bridge tooling.

RT comes basically in two types. In 'Direct RT', the shapes of the tooling components are obtained directly by means of an additive/stacking process followed by certain machining and other post-processing operations. Often, tools so produced are not durable enough to survive large volume production, thus restricting their application to soft or bridge tooling. However, some direct hard tooling methods have started appearing in recent times.

Indirect RT does not aim at producing the final tooling components. Rather it aims at producing the essential shape-generating element in a conventional tool-fabrication technology. For instance, the pattern is the main shaping element in conventional casting and investment casting. The patterns are usually made from wood, plastic, aluminum, steel, or cast iron. The conventional approaches to the production of these patterns can be quite tedious and time consuming. A faster approach would be to produce the pattern via an SFF process and then coat it with silicone rubber or ceramic slurry.

1.3 HISTORY OF RP

The idea of layered manufacturing (LM) actually precedes the modern era of RP. Since several of the ideas contained in these early developments have parallels in modern RP methods, it is useful to briefly recount them. We will base our discussion of the early developments on the excellent illustrated summary presented by J.J. Beaman as a part of the report on RP in Europe and Japan prepared by the World Technology Evaluation Center, Baltimore, MD (Anon. 1997).

As early as in 1892, Blanther used a primitive LM method to make molds for topographical relief maps (Blanther 192). He impressed topographic contour lines on a series of wax plates, so they could be cut and stacked along the contour lines. In 1940, Perera used a similar approach using cardboard (Perera 1940). In 1964, Zang obtained a more refined effect by using transparent plates inscribed with topographic detail (Zang 1964). Subsequently, such 3D relief maps were used as teaching devices (Gaskin

1973). In 1976, DiMatteo stacked and bolted together contour milled metallic sheets to produce complex surfaces that could not be conveniently produced via machining (DiMatteo 1976). Examples of objects thus produced include 3D cams, air-foils, and punching dies. Later Nakagawa of Tokyo University, Japan, produced laser-cut laminated sheet tools for drawing dies (Kuneida and Nakagawa 1984) and injection molding dies (Nakagawa *et al.* 1985).

In 1974, Matsubara of Mitsubishi Motors, Japan, tried to avoid the mechanical stacking approaches described above by a process that took advantage of photo-polymerization. He coated a photopolymer resin onto graphite powder or sand, spread the coated particles into a layer, and heated the layer to form a thin sheet (Matsubara 1974). He then put the sheet under a mercury vapor lamp so as to harden the desired parts of the sheet though light projection or scanning. After dissolving away the unhardened portions, he stacked the layers to form the desired relief map.

An alternative to the use of photopolymers was also conceived in the early 1970s. In 1972, Ciraud proposed applying a partially meltable material in powder form to a matrix (Ciraud 1972). He suggested that particle application could be carried with the aid of gravity, magnetostatics, electrostatics, or a nozzle whereas particle adhesion could be achieved through local heating effect achieved by a laser, electron beam, or plasma beam. A similar process is used in contemporary RP under the name of selective laser sintering (SLS).

Another early LM development that also led to photopolymer-based prototyping eventually is photo-sculpture. In 1860, Francois Williams, a Frenchman, simultaneously photographed a human subject by 24 cameras placed at equal distances along a circle centered at the subject. Next an artisan carved out 1/24th of a cylindrical portion of the figure on the basis of the information contained in the silhouettes of the photographs. In 1904, Baese obtained a similar effect by using a more advanced method involving exposure of photosensitive gelatin to graduated light (Baese 1904). The gelatin had the interesting property of expanding in proportion to exposure after being treated with water. Subsequently Moroika used structured light consisting of black and white bands of light to create the contour lines needed for the stacked sheets forming the photo-sculpture (Moroika 1935 and Moroika 1944). By the 1950s these techniques culminated in a technology somewhat similar to present day SL. A notable effort in this direction involved selective exposure of a transparent photo-emulsion to light (Munz 1956). The layers were created by lowering a piston in a cylinder and adding appropriate amounts of the photo-emulsion along with a fixing agent. The resulting solid transparent cylinder contained an image of the object that could be carved or photo-chemically etched out to create a 3D replica. Subsequently Swainson proposed using two laser beams such that selective 3D polymerization occurred at the intersection of the beams (Swainson 1977). A similar photochemical etching process was also developed at Battelle Laboratories, USA (Schwerzel 1984).

However, it was left to A. Kodama of Nagoya Prefecture Research Institute, Japan, to develop the first functional RP system using a photopolymer exposed to laser radiation via masks and/or an optical fiber (Kodama 1 981). M eanwhile A . H erbert o f 3 M i n M inneapolis, U SA, w as working on a single laser beam directed on to an x-y plotter to achieve selective photo-polymerization. Likewise C. Hull and A. Reed were trying to develop an SL system at Ultra Violet Products, Inc. (UVP), CA. However, only Hull continued the work beyond patenting to the stage of commercial exploitation by founding 3D Systems, Inc. In 1987, this company unveiled the first ever commercially available RP machine, SLA-1. Since then several larger and more advanced models have appeared. Further 3D Systems and Ciba-Geigy, Ltd. partnered to produce advanced photosensitive polymers. However, it is generally recognized that the launching of SLA-1 represents the beginnings the modern RP industry.

Why d id R P h ave t o w ait s o l ong t o b e c ommercially e xploited? The answer is in the availability of 3D computer models that are crucial to realizing the concept of layered object creation, as well as of other essential technologies such as affordable laser systems, photo-curable materials, and powerful personal computers. Indeed, RP is an interdisciplinary engineering field (see Figure 1-8) and RP machines are truly mechatronic in nature.

Table 1-1 outlines the chronological development of RP since 1987. It is clear that the field of RP has been attracting intense innovative efforts from several companies distributed mainly in the USA, Europe, Japan and Israel. Interestingly, several of these developments were based on the work conducted by university academics. For instance, selective laser sintering (SLS) was first developed by Carl Deckard at the U niversity of Texas at Austin, USA, to build plastic parts. Likewise, CMB (Controlled Metal Buildup) was developed at Fraunhofer Institute for Production Technology (IPT) in Germany; 3DP (3-dimensional Printing) at Massachusetts Institute of Technology (MIT), MA; Shape Deposition Manufacturing (SDM) at Carnegie Mellon University, PA; and Contour Crafting (CC) at the University of Southern California, CA.

Table 1 -1 r eveals that t he v arious R P d evelopments c an b e subsumed under categories c haracterized b y a f ew k ey R P p rocesses. We w ill study these processes in detail in Chapters 7, 8, and 9. However, it is useful at this stage to outline the working principle of each key process, so its significance within the overall development of RP can be appreciated. We will take the key processes in the order they appear for the first time in the last column of

Table 1-1. (We will not recount the working principle of SL since we have already outlined it.)

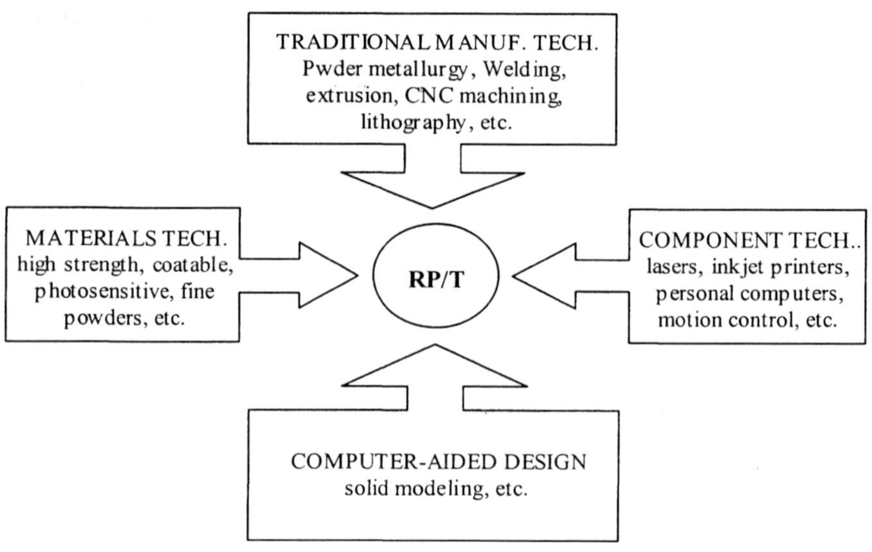

Figure 1-8. RP/T enabling technologies.

Table 1-1: Chronological development of RP/T.

Year	Company/Org.	Country	RP system	Key Process
1987	3D Systems	USA	SL Apparatus SLA-1	SL
1988	NTT Data CMET	Japan	Solid Object Ultraviolet Plotter (SOUP)	SL
1989	Sony/D-MEC	Japan	Solid Creation System (SCS) based	SL
1989	MIT	USA	3D Printing	Inkjet
1990	EOS	Germany	Steros SL system	SL
1991	Stratasys	USA	FDM^{TM}	FDM
1991	Cubital	Israel	Solid Ground Curing (SGC)*	SL
1991	Helisys	USA	Laminated Object Manufacturing (LOM)	LOM
1992	Teijin Seiki[1]	Japan	Soliform SL system based on DuPont SOMOS	SL
1992	DTM[2]	USA	Selective Laser Sintering (SLS)	SLS
1993	Denken	Japan	Solid-state SL: 1st Desktop RP system	SL
1993	Soligen1	USA	Direct Shell Production Casting (DSPC)	Inkjet
1994	Sanders[4]	USA	ModelMaker	Inkjet
1994	Meiko	Japan	Small SL machine for jewelry	SL

Table 1-1 (continued)

Year	Company/Org.	Country	RP system	Key Process
1994	Kira	Japan	Solid Center	LOM
1994	Fockel & Schwartz	Germany	An SL machine on a limited basis	SL
1995	EOS	Germany	EOSINT	SLS
1995	Ushio	Japan	Unirapid machine	SL
1996	BPM Tech.[5]	USA	Personal Modeler 2100*	Inkjet
1996	Kinergy	Singapore	Zippy paper lamination systems	LOM
1996	Stratasys	USA	Genisys system based on technology developed by IBM's Watson Res. Center	FDM
1996	3D Systems	USA	Actua 3D printer (multiple jets)	Inkjet
1996	Z Corp.	USA	Z402 concept modeler	Inkjet
1996	Schroff Dev.	USA	Less expensive paper lamination machine	LOM
1997	Aeromet[6]	USA	Laser Additive Processing	SLS
1998	Optomec	USA	Laser Engineered net Shaping (LENS)	LC
1999	3D Systems	USA	ThermoJet (less expensive than Actua)	Inkjet
1999	Extrude Hone	USA	RTS-300 metal part builder	Inkjet
1999	Fockel & Schwartz	Germany	Powder-based selective laser melting	SLS-like
1999	Röders	Germany	Controlled Metal Buildup	LC
2000	Objet Geometries	Israel	Quadra multi-jet machine	Inkjet
2000	SDI	USA	PatternMaster	Inkjet
2000	POM	USA	Direct Metal deposition (DMD)	LC
2000	Z Corp.	USA	1st color 3D printer	Inkjet
2000	Stratasys	USA	Prodigy for ABS parts	FDM
2001	Solidimension	Israel	PVC sheet based desktop m/c	LOM
2001	Solidica	USA	Combination of ultrasonic welding and CNC machining	
2001	Stratasys	USA	Titan machine for ABS, etc.	FDM
2001	Envision Tech.	USA	Perfactory m/c based of Digital Light Technology (DLT) of Texas Instruments	
2001	Concept Laser GmbH	Germany	A m/c combining of laser sintering, marking and machining	SLS
2001	RSP Tooling	USA	Rapid Solidification Process	Spraying
NOTES				
* Not available anymore. [1] Now a part of CMET . [2,3] Now parts of 3D Systems. [4] Now called Solidscape. [5] Ceased operations in 1997. [5] Subsidiary of MTS Sys.				

Inkjet printing comes from the 2D printer and plotter industry where graphic images are printed on paper by shooting tiny droplets of ink. RP inkjet techniques shoot droplets of liquid-to-solid compound and form a layer of an RP model. For instance, Multi-Jet Modeling™ offered by 3D Systems uses a 96-element print head to deposit molten plastic for layering. Since the system is small and produces good appearance models with minimal operator effort, it is mainly targeted at the engineering office where the system must be non-toxic, quiet, small, and with minimal odor. Many desktop RP systems are based on the inkjet principle.

Fused Deposition Modeling (FDM) extrudes a thin bead of plastic, one layer at a time. A thread of plastic is fed into an extrusion head, where it is heated into a semi-liquid state and extruded through a very small hole onto the previous layer to build the part as well as the support material. Because it can use high strength ABS plastic, it is the favored technology for prototyping plastic parts requiring strength.

In laminated object manufacturing (LOM), a laser traces the outline of the layer on a sheet of paper. The sheet is then adhered to a substrate with a heated roller. Non-part areas are crosshatched to facilitate removal of waste material. Once the laser cutting is complete, the platform holding the part is moved down and out of the way so fresh sheet can be rolled into position. The platform then moves back up to one layer below its previous position. The process is repeated until the part is fully built. The prototype part has a wood-like texture composed of the paper layers. Most models are sealed with a paint or lacquer in order to avoid moisture being absorbed by the paper.

In selective laser sintering (SLS), the basic material consists of plastic, metallic, and/or ceramic powder with particle sizes in the order of magnitude of 50µm. Successive powder layers are spread on top of each other. After deposition, a computer-controlled CO_2 laser beam or a pulsed Nd-YAG laser scans the surface to selectively bind together the powder particles over the desired regions of the layer. There are two kinds of SLS, indirect and direct. Powders for indirect SLS are coated with a thermoplastic so that, during laser exposure, the powder temperature rises above the glass transition point of the coating after which adjacent particles flow together, i.e., get sintered. Direct sintering typically uses a powder consisting of the primary material with a secondary material exhibiting a lower melting temperature, so the primary materials can bind together through liquid phase sintering. Many industries have used SLS for a broad range of applications, including power tool housings for functional testing, engine components subjected to high temperatures, pumps and valves that transmit harsh chemicals, and patterns for complex investment castings. The US space program has even sent the shuttle and the space station into orbit with custom-made components built via SLS.

In l aser c ladding (LC), p owder m aterial is m elted b y f ocusing a l aser beam. The melt is then distributed layer by layer over the partly built part and allowed to solidify within an inert environment. A range of materials including aluminum, copper, Inconel, stainless steel, tool steel, and titanium can be deposited. Since p owder composition can be changed dynamically and continuously, it is possible to build multi-material and functionally graded parts.

In the spraying process, a high-velocity jet of inert gas is made to spray tiny droplets of molten metal alloy onto a plastic or ceramic pattern The rapid solidification process circumvents the majority of the fabrication steps in c onventional m ethods. B ecause t he s urface a rea of t he d roplets i s v ery large compared to their volume, the droplets cool at a rate somewhere between 100 t o 100,000°C per second. Such a fast c ooling rate r esults i n unusual beneficial characteristics of the alloy. The system is capable of building full-size dies.

The n umber o f R P t echnologies h as n ow g rown t o s uch a l evel a s t o prompt researchers to start classifying them. The major classification criteria suggested so far include the state of the raw material, the kind of raw material, and the chemical and physical transformations involved in material conversion (Kochan *et al.* 1999 and Kulkarni *et al.* 2000). Table 1-2 classifies the key RP processes according to these criteria.

A benchmarking exercise conducted as long back as in 1995 has revealed that the dimensional precision provided by R P processes such as FDM, SGC, SL, SLS, and LOM are much the same and are on a par with those associated with casting and forging. Likewise the surface finish is comparable to that obtainable by machining (Ippolito and Luliano 1995).

Table 1-2: Classification of RP processes.

RP process	State of raw material	Kind of raw material	Chemical or physical transformation
3DP	Powder	Ceramic and wax	Adhesive bonding
FDM	Solid	Wax and polymer	Solidification
LOM	Solid	Paper	Cutting and bonding
SL	Liquid	Polymer	Photo-polymerization
SLS	Powder	Ceramic, wax and alloy	Sintering

Although RP/T is still a young field, many books have already appeared on the subject (see Table 1-3 for a list of major books published upto 2002). However, owing to the r apid progress, several of t hese b ooks are a lready partly outdated. Further some books are biased towards the particular RP techniques that the author(s) had helped develop. Lastly, it is only recently that RP/T is being included as a distinct course in many industrial, manufacturing, and mechanical engineering curricula. Consequently, there are very few books providing a balanced and updated understanding of the wide range of RP/T technologies available today. This book aims to fill the gap.

Table 1-3: Books on RP/T

Year	Author(s)/Editor(s)	Book Title	Pages
1992	Bjørke, Ø.	*Layer Manufacturing*	182
1992	Jacobs, P.F. (Ed.)	*Rapid Prototyping & Manufacturing: Fundamentals of Stereolithography*	434
1993	Burns, M.	*Automated Fabrication: Improving Productivity in Manufacturing*	370
1993	Kochan, D. (Ed.)	*Solid Freeform Manufacturing*	212
1993	Wood, L.	*Rapid Automated Prototyping: An Introduction*	140
1995	Johnson, J.L	*Principles of Computer Automated Fabrication*	156
1996	Jacobs, P.F. (Ed.)	*Stereolithography and Other RP&M Technologies: From Rapid Prototyping to Rapid Tooling*	396
1997	Beaman, J.L. *et al.*	*Solid Freeform Fabrication: A New Direction in Manufacturing*	330
1997	Bennet, G.	*Rapid Prototyping Casebook*	212
1997	Chua, C.K. and L.K. Fai	*Rapid Prototyping: Principles & Applications in Manufacturing*	318
1998	Bernard, A. and G. Taillandier	*Le Prototypage Rapide*	256
1999	Binnard, M.	*Design by Composition for Rapid Prototyping*	139
2000	Hilton, P.D. and P.F. Jacobs (Ed.)	*Rapid Tooling: Technologies and Applications*	270
2000	Leu, D.	*Handbook of Rapid Prototyping and Layered Manufacturing*	800
2001	Pham, D.T. and S. S. Dimov	*Rapid Manufacturing: The Technologies and Applications of Rapid Prototyping and Rapid Tooling*	214
2001	Cooper, K.G.	*Rapid Prototyping Technology: Selection and Application*	226
2001	Lu, L. *et al.*	*Laser-Induced Materials and Processes for Rapid Prototyping*	268
2001	McDonald, J.A. *et al.*	*Rapid Prototyping Casebook*	260
2001	Varadan, V.K. *et al.*	*Microstereolithography and Other Fabrication Techniques for 3D MEMS*	274
2002	Gibson, I. (Ed.)	*Software Solutions for Rapid Prototyping*	380
Note: For full details, see References section for the given author(s)/editor(s) and year.			

1.4 THE STATE OF RP/T INDUSTRY

The annual reports provided by Wohlers Associates, Inc., provide a good source to study the trends in RP industry. This section will summarize the major trends as reported in (Wohlers 2002).

That the RP market is highly competitive can be surmised from the fact that "many companies have emerged and submerged over RP's relatively short h istory (Wohlers 2 002)." F or i nstance, w hen 3D S ystems a nd D TM announced in 2001 that they would be merging, the US Department of Justice Antitrust Division approved the deal only on the condition that 3D Systems licenses either stereolithography (of 3D Systems) or laser sintering (of D TM) t o a t hird p arty. E ventually, t he licensee happened to b e Teijin Seiki, Tokyo, Japan.

In 2002, some 3.55 million prototypes were produced worldwide using RP technologies. Since this represents an 18.3% growth over the previous year's figure, one may assume that RP activity worldwide is expanding. This appears to be mainly because of increasing throughput of the latest generation of RP systems and upgrades of installed equipment. Of the prototypes produced, about 22% were provided by service bureaus. At the same time, the proportion of RP revenue generated by them remained steady around 45%. These results show that service bureaus play a significant role in the RP industry.

Figure 1-9 shows the trends in annual number of RP units sold worldwide and annual revenue (figures for 2002 and 2003 are estimates). It is clear that RP usage has grown substantially and fairly rapidly in the last fifteen years. The rate of growth has, however, slowed down in the last few years. It is not immediately clear whether this saturation in penetration is an indication of the maturity of RP whether it is just a result of the recent economic downturn worldwide. However, interestingly, the growth of RP in its f irst 13 years s eems to be faster than that of CNC market in its early years. RP revenues grew by an average annual rate of 37.3% from 1970 to 1981, which is significantly higher than the rate of 22% experienced by CNC market in the period 1970 to 1981.

Nearly 81% of the systems sold in 2001 came from U.S. companies, so the dominant producer and seller of RP systems in the world is U.S. The rest 19% consisted of 10.9% from Japan, 4.6% from Europe and 3.8% from other regions. With 39.6% of the units sold within the US, 3D Systems dominate the RP scene. The other US companies with notable output are Stratasys (26.5%), Z Corp. (18.0%), and Solidscape (9.8%). The major RP vendors from Japan are Autostrade (21.1% of units sold by Japan in 2001), CMET (21.1%), Sony/D-MEC (19.0%), Kira (13.4%), Meiko (11.3%), Denken (7.0%), and Unirapid Inc. (4.9%).

What are the currently favorite RP t echnologies? T o answer, we may look at the mix of systems sold by 3D Systems. In 2001, the company had sold 194 SLAs, 182 ThemoJet printers, and 39 SLS machines. These figures suggest that SL is the favorite technology although desktop manufacturing systems carrying much lower prices are catching up. Other statistics though lead to mixed conclusions regarding desktop manufacturing. For i nstance, some 1 800 3 D p rinters (desktop m achines) w ere s old i n t he p eriod 1 996-2001. The growth rate was at the exciting level of 52.8% in 1999-2000. However, there was negative growth (-4.3%) in the following year. These figures suggest that the shift from high-end RP to desktop RP has begun, but factors impeding its growth continue to exist. Despite the lure of an affordable price, consumers are not (yet?) finding sufficient value. Another plausible explanation is that the desktop wave arrived exactly when the world had entered a new cycle of recession.

The relative degree of usage of RP in a country can be used as an indicator of the relative degree of innovation (includes design) in that country. From this viewpoint, it is worth noting the proportions of RP systems sold in different countries during 2001: USA (35.4%), Japan (16.6%), Germany (10.2%), China (6.1%), UK (4.7%), Italy (4.6%), France (3.8%), Korea (2.2%), Taiwan (2.0%), India (1.5%), and others (12.9%). It can be seen that US, Europe and Japan are the big users—in that order. However, with a proportion exceeding 10.3%, the Asia-Pacific region is not far behind. This indicates that this region is progressing beyond OEM into ODM.

Which industry-sectors are the major users of RP? The data reported in (Wohlers 2002) indicate the following pattern for 2001: consumer products (25.5%), motor vehicles (23.8%), business machines (9.6%), medical (10.1%), aerospace (8.6%), government/military (6.9%), academic institutions (6.7%), and other (8.7%). The high figures for consumer products, motor vehicles and business machines should not be surprising because these industries mainly compete through rapid designs. The fairly high figure for academic institutes is probably because a good proportion of RP-related research and development is being undertaken by academia and that RP is now being seen as a useful part of industrial, manufacturing, and mechanical engineering curricula. In particular, in Europe and Japan, educational infrastructure and computational environment have been recognized as key factors for broadening the use of RP in industry.

What is RP being used for? An idea can be obtained from the data illustrated in Figure 1.10. Note that the main use is for producing models for the purpose of design evaluation. The figure of 60.4% for this application is made up as follows: functional models (19.8%), visual aids for engineering (16.5%), fit/assembly (15.9%), visual aids for tool makers (4.5%), and models for ergonomic studies (3.7%). Interestingly, RP usage for creating models for the evaluation of function, fit, and assembly has surpassed that for producing visual aids. Indeed, RP has come a long way from the days when it could produce mere look-alikes of products.

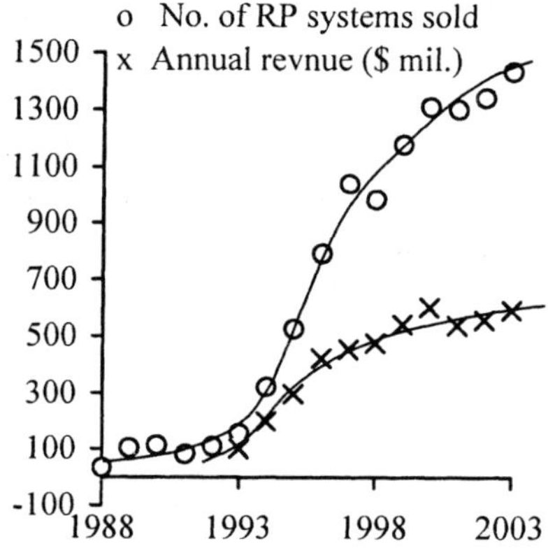

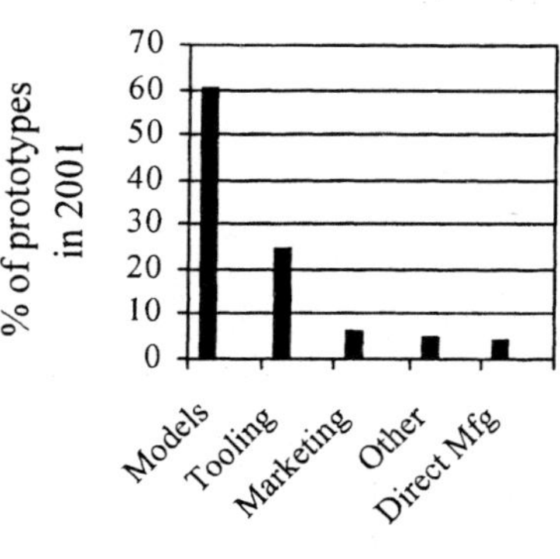

Figure 1-9. Growth of RP industry (Wohlers 2002).

Figure 1-10. RP usage in 2001 (Wohlers 2002).

The next major use of RP is for tooling (24.8%): patterns for prototype tooling (13.6%), patterns for cast metal (13.6%), and tooling components (4.0%). It appears that rapid tooling (RT) has not yet progressed beyond soft or bridge tooling into hard tooling capable of supporting mass production. This conclusion is reinforced by the fact that only 3.8% of RP usage is aimed at direct manufacture. This indicates that the dream of RM is still a long way off. A less expected observation is that RP is also being used for marketing purposes (proposals: 5.0%, and quotes: 3.7%).

Chapter 2

MATERIALS BASICS

Much of the progress in RP processes such as laser stereolithography, selective l aser s intering, f used d eposition m odeling a nd 3 D p rinting h as occurred owing to the development of new materials. Likewise, the progress from mere 'look alike' to functional parts has required the development of appropriate materials. This chapter provides a basic introduction to classes of materials that are of particular importance in gaining a deeper understanding of the concepts covered in subsequent chapters. In particular, ceramic, polymeric, powdered, and composite materials are covered in view of their wide use in RP. Although of equal importance, a discussion of metals is not included in the expectation that engineering students and professionals would already be quite familiar with them.

2.1 ATOMIC STRUCTURE AND BONDING

Molecules are the building blocks of any material. Molecules can be broken down further only to the level of atoms without changing the chemical properties of the substance. A molecule may comprise atoms of the same type (belonging to the same element) or of different types. The way the atoms within and between molecules are bonded (held together) can significantly influence the chemical (affinity for oxidation, etc.), physical (density, melting temperature, thermal conductivity, electrical conductivity, opaqueness to light, etc.) and mechanical properties (strength, hardness, brittleness, toughness, ductility, fatigue resistance, shrinkage, etc.) of the material.

An atom of any element has a nucleus made up of electrically neutral neutrons and positively charged protons of approximately the same mass (1.67×10^{-27}). In the periodic table, elements are arranged in the order of number of protons in the element's nucleus. This number is called the atomic number. The nucleus occupies less than one trillionth of the atom's volume but still accounts for over 99.9% of its mass. Much of the rest of the atom's space is empty at any time except for the momentary space taken by the atom's electrons.

Irrespective of the nature of the element in question, an electron carries a negative charge equal to 1.602×10^{-19} coulomb whereas a proton carries a positive charge of equal magnitude. An atom in its natural state is electrically neutral, so the numbers of electrons and protons are equal. Given the presence of opposing charges in close vicinity, one would expect the electrons to collapse into the nucleus. To explain why this does not happen, initially electrons were viewed as particles 'orbiting' around the nucleus in a hierarchically organized series of shells. However, with a rest mass of 9.109×10^{-31} kg an electron is 1836 times lighter than a proton. At this scale, matter exhibits a dual nature, i.e., it can be viewed as a particle as well as a wave. The particle view is simpler to visualize but cannot explain all aspects of electron behavior. A more complete and rigorous understanding requires the wave view consistent with modern concepts of quantum mechanics.

There are only a finite number of elements in nature (so far, only 114 have been discovered) and these few exhibit certain regularities with respect to their chemical properties and spectral characteristics. The periodic table captures many of these regularities. Thus, ordered discreteness seems to be the nature of matter. Modern quantum mechanics has arrived at a set of rules consistent with this finite and ordered nature of matter.

According to modern quantum theory, the orbiting electrons may be viewed as a series of stationary (standing) waves surrounding the nucleus. At any given location and time, the absolute value of the square of the amplitude of an electron-wave is a measure of the probability that the electron will be found there. The state of an atom can be modeled through four discrete quantum numbers *(QN)*. Each of these numbers is governed by certain rules. The more important of these are outlined below.

The first is the principal *QN*, *n* (=1,2,3...7). These represent the average distance of the electron from the nucleus and, hence, the principal energy state analogous to the shell in the particle view. Larger the value of *n*, higher is the electron's energy level. Shell *n* can accommodate a maximum of $2n^2$ electrons. Thus the first shell (*n*=1) can have no more than 2 electrons, the second (if it exists) 8, and so on.

The second is the azymuthal *QN* (or, orbital *QN*), *l* characterizing the electron's angular momentum (the product of the mass, velocity and radius of the orbiting electron) Orbitals (*l*-values) can be viewed as energy sublevels of associated with a shell. Within a shell, they can take up values 0 to *n*-1. Chemists use letters in place of numbers for *l* (s, p, d, f, g...). Thus, an $n=3$, $l=1$ electron would be denoted "3p". The number of sublevel types associated with a shell is equal to the shell number. Thus, shell 1 has only the s type of sublevel associated with it, shell 2 has s p, shell 3 has s p d, and so on. Each orbital has a limited capacity for accommodating electrons: s can take up to two electrons, p up to six, d up to ten, up to fourteen, and so on. While all electrons within a sublevel have the same energy, there is an increase in energy up the sequence s p d f within a shell. The energy level of an orbital is approximately equal to the sum of n and l. This means that orbitals may be ordered according to their $n+l$ values. They are filled lowest energy level first. If the $n+l$ values are equal, those with lower n-values are filled first.

The third is the magnetic QN, m. This number describes the spatial orientation of the 'orbiting' electron and can take on values from $-l$ to $+l$. In other words, for a given l, there are $2l+1$ values of m_l. The magnitude of this number does not affect electron energy

The final number is the spin QN, m_s, which is a fundamental property of all elementary particles including electrons. Electrons can be viewed as spinning toy tops. They may spin clockwise (up-spin) or counter-clockwise (down-spin). The rotation of the electric charge causes a local magnetic field. Spin is the angular momentum present even when the particles are not orbiting. The spin of subatomic particles (including electrons) can only be expressed in a discrete fashion as multiples of $h/(2\pi)$ where h is Planck's constant ($= 6.6\times10^{-34}$ Js). A neutron, proton, or electron can have a positive ('up') or negative ('down') spin of magnitude equal to ½.

The manner in which electrons are organized in an atom in its ground state (i.e., in the unexcited state) may be understood by referring to three well-known principles. The first is the Pauli exclusion principle that states that no two electrons in an atom can have identical set of quantum numbers. This means that there can only be zero, one, or two electrons with the same combination of n, l and m_l. Alternatively, the number states of an electron with a specific combination of n and l is equal to the number of possible m_l values, i.e. equal to $2l+1$. Thus, there can only be one type of s-orbital ($l=0$), 3 types of p-orbitals ($l=1$), five types d-orbitals ($l=2$), and so on. The principle also implies that two electrons that fill any particular orbital must have opposite spins. If an orbital contains only one electron, the electron's magnetic field is not cancelled out and it can be attracted to other external magnetic fields. Atoms having at least one unpaired electron are paramagnetic and can be attracted to magnetic fields. Atoms with no unpaired electrons are not magnetic.

The second is the Aufbrau principle that states that each electron will assume the lowest possible energy first, i.e., higher levels can be filled only after the lower levels have been filled. Following this rule, it can be shown that electrons will fill orbitals according to the following sequence: $1s^2\ 2s^2$ $2p^6\ 3s^2\ 3p^6\ 4s^2\ 3d^{10}\ 4p^6\ 5s^2\ 4d^{10}\ 5p^6\ 6s^2\ 4f^{14}\ 5d^{10}\ 6p^6\ 7s^2\ 5f^{14}\ 6d^{10}\ 7p^6$. The number in superscript for each orbital is the number of electrons needed to completely fill it. The seemingly odd behavior in some sections of this sequence arises from the fact that the energy associated with certain orbitals in some higher shells can be smaller than that associated with certain orbitals in a lower shell. Given the above filling sequence, one can easily identify the orbital structure of an element from its atomic number. For instance, given the fact that carbon has an atomic number of 6, we can determine its electron configuration to be $1s^2\ 2s^2\ 2p^2$. Likewise, the configuration of Fe (atomic number 26) is determined as $1s^2\ 2s^2\ 2p^6\ 3s^2\ 3p^6\ 4s^2\ 3d^6$.

The electron configurations of He (2), Ne (10), Ar (18), Kr (36), and Xe (54) are particularly interesting because all their shells (including the outermost shell, i.e., the shell with the largest principal QN) are full. It is known that these elements show little tendency to react with other elements—hence they are said to be 'noble'. The atomic structures of these elements are stable and the electrons in these structures are said to be core electrons. With the exception of hydrogen, it is conventional to represent the structures of other elements in a compressed form as '[preceding noble gas in terms of atomic number] followed by the rest of the structure'. Following this convention, C and Fe have configurations [He] $2s^2\ 2p^2$ and [Ar] $4s^2\ 3d^6$ respectively.

It must be noted that a major exception to the filling sequence occurs within the transition elements of the periodic table. With large sets of orbitals like the d orbitals there is a greater stability associated with half filled, d^5, and totally filled, d^{10}, orbitals. This is the reason for Cr and Cu showing the seemingly anomalous configurations [Ar] $4s^1\ 3d^5$ and [Ar] $4s^1$ $3d^{10}$ respectively. Further such exceptions occur when the atomic configurations contain 4f and 5d orbitals that are close enough in energy.

The electrons beyond core electrons are called valence electrons. Thus, carbon with atomic number equal to 6 has two core electrons corresponding to the structure $1s^2$ of the preceding noble gas, He (atomic number 2), and 4 valence electrons corresponding to the rest of the structure, $2s^2\ 2p^2$. On the same basis, Fe has 6 valence electrons. The number of valence electrons of an element is called the valence of the element. Thus, carbon is a tetravalent (valence = 4) element.

The valence of an atom determines its ability to gain, lose, or share electrons, which in turn determines its chemical and electrical properties of the atom. An atom that is lacking only one or two electrons from its outer shell will easily gain an electron to complete its shell, but a much larger amount of energy is required to free any of its electrons. An atom having a relatively small number of electrons in its outer shell in comparison to the number of electrons required to fill the shell will easily lose or share these valence electrons. The valence shell always refers to the outermost shell. However, it must be noted that the notion of single-valued valence does not work in a reliable way in the case of transition metals.

Attractive forces can arise between atoms when they get close enough. A chemical bond is said to have formed if the forces are large enough to keep the atoms together. The formation bonds usually results in energy being given out so that the resulting compound is at a lower energy level. Conversely, the bond can be broken only by inputting energy through heating, passage of electric current, etc. The energy required to break a bond is called the bond energy, or bond enthalpy and is usually specified as kJ/mol where a mol is the number of bonds equal to Avagadro number, 6.02×10^{23}.

Atomic bonds can be classified into two broad types depending on their strength: primary and secondary. With both types, atoms tend to have full valence shells. Different types of such bonds can be formed depending on whether the atoms forming the bond are transferring or sharing electrons. Bonding results in molecules that have properties completely different from those of the parent atoms. Bonds can exist within as well as between molecules. Primary bonds are significantly stronger than secondary bonds.

There are three types of primary bonds: ionic, covalent, and metallic. Depending on the substance, these three types of bonds can occur exclusively or in some degree of combination Broadly speaking, bonding turns progressively from ionic to covalent to metallic as we progress up the sequence from ionic ceramics (mostly ionic), complex salts, ionic glasses, H-bonded polymers, polymers, diamond (purely ionic), graphite, liquid crystals, doped semiconductors, transition metals, alloys, to alkali metals.

Bonding between a metal and nonmetal usually occurs through transfer of electrons between bonding atoms. To illustrate this notion, consider the bonding between Na, a metal with atomic number 11 and electron configuration) and Cl (a nonmetal). The atomic number of Na is 11, so its electron configuration is [Ne] $3s^1$. The outermost shell (shell 3) has 1 electron whereas its capacity is 8 ($=2 \times 3^2 - 2 \times 2^2 - 2 \times 1^2$) electrons. Hence, to complete its 3rd shell, an Na atom needs to either gain 7 electrons or lose 1 electron. Doing the former is energy intensive, so giving up 1 electron to a recipient atom will be preferred. Now, Cl has an atomic number of 17 so that its configuration is [Ne] $3s^2 3p5$. Clearly, this atom would like to receive 1 electron. Given this match, one Na atom will readily react with one Cl atom to form an NaCl (common salt) molecule. While bonding, Na will lose an electron, so it becomes a positively charged atom (a cation, i.e., a charged particle, or ion with tendency to be attracted to a cathode). Likewise, Cl becomes a negatively charged ion (anion). The resulting electrostatic attraction between the oppositely charged ions, results in ionic bonding.

The energy needed to pull one electron off the outer shell of an atom is usually of the order of a few electron volts (eV). It becomes progressively harder to remove further electrons. When an electron is pulled off, the positively charged nucleus pulls the remaining electrons more tightly closer so that the resulting cation is smaller than the neutral atom. In contrast, the anion resulting from the receipt of an electron swells considerably owing to the fact all the core electrons as well experience less attraction to the nucleus.

Compounds formed through ionic bonding usually are solids with crystalline structures whose nature depends on the sizes and shapes of the participating ions. Lattice (or, bond) energy typically ranges from 600 to 1,000kJ/mol, so ionic compounds have high melting and boiling temperatures. They dissolve easily in water and the solutions are electrically conductive, i.e., the ions can move about in the solution. Likewise, molten ionic compounds are electrically conductive.

Atomic bonding through sharing of electrons is called covalent bonding. Viewed as a particle, an electron in the outer shell of an atom spends most of its time at locations far from the atom's nucleus. While out there, the orbit can sometimes overlap a similar orbit of a similar or different atom. If that orbit is not full, the electron can penetrate deep into the second atom. An electron thus shared between two atoms will have a lower energy level. Two electrons (from disparate atoms) doing the same leads to even lower energy and the modified orbit can accommodate a spin pair. This, however, is a simplistic view and the reality in many cases (hybrid orbits) can be more complex. Whatever the nature, the essence of covalent bonding is in the sharing of electrons.

Covalent bonding can occur between like as well as unlike atoms. For instance, two Cl atoms can covalently bond to form a chlorine molecule, Cl_2. Likewise, two H atom can share their two lone electrons with an oxygen atom to form a water molecule, H_2O. Sometimes, covalent bonding can lead to giant molecules as in diamond where each tetravalent C atom can form single covalent bonds with 4 neighboring C atoms.

Each pair of shared electrons, one from each contributing atom, represents one bond between the atoms. Atoms can share multiple bonds. For instance, a carbon atom can achieve stability mimicking noble gases through sharing of its unpaired atoms with specific other atoms. Since carbon is tetravalent, it can share only one, two, or three electrons (single, double, or triple bonds). For instance, a molecule of ethylene consists of two carbon atoms each linked to two hydrogen atoms through single bonds and to the other carbon atom through a double bond (see Figure 2-1a).

In simple molecules, covalent bonds are quite strong (bond energy usually ranges from 100 to 500kJ/mol) but highly directional. The directionality results in the material assuming a spatial network that is not closely packed. Because of the high bond strength, the material's strength, hardness, brittleness, electrical resistance and melting point tend to be higher whereas thermal expansion tends to be lower. However, many covalent compounds are liquids or gases at room temperature since the bonds between different molecules are relatively weak. Many covalent bonds (except those forming hydrogen bonds) are insoluble in water. Since no ions are present, solutions of covalent compounds do not conduct electricity. Covalently bonded giant molecules are directional.

In many metals, r oughly o ne o uter e lectron p er a tom g ets d elocalized and becomes capable of roaming freely about the metal (diffuse electrons) so that we have, in effect, an array of positive ions attracted to a sea of electrons. The i on lattice provides a s pecific shape to m etallically b onded molecules, so they are crystalline with high melting and boiling points. The material is electrically conductive because the delocalized ions move about easily when an electric potential is applied.

Three types of secondary bonds are of particular importance. The weakest (2 to 8kJ/mol) of these are bonds formed due to Van der Waals forces as in polyethylene. In fact, these forces are always present because even neutral (uncharged) molecules are usually electric dipoles, i.e., pairs of equal and opposite electric charges with noncoincident centers. These dipoles tend to align with each other and induce further polarization in neighboring molecules so that a net attractive force is always present.

The next level of secondary bond is the dipole bond that arises because the sharing of electrons in a covalent bond results in the atoms losing electrons appearing to be positively charged and *vice versa*. Thus, a permanent dipole is set up. Such polar molecules containing Cl, F, or O valences make stronger bonds (6 to 13kJ/mol) as in PVC (polyvynil chloride).

The final level of secondary bond in terms of strength (13 to 30kJ/mol) is the hydrogen bond that is a special case of dipole bond occurring between hydrogen and F, O, or N atoms. An example of a polymer with hydrogen bonding is Nylon 66.

2.2 CERAMICS

The word ceramics is rooted in the Greek word *keramos* meaning potter's clay. Ceramics date back to 4000B.C. when they were used to produce pottery and bricks. Today, different types of metallic and nonmetallic elements are combined to produce a great variety of monocrystalline and polycrystalline ceramics.

Both natural and manufactured ceramics are of industrial importance. For instance, SiO_2 is abundant in nature. Like carbon, silicon is tetravalent and forms a spatial network of hexagonal form when combined with O. This hexagonal form, quartz, may be used as a force transducer. Likewise, SiO_4 is tetrahedral, so it can join into sheets to result in asbestos. Examples of manufactured ceramics of industrial importance are SiC that is useful as an abrasive or heating element, Al_2O_3 that is used in grinding wheels, and calcium silicate/aluminate that is useful as hydraulic cement

Primary bonding between atoms in a ceramic can be at once covalent or ionic (dual bonding). For instance, the degree of covalent bonding of MgO is 0.25 (compare to carbon for which the degree is 1). Sometimes different constituents of a molecule can exhibit different bonding mechanisms. For instance, in CaSO4 (gypsum), Ca^{2+} is ionically bonded to SO_4^{2-}, while the S within SO_4 is covalently bonded owing to the lower difference in electronegativity (affinity of atoms to attract electrons) between the atoms being bonded. Thus, while the degree of covalent bonding in C is 1, those in SiC, Si_3N_4, SiO_2, and MgO are >0.9, 0.7, 0.5, and 0.25 respectively.

Secondary bonding is important in ceramics forming platelets. While the bonding of atoms within platelets, the bonding between platelets is secondary. Secondary bonds are influenced by the addition of gas or liquid molecules, so they can be controlled to facilitate processing.

Crystals are low energy atomic structures. The structures assumed by the same material can change (polymorphism) as temperature and pressure are varied. Such changes in ceramics can lead to volume changes that are more significant than in metals, thus leading to part fracture. Crystal structures may change through distortion or breaking of existing bonds (displacive or reconstructive bonds respectively). When a molten ceramic is cooled at a sufficiently fast rate, there is no time for the crystals to form so that the end product is amorphous. Such ceramics are said to be in the vitreous state, or just glassy. However, a glass can get devitrified if it is held at high enough temperature for a long time.

Covalently bonded ceramics are not closely packed, so they can accommodate higher vibrational amplitudes without exhibiting dimensional expansion. Hence, ceramics have low thermal conductivity. In fact, some ceramics such as $LiAlSi_2O_6$ exhibit zero thermal expansion. With regard to electrical conductivity, ceramics range form insulators to semiconductors to, even, good conductors. Some ceramics are piezoelectric, i.e., they generate a potential difference in a plane normal to mechanical loading. The effect is reversible, so a dimensional change occurs when a potential difference is applied. Ceramics can also exhibit a wide range of magnetic and optical properties. Single crystals of ionically bonded nonporous and isotropic ceramics can be transparent whereas covalently bonded ones may range from transparent to opaque. Through the use of appropriate additives, certain ceramics can be made to absorb selected segments of light spectrum. Thus, ceramics can come in different colors. The diffraction index can be controlled likewise. Ceramics can also exhibit phosphorescence, i.e., emit light of a characteristic wavelength when stimulated by an electron beam or electrical discharge. Some lasers use single crystal ceramic rods. Many ceramics do not chemically react easily with gases, liquids and, even, high-temperature melts, so they can be used to form furnace-linings, turbine blades, etc.

Ceramics usually contain many micro-cracks, so they are brittle and can only be used in compression loading. Since the cracks are contained by grain boundaries, fine-grained ceramics are tougher. Thus, sub-micron particle size is often adopted for high-technology applications. Toughness may also be improved by incorporating fine fibers of graphite or silicon carbide. Ceramics remain strong even when they approach temperatures close to the melting point. However, they may experience creep in high temperature states. The strength properties of ceramics may also be improved by imparting compressive residual stresses.

2.3 POLYMERS

2.3.1 Nature of Polymers

The word 'polymer' was first used in 1866. It has two Greek roots: *poly* (many) and *mers* (parts). Polymers consist of macromolecules formed through repeated linking of mers.

Polymers belong to the general class of plastics (from the Greek word *plastikos*), i.e., they can be molded and shaped relatively easily. The earliest polymers w ere m ade o f n atural m aterials s uch a s c ellulose. S ubsequently, other types of polymers were produced by subjecting natural polymeric materials to certain chemical reactions. For instance, cellulose was modified into cellulose acetate to make photographic film (celluloid). The earliest man-made (synthetic) polymer was phenol-formaldehyde developed by Baekaland in 1906 under the trade name of Bakelite. However, the development of commercial synthetic polymers began only in the 1920s when it had become possible to extract the raw materials needed from coal and petroleum products.

All polymers are organic in nature, i.e., they contain carbon as an essential element Most polymer molecules also contain hydrogen atoms. In addition, they may contain atoms of Cl, F, O, or N and 'pendant groups' such as CH_3. Organic compounds with no double or triple bonds are called saturated compounds.

A polymer consists of repeated units called monomers (from Greek *mono* "one" and *meros* "part"). A monomer is a simple molecule of a compound of relatively low molecular weight and consisting of simple unrepeated structural units, but capable of reaction to form a polymer. An example of a monomer is ethylene (C_2H_4) that contains two carbon atoms joined together by double bond and each carbon atom is connected to two hydrogen atoms through single bonds. Other examples of monomers include styrene, methyl methacrylate, vinyl acetate, butyl acrylate, butadiene, acrylonitrile.

Most polymers have hydrocarbons (such as ethylene) as their basic building blocks. Any of a class of organic compounds composed only of carbon and hydrogen is a hydrocarbon. Two types of hydrocarbons are of particular importance: aliphatic and aromatic. Aliphatic hydrocarbons have their carbon atoms arranged in linear or branched chains and may be saturated (alkanes or paraffins—Figure 2-1a) or unsaturated (alkenes or olefins). Aromatic compounds exhibit delocalization of electrons in a ring system containing multiple conjugated double bonds (as in benzene shown in Figure 2-1b).

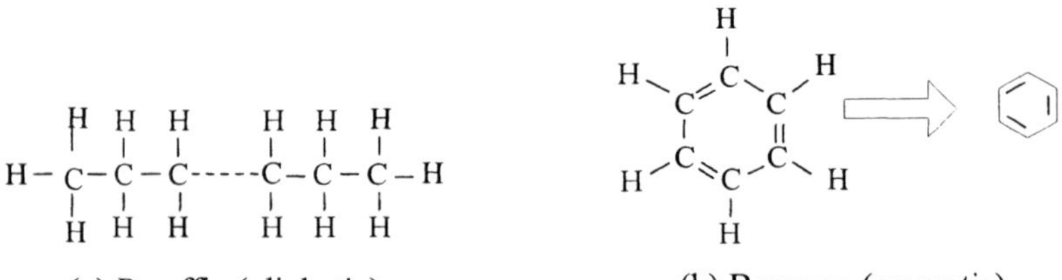

(a) Paraffin (aliphatic)   (b) Benzene (aromatic)

Figure 2-1. Types of Hydrocarbons.

A carbon double bond contains four electrons of which only two are sufficient to hold the two carbon atoms together in what is called a sigma bond. The other two are called pi electrons and are free to bond with initiator or monomer molecules. A radical is a group of atoms that can exist only in combination with other molecules to form a free radical, i.e., a chemical component containing a free electron that forms a covalent bond with an electron on another molecule.

Not all single bonds need occur with H atoms. They may form with 'pendant groups' such as CH_3 in the case of polypropylene (PP). Some hydrocarbons are aromatic, i.e., they contain benzene rings (e.g., polystyrene). Polymerization may also proceed with non-hydrocarbon molecules such as those containing N, O, S, P, or Si while H may be replaced by monovalent Cl (e.g., polyvinylchloride), F as (e.g., polytetrafluoroethylene, or Br.

The ordering (*taktika* in Greek) of pendant groups can significantly influence the properties of the resulting polymer. If all the groups are on one side of the carbon backbone, the polymer is said to be isotactic (Figure 2-2a). Syndiotactic polymers have the groups have groups alternate on both sides of the backbone (Figure 2-2b). Atactic polymers have randomly arranged groups (Figure 2-2c). Some benzene-based polymers can have backbones of double strand (Figure 2-2d). Such ladder polymers are usually highly temperature resistant because two bonds instead of one must be broken before degenerating into products of lower molecular weight.

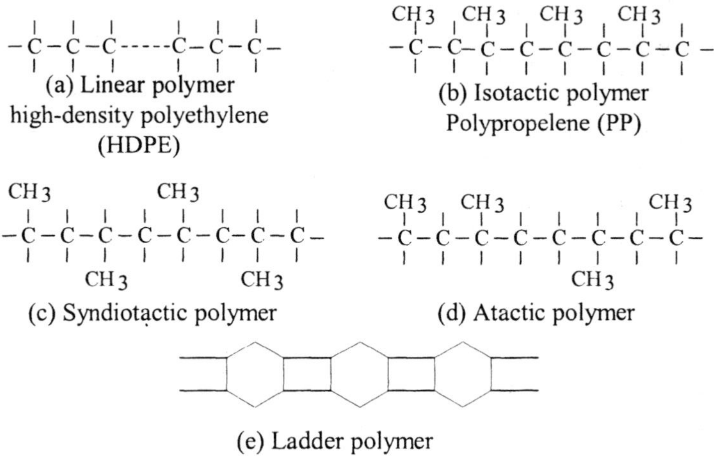

(a) Linear polymer
high-density polyethylene
(HDPE)

(b) Isotactic polymer
Polypropelene (PP)

(c) Syndiotactic polymer

(d) Atactic polymer

(e) Ladder polymer

Figure 2-2. Some classes of polymer structures.

All the polymers discussed so far have backbones made up of a single type of mer, i.e., they are monomers. These are analogous to pure metals. However, just as two or more metals can be combined to form an alloy, it is possible to have polymers with two (copolymers) or three (terpolymers) mers. For instance, ethylene-propylene is a copolymer whereas ABS (acrylonitrile butadiene styrene) is a ternary polymer.

The repeating units in a copolymer or terpolymer may be linked in different w ays. F or i nstance, a c opolymer o f m ers A a nd B c ould e xhibit random (BBBABABABBA), alternating (ABABAB), or block (AAAAABBBAAABBBBBBAA) sequences. Some times B-chains can branch off from A-chains producing a graft polymer.

Not all polymer pairs can form joint chains. Such polymers are said to be incompatible. However, it might be possible to produce polymer alloys (or, polymer blends) by mixing incompatible polymers with one of the polymer serving as the matrix.

2.3.2 Free Radical Polymerization

Polymerization is the process of forming high molecular mass molecules from simple organic molecules called monomers. There exist several basic types of polymerization. Among these, chain reaction (addition) polymerization and cationic polymerization are particular interest in RP.

Chain reaction polymerization occurs in three steps and involves two chemical entities. The first entity is a monomer or an oligomer that, in most cases, has at least one carbon-carbon double bond, e.g., ethylene (2-3a). An initiator is a catalyst capable of forming a free radical (Figure 2-3b).

Figure 2-3c illustrates free radical formation occurring when a common initiator called benzoyl peroxide is heated. Note that two free radicals have been formed from the one molecule of benzoyl peroxide. During initiation, the first step in the polymerization process, the free radical catalyst reacts with one of the monomer molecules and breaks apart the carbon double bond. As a result, the monomer molecule bonds to the free radical, and the free electron is transferred to the outside carbon atom (Figure 2-3c). The initiation rate is usually taken to be given by $k_i[\text{M}]\text{R}^\bullet$ where k_i is the initiation rate constant, M is the concentration of monomer, and $\text{R}^\bullet$ is that of the free radical.

The next step in the process is propagation of the chain through opening up of successive monomer molecules such that the free electron is progressively passed down the line of the chain to the outside carbon atom (Figure 2-3d). The chain reaction can take place continuously because sum of the energies of the polymer is less than the sum of the energies of the individual monomers. In other words, the single bonds in the polymeric chain are more stable than the double bonds of the monomer. The rate of propagation is usually given by $k_p[\text{RM}^\bullet][\text{M}]$ where k_p is the propagation constant. When the reactive monomer is difunctional (i.e., it has two different types of reactive chemical groups), the polymerized m aterial can become a superviscous solid. If the monomer molecules have three or more reactive chemical groups, significant cross-linking by strong covalent bonds may result. When the reaction is sustained long enough, the resulting molecular weight can be quite high.

The final phase is 'termination'. Three types of termination are possible: coupling (or, 'recombination'), disproportionation, and occlusion. Recombination involves joining of two radicals to form a non-reactive molecule. As illustrated in Figure 2-3e, the free radical (R-O) that was left over from the original splitting of the organic peroxide can link with the last CH_2 component of the polymer chain. Alternatively, termination can occur when two unfinished chains bond together. Disproportionation occurs when β-hydrogen gets transferred from one radical to another thus resulting in a pair of polymer molecules, one with a saturated end and the other with an unsaturated end. Occlusion or frozen mobility of the radicals can also terminate the chain reaction. The rate of termination is usually expressed as $k_t[\text{Pol*}]^2$ where k_t is the termination constant and $[\text{Pol}^*]$ is the concentration of the polymer. Further, k_t can often be expressed as $k_t = k_{tc} + k_{td}$ where are the components due to combination and disporoprtionation respectively. The entire reaction is highly exothermic. When the monomer is ethylene, the product of the chain reaction is polyethylene and the reaction can take place in under 0.1 second.

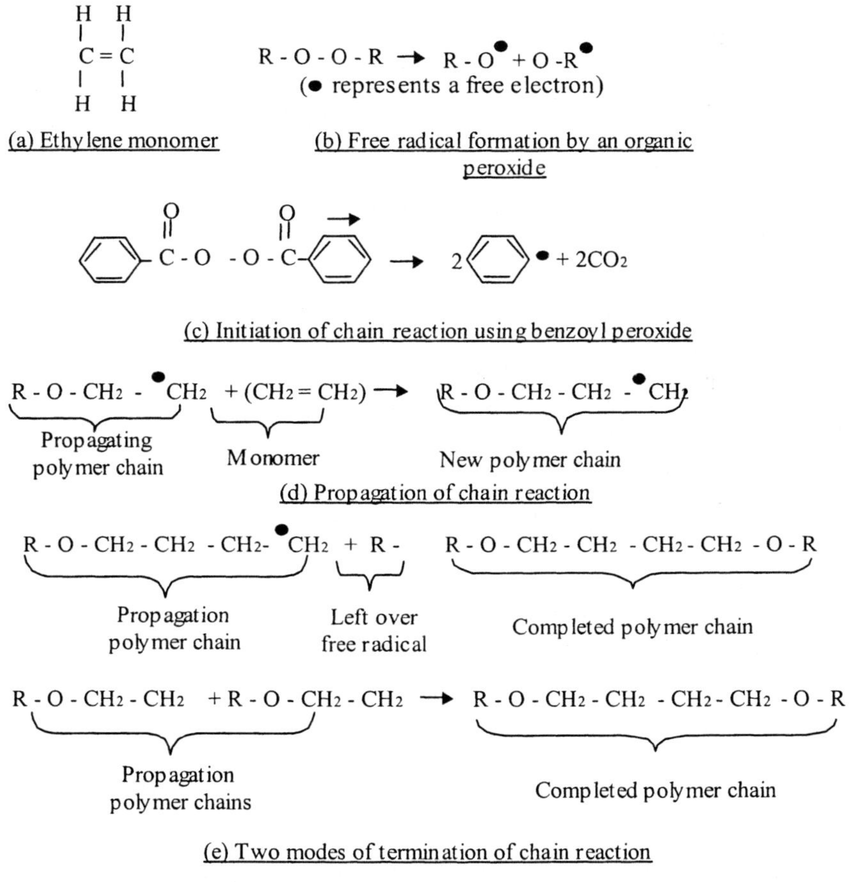

(a) Ethylene monomer

(b) Free radical formation by an organic peroxide

(c) Initiation of chain reaction using benzoyl peroxide

(d) Propagation of chain reaction

(e) Two modes of termination of chain reaction

Figure 2-3: Chain reaction (addition) polymerization.

Under appropriate conditions, it is possible to open up the multiple-bond between the carbon atoms of a monomer's backbone with the aid of an initiator such as heat, pressure, ultraviolet (UV) energy input, or a catalyst. Of these, the UV approach that is most prevalent in RP. There are several other hydrocarbons containing double or triple bonds and hence capable of producing similar chain reactions.

2.3.3 Cationic Polymerization

Cationic polymerization is of more recent origin (Pappas 1985). The first patent related to the idea was issued in 1965 (Lacari 1965). The first commercial applications based on the principle were produced in the late 1970s (Crivello and Lam 1976a, Crivello *et al.* 1977, Crivello and Lam 1979).

In cationic vinyl polymerization too, the pi electrons of the carbon-carbon double form the new linkages. However, instead of splitting up as in free radical polymerization, the pi electrons move together to form the new bond. The process is called cationic polymerization because the initiator is a cation (an ion with a positive electrical charge) shown as 'Init+' in Figure 2-4a. The cation attracts a pair of electrons from the carbon-carbon double bond thus leaving the carbon-carbon double bond to form a single bond with the initiator. Thus electronic donating groups (shown as 'R' in 2-4a) are needed. These stabilize the propagating species by resonance. In the process, as a result of the loss of electrons, one of the former double bond carbons becomes a cation by carrying a positive charge. This new cation reacts with a second monomer molecule in the same manner as the initiator had reacted with the first monomer molecule. This process is repeated and the molecular weight of the polymer increases progressively. Termination of cationic polymerization occurs mainly through the formation of impurities such as water and other compounds containing hydroxyls. In contrast to radical polymerization, the presence of oxygen does not inhibit cationic polymerization. Figure 2-4b illustrates the propagation phase with respect to vinyl ether.

The simplest initiator of cationic polymerization is a proton from a strong acid such as H_2SO_4 (sulfuric acid), CF_3SO_3H (triflic acid), or $HClO_4$ (perchloric acid). However, in laser-induced RP, we need photochemical acid generating compounds. In this context, HCl and similar acids do not work because the counterion (Cl^- for HCl) reacts with the initial cation. The result is not polymerization but addition of the Markovnikov type. This problem may be solved by a Lewis acid since it can generate cations *in situ*. A Lewis acid has molecules comprising less than an octet of electrons in the valence shell. Hence it can act as an electron-pair acceptor. Many metal cations (e.g., Fe^{3+}) can act as Lewis acids. Compounds with multiple bonds can also behave as Lewis acids (i.e. they can accept pairs of electrons and form new bonds. For instance, the carbon atom in a molecule of CO_2 is bonded to the two oxygen atoms through a pair of double bonds.

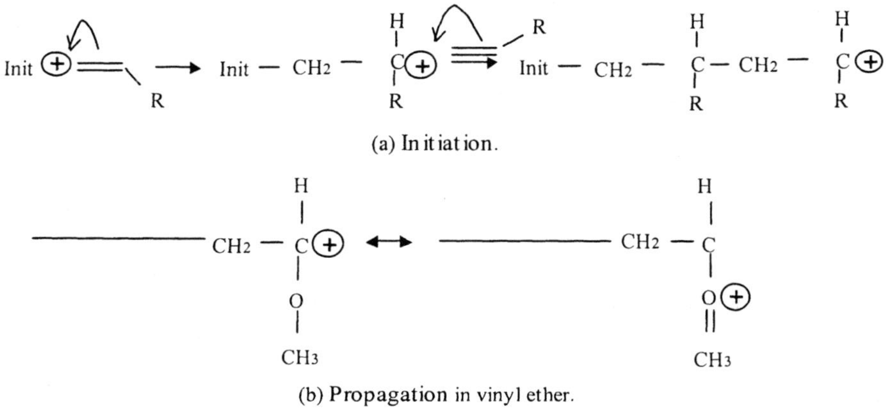

(a) Initiation.

(b) Propagation in vinyl ether.

Figure 2-4. Cationic polymerization.

Cation propagation is usually quite rapid. However, it is quite difficult to grow long chains using cationic polymerization because of the incidence of competing chain transfer reactions. After some time, the currently active chain(s) stop growing, and others start so that the monomer being consumed gets divided up among many chains. As a result, the achievable molecular weight can be low.

Just as radical polymerization, cationic polymerization is an exothermic process. It is also accompanied with volumetric shrinkage. However, the exothermic reactions resulting from cationic polymerization continue for quite some time after the energy source activating the reaction (the laser beam in many RP processes) has been turned off.

2.3.4 Thermoplastic and Thermosetting Polymers

When primary covalent bonds occur within polymer molecules, the bond energy is of the order of 350 to 830kJ/mol, so the molecule itself does not break easily. However, the strength of a macromolecule is not determined mainly by the primary bonds but by the secondary bonds present between pairs of molecules. Since secondary bonds are much weaker, they can be easily broken and reformed during deformation.

It is possible to control the average length of the macromolecules by terminating the chain reaction at the appropriate time. Thus, different polymers formed from the same monomers can have a different molecular weight (the number of grams equal to Avagadro's number) or a different degree of polymerization (the number of mers in the average molecule).

The number of secondary bonds and hence the strength of a polymer increases with increasing chain length. At the same time, if thermal excitation is high, secondary bonds are easily broken and reformed, i.e., there can be relative motion between molecules. However, when the thermal excitation is removed (i.e., the heated polymer is cooled), the polymer practically returns to its original hardness and strength. Polymers exhibiting such reversible behavior are called thermoplastics. Thermoplastics are extensively used in industry and are shaped using processes such as injection molding. Such plastics are also used in rapid prototyping processes such as laser stereolithography. However, very long-chain plastics such as UHMWPE (ultrahigh molecular weight polyethylene) and PTFE may char before reaching the moldable state.

In some polymers, covalent bonds (primary bonds) can also exist between molecules. Such links are called crosslinks. Crosslinking was accidentally discovered first by Charles Goodyear in 1839 when he managed to vulcanize rubber with the aid of sulfur. The degree of crosslinking can vary in a wide range, from branching as in low density polyethylene (LDPE) to networked structures as in vulcanized rubber. Crosslinking interferes with the relative movement between molecules. Hence, greater the degree of crosslinking greater is the resistance to deformation and stress cracking. Crosslinking also imparts hardness, strength, brittleness, stiffness, and dimensional stability to the material.

When the long-chain molecules in a polymer are crosslinked in a spatial network, the structure has in effect become a giant molecule with its shape permanently set. Such polymers are called thermosetting polymers or, simply, thermosets. In contrast to thermoplastics, the curing reaction (crosslinking) in thermosets is irreversible.

Already formed thermoplastics may be cured by subjecting them to high-energy radiation such as UV radiation, electron beams (EB), gamma-rays, x-rays, and particle beams. However, the material may degrade when the radiation is high.

Within the realm of rapid prototyping, selective laser sintering (SLS) and fused deposition modeling (FDM) utilize either linear or branched thermoplastic materials. In contrast, stereolithography (SL) utilizes low molecular weight liquids capable of exhibiting thermosetting behavior under ultraviolet radiation.

2.3.5 Polymer Structures

That a macromolecule is linear does not necessarily mean that it is straight. For instance, as is typical of covalent bonding, the C-C bond in polyethylene has a spacing of 0.154nm at a fixed bond angle of 109.5°. Hence, the spacing in a straight line is approximately 0.126m $\{=0.154\sin(109.5°/2)\}$. Thus, a macromolecule of ultrahigh molecular weight polyethylene (UHMWPE) macromolecule consisting of 150,000 mers will be 18,900nm long when fully stretched. However, the macromolecule often becomes coiled in practice so that the end-to-end distance is much smaller. Such coiled macromolecules will not be able to exhibit any long-range order, i.e., they will be amorphous (having the appearance of spaghetti). Amorphous polymers do not exhibit a sharp melting temperature, T_m. All polymers exhibit an amorphous structure when they a re h eated t o some h igh t emperature but not s o h igh a s to b reak t he primary bonds. The state is, however, not rigid and fixed. Owing to the elevated temperature, the polymer molecules are in constant motion and the fee volume is large.

What happens when a polymer heated to amorphous state is cooled depends on the polymer type. Some long-range order may develop if the polymer molecules exhibit chemical regularity along the chain and the conditions are right. However, in contrast to metals, crystallization in polymers occurs only in certain regions. Crystalline regions are called crystallites. A crystallite may look like a spherulite, crystal or *shish-kebab*. The specific volume in the ordered regions will drop, i.e., the polymer becomes denser (Figure 2-5). As crystallinity increases, the polymer will become harder, stiffer, less ductile, less rubbery and more chemically resistant. More crystalline polymers have more boundaries between crystalline and amorphous regions so that the polymer becomes more opaque (less transparent). Partly crystalline polymers exhibit a distinct melting temperature, T_m. A greater degree of crystallinity is usually achieved if the polymer has no side branches and pendant groups. Isostactic forms crystallize more readily than atactic forms. Copolymers and random or graft polymers often find it difficult to crystallize. Slower cooling promotes crystallization by providing the time to create the folds needed.

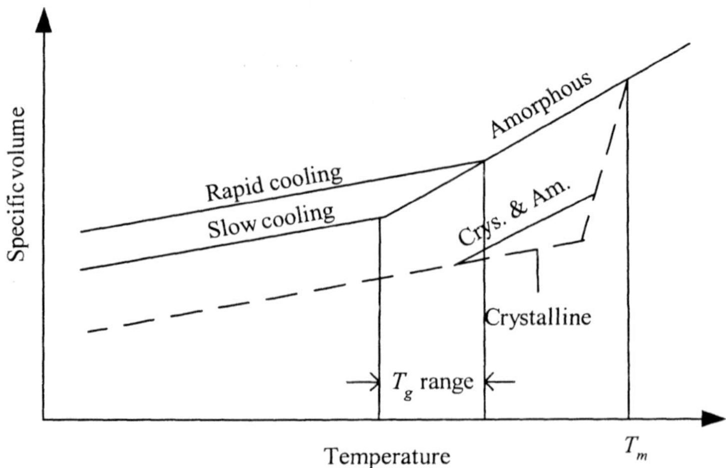

Figure 2-5. The Effect of temperature on polymer structure.

If the conditions are unfavorable for the development of some long-range order, the amorphous state will continue. Specific volume will drop at a rate typical of molten state. The relative motion between molecules drops as temperature falls. At some typical temperature the relative motion may be completely arrested. This temperature is called the glass-transition temperature, T_g, of the polymer. The reference to glass is because glass also exhibits such a temperature behavior. Below T_g, there usually is a sharp drop in the sensitivity of specific volume to temperature. T_g may be determined by noting the temperature at which this sharp drop occurs. The sharp drop is the reason for semicrystalline polymers exhibiting 4-8% shrinkage relative to that at the freezing point, T_m. This shrinkage adversely affects the accuracy of parts produced by RP processes such as stereolithography, selective laser sintering, and fused deposition modeling.

2.3.6 Properties of Polymers

Polymers are viscoelastic materials, so their properties are time and temperature d ependent. H ence t he t esting r ate a ffects t he r esults. P olymer chain alignment can be influenced by shear flow. Oriented polymers exhibit higher stiffness values than unoriented polymers. Polymer failure may occur through ductile as well as brittle fracture. The nature of failure depends on whether primary or secondary bonds are broken.

The molecular weight of a polymer is simply the molecular weight of the repeat unit times the number of repeat units in the chain. However, polymers are normally made up of macromolecules exhibiting the same structure but different chain lengths, i.e., they are polydisperse. Hence, one needs to determine an average molecular weight. While several types of average can be calculated (Billimeyer 1962), two are most commonly used. The number average molecular mass, M_n, is obtained by weighting each molar mass by the number of molecules present in that molar mass: $M_n = \sum(M_i N_i)/\sum(N_i)$. Likewise, the weight average molecular mass is $M_w = \sum(M_i M_i N_i)/\sum(M_i N_i)$. The average MW decreases in proportion to increasing ratio of the concentration of the chain transfer agent to that of the monomer (Vail 1994). Thus, the magnitude of M_n can be controlled between 5000 (below which they are called oligomers) and infinity through appropriate chemical synthesis. For instance, processes such as stereolithography start with low MW material and build the MW through the chemical reactions arising from exposure to UV light.

Products with widely different mechanical and rheological (flow) properties can be produced by controlling M_n (see Figure 2-6). For example, whereas 5000 MW cis-polybutadiene is an easily poured viscous liquid at room temperature, the same material is nonpourable and rubbery when the M_n is raised to 2 million. Finally, note that the ratio M_w/M_n (called the polydispersion index, Pdi) characterizes the degree of polydispersion.

Several properties (e.g., T_g, modulus, and tensile strength) of polymers follow a peculiar pattern with increasing molecular weight. Small molecules have small values. As the molecules grow to intermediate size (oligomers), the properties rise sharply (Figure 2-7). Finally, the properties level off as the chains become long enough to be full polymers. In contrast, a few properties like melt viscosity and solution viscosity, increase monotonically with MW. Hence, the goal of polymer synthesis should not be to make the largest possible molecules, but, rather, to make molecules large enough to get onto the plateau region. However, properties such as color, dielectric constant, and refractive index are not changed by MW because they are dictated by the repeat units alone.

The rheological properties of liquid polymers are of particular importance in many manufacturing processes using polymers. For instance, in stereolithography, the effectiveness of the recoating cycle is strongly influenced by the viscosity of the liquid resin. Polymer melts exhibit much higher viscosities (in the range 100-10,000poise) than that of water (0.1poise). Further, unlike water, polymers exhibit non-Newtonian behavior by exhibiting decreasing viscosity with increasing flow rate (increasing shear rate). However, at very low shear rates, the viscosity remains essentially constant. It has been found that this apparent viscosity plateau, η_0, is proportional to $(M_w)^{3.4}$ (Graessley 1974). This is why the material appears to be less viscous and much harder as polymerization proceeds.

Polymer viscosity is also sensitive to temperature and, in the case of some polymers, can even become twice as large following a temperature reduction of just 15-25°C (Tadmore and Gogos 1979).

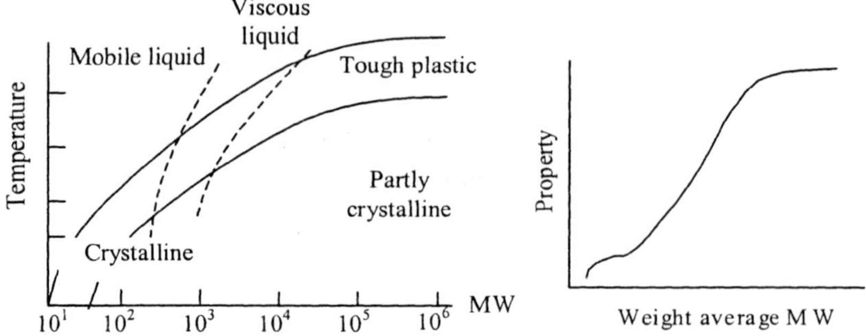

Figure 2-6. Effect of MW on polymer behavior (Billimeyer 1962).   *Figure 2-7.* Influence of MW on polymer properties.

The glass transition temperature, T_g, value of a copolymer consisting of two components with mass fractions, M_1 and M_2, and having glass transition temperatures, T_{g1} and T_{g2} respectively, can be estimated by means of Gordon-Taylor equation (Olabisi *et al.* 1979):

$$T_g = M_1 T_{g1} + M_2 T_{g2} \tag{2.1}$$

The magnitude of T_g of a polymer or copolymer usually decreases with average molecular weight as (Wu 1981)

$$T_g = T_{g\infty} - \frac{K'}{M_n} = T_{g\infty} - \frac{K''}{M_w} \tag{2.2}$$

where $T_{g\infty}$ is the glass-transition temperature at infinite molecular weight, and K' and K'' are constants.

Polymers are viscoeleastic materials, i.e., they simultaneously exhibit the viscous properties of fluids and the elastic properties of solids. When a viscoelastic material is deformed to a fixed value and released, the retractive forces within the material diminish over time owing to volume change. This stress relaxation can be modeled as a Maxwell fluid in which the elastic response is characterized by a spring with shear modulus, G, connected in series with a 'dashpot' containing viscous fluid with viscosity, η_0. The response of the dashpot under external load is

$$\tau = \eta_0\, \dot{\vartheta} \tag{2.3}$$

where τ is the shear stress and $\dot{\vartheta}$ is the strain rate, so that the system exhibits a time constant:

$$\zeta = \eta_0 / G \tag{2.4}$$

Thus, the time dependent relaxation modulus can be expressed as

$$G(t) = G(0)(-t / \zeta) \tag{2.5}$$

When such a viscoeleastic material is heated to temperature T, viscosity decreases according an Arhenius-type equation

$$\eta_0(T) = A \exp\left(\frac{\Delta E}{RT}\right) \tag{2.6}$$

where A is a prefactor, ΔE is the activation energy of viscous flow, and R is the universal gas constant (McKelvey 1962). Since ζ is proportional to η_0, it follows that the same value of $G(t)$ can be obtained by adjusting time, t, or temperature, T. This is known as the time-temperature superposition principle (Ferry 1961). The principle enables one to model material behavior over longer times and different temperatures on the basis of data gathered over shorter times.

Many practical materials, however, exhibit a more complicated behavior that cannot be explained by a single time constant ζ. Such materials can be modeled in terms of n_M parallel Maxwell units (Tobolsky 1960) so that

$$G(t) = \sum_{i=1}^{n_M} (t - t / \zeta_i) = \int_{-\infty}^{+\infty} G(\zeta) \exp(-t / \zeta) d\zeta \tag{2.7}$$

where the various time constants can be determined by fitting of experimental data.

The specific heat capacity of a solid or powder can be measured by a variety of methods. One common method is Differential Scanning Calorimetry (DSC) (Anon. 1976). This technique enables one to study what happens to a polymer when it is heated with a view to achieving thermal curing. For instance, it can detect the melting or glass transition.

A DSC apparatus consists of a pair of identically positioned platforms in good thermal contact placed in a furnace by a common heat flow path (Figure 2-8). One of the pans, called the reference pan, is kept empty or contains a standard material of known heat capacity such as sapphire (Ginnings and Furukawa 1953). The two pans are placed on a metal disc to ensure good thermal contact. Two independent thermocouples measure the temperatures of the two pans. There is a computer controlled cylindrical heating element surrounding the pans. The system is capable of heating both the pans at an identical rate, usually something like $10^{\circ}C$ per minute. However, since one pan is empty and the other holds the polymeric sample, the heat capacities of the two pans are different, so heat must flow between the two pans. Hence the temperatures of the two pans can be different at a particular moment. This temperature difference is easily calculated from the pair of thermocouple readings. Procedures exist for calculating the differential heat flow for a given sample mass. The computer monitoring and controlling the apparatus makes these calculations and outputs a plot of sample temperature versus differential heat flow into the sample is prepared. The plots are used to study thermally induced transformations in the sample material. For instance, if the sample is a polymer, one can expect to see a dip in the plot when the polymer has reached the glass transition temperature, T_g, a peak when an amorphous region has turned crystalline (temperature T_c), and a dip again upon reaching the melting temperature, T_m. The areas under the dips or peaks can be used to calculate the total heat involved in the corresponding transitions. The amount of crystallization associated with an amorphous to crystalline transition can be estimated from the total heat calculated corresponding to the T_c-peak.

DSC is useful mainly for measuring the amount of heat released or absorbed in a reaction during thermal cycling. However, while investigating reactions associated with curing by means of ultraviolet light or lasers (as in stereolithography), one needs to use differential scanning photo-calorimetry (DSP). The technique is essentially similar to DSC.

Another technique useful in studying thermal or UV curing of polymers is Raman spectroscopy. The technique exploits Raman scattering (or Raman effect) discovered in 1928 by C.V. Raman. When electromagnetic radiation is absorbed (scattered) by a molecule or by a crystal, one photon of the incident radiation is annihilated and, at the same time, one photon of the scattered radiation is created. There exist two possibilities concerning the relative values of energy of the incident and scattered photons. If the energy of the incident photon is equal to that of the scattered one, the process is called Rayleigh scattering. Otherwise, it is called *Raman scattering*. If the substance being studied is illuminated by monochromatic light of certain frequency, the spectrum of the scattered light consists of a strong line (the exciting line) of the same frequency as the incident illumination together with several weaker lines on either side. The lines of frequency less than the exciting lines are called Stokes lines, the others anti-Stokes lines. The spectral lines differing in frequency from the exciting line are called Raman lines. Raman lines appearing close to the exciting line are correlated with changes in the rotational energy states of the molecules without changes in the vibrational energy states. Bands of lines farther from the exciting line are associated with simultaneous changes in the vibrational and rotational energy states. However, interestingly, the differences between the frequencies of the Raman lines and the exciting line (called frequency shifts) are independent of the frequency of the exciting line. Usually, these frequency shifts are studied by plotting the delta wave-numbers (frequency/speed of light) on the x-axis and intensity (particle counts per second) on the y-axis.

Further details concerning polymer synthesis and polymer properties can be obtained from (Billimeyer 1962, Tadmore and Gogos 1979, Allcock and Lampe 1990, and Stevens 1990).

2.3.7 Degradation of Polymers

Polymers can degrade owing to oxidation, random chain scission and depolymerization owing to exposure to heat and radiation. Most degradation mechanisms entail a loss in molecular weight and the associated properties. In most situations, polymer degradation is undesirable. However, in some processes such as selective laser sintering, the polymer binders are deliberately made to degrade in order to get rid of them from the final product. In either case, it is important to understand the degradation characteristics of the polymer in use.

Albeit slow, oxidation of polymers can occur even at apparently low temperatures (<100°C) (Beaman *et al.* 1997). In most cases, ambient oxygen reacts with hydrogen in the polymer to form hydroperoxides (-OOH) that decompose producing free radicals. The result is a reduction in the molecular weight. The oxidation rate is proportional to the concentration of easily abstractable hydrogen in the polymer whereas the reaction propagation is more rapid than initiation and termination occurs through bimolecular reactions leading to nonradical products (Shalaby 1979). Generally, polymers with aliphatic backbones tend to oxidize more readily than those with aromatic backbone

Thermal depolymerization, sometimes called 'unzipping', resulting in a monomer can occur in some polymers such as poly(methyl methacylate) and poly(a-methyl styrene).Random chain scission occurs in almost every polymer to some degree. It is usually higher at higher temperatures and when subjected to radiation.

2.4 POWDERED MATERIALS

2.4.1 Types of Powders

Many conventional manufacturing processes use powders of metals, ceramics, polymers, etc. for producing parts. In the early 1900s, tungsten filaments used in incandescent light bulbs were made from tungsten powder using powder metallurgy. Powder metallurgy refers to manufacturing processes involving blending (mixing) the desired powders with a lubricant/binder, compacting (or, consolidating) the paste to form green parts with the help of suitable dies, and sintering the green parts. The green parts may exhibit some porosity but are usually strong enough to permit gentle handling. Sintering involves heating of the green part without total melting (there can be localized melting).

Mixtures of pure metals and alloys may be subjected to powder metallurgy. Aluminum, copper, iron, tin, nickel, titanium, and refractory metals are among the metals used. Prealloyed powders are needed for producing brass, bronze, steel and stainless steel parts.

Ceramic powders may be made through comminution (milling), reaction in the vapor phase, spray drying and granulation, freeze drying (sublimating from a w ater s olution), e tc. M etal p owders a re p roduced b y c omminution processes such as roll crushing, ball milling, vibratory ball milling, attrition milling, rod milling, hammer milling, impact milling, and fluid energy milling. Comminution is continued until the desired particle size is achieved. Over-comminution of the metal to be powdered on its own can result in agglomeration (particle clustering) owing to Van der Waals forces. Agglomeration c an b e p revented t hrough t he u se o f a ppropriate additives. For instance, in dry milling, lubricant-type compounds such as stearic acid may b e u sed. I n w et m illing, c hemicals i mparting a s urface charge t o t he particles may be used. Using such technologies, particle sizes from $0.1 \mu m$ to $1000 \mu m$ can be realized. Much finer metal particles may be produced through atomization where molten metal s tream is p roduced by forcing it through a small orifice and then interrupting the stream by jets of inert gas, air, or water. Other methods of producing powders include reduction of oxides using hydrogen, electrolytic deposition, use of appropriate chemical reactions, p recipitation from a c hemical s olution, v apor c ondensation, a nd reacting volatile halides with liquid metals. Powders may be blended by placing them in a rotating cylindrical, cubic, double-cone, or twin-shell container.

2.4.2 Compaction and Sintering of Powders

Compaction aims to obtain the required shape, density, particle contact conditions, and strength in the green part. The density of the green part increases with the compacting pressure and approaches the theoretical density of the material in bulk at high pressures. If all particles are of uniform size, the void volume will be high and the density low. Density will be higher when smaller particles are introduced to fill up the voids. Higher density indicates lower porosity and hence leads to a stronger green part. Local density can vary across the green part owing to friction conditions between particles and between the powder and the die.

Compacted green parts may form in the presence of water vapor simply through the tendency of liquid to condense in negative curvature regions (Figure 2-9). The vapor phase forms a neck between adjacent particles through material transport. The particles are bonded while keeping the center distance between them constant. Hence no shrinkage occurs. One may also apply pressure to mechanically interlock the particles. The required compaction pressure for metal powders may range between 70 to 800MPa. Compaction may be carried out using conventional pressures (typically, 1.8 to 2.7MN), cold isostatic pressing (CIP), or hot isostatic pressing (HIP). CIP involves placing the powder in a flexible mold and subjecting the mold to hydrostatic pressure in a water chamber. HIP involves a similar process but the pressurizing medium (100 to 300MPa) is a hot (500 to 2000°C) inert gas or a glass-like fluid. HIP is gaining popularity in the context of some rapid prototyping processes. Because of the absence of die-wall friction, isostatic pressing produces more uniformly dense parts irrespective of part shape.

Sintering involves prolonged heating of the green parts in a neutral (nitrogen, vacuum, etc.) to reducing (hydrogen) environment to obtain the desired final part properties. Sintering temperatures range from $0.7T_m$ to $0.9T_m$. Sintering times range from a minimum of 10 minutes for iron and copper alloys to 8 hours or more for tungsten and tantalum. Sintering furnaces typically consist of three chambers, one for volatilize lubricants in the green compact, one with high-temperature for sintering, and one for controlled cooling.

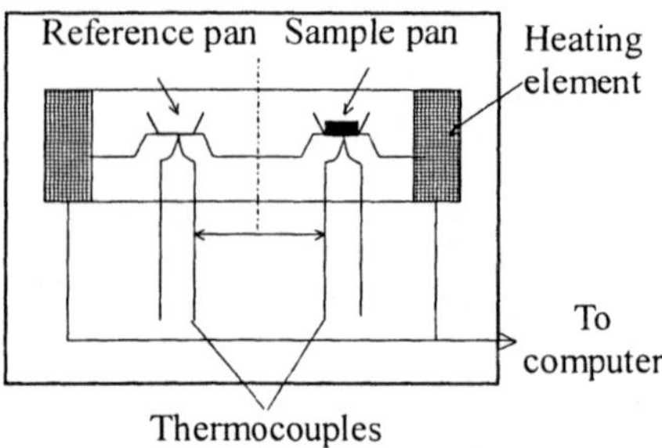

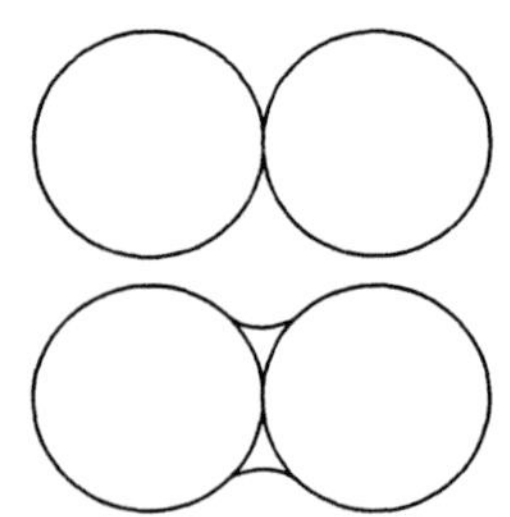

Figure 2-8.
Differential Scanning Calorimetry.

Figure 2-9.
Vapor phase neck formation.

Sintering can occur through several mechanisms: solid phase sintering, liquid phase sintering, and reaction bonding. The first two are of particular importance in RP. During solid phase sintering, the green body is converted into a solid owing to heating. As temperature increases, two adjacent particles begin to form necks due to diffusion. The process is thermally activated and maintained by diffusion. Void surfaces tend to coalesce in order to decrease total surface energy. As a result, this form of sintering is associated with volume shrinkage, hence to densification. At the same time the strength and ductility increase. The driving force behind sintering can be characterized by the ratio of the surface energy, E_s, to the volume, V_p, of the particles. Assuming that the particles can be approximated as spheres of radius, r

$$E_s / V_p = \gamma(4\pi r^2)/(4\pi r^3/3) = 3\gamma / r \tag{2.8}$$

where γ is the surface tension. This equation suggests that the driving force for sintering increases with decreasing particle size. A typical ceramic has surface energy of the order of $1 Jm^{-2}$. Thus the driving force for a 1mm ceramic powder is approximately equal to $3 Jm^{-3}$.

In liquid phase sintering, the adjacent particles are of different materials with one particle of significantly lower melting temperature than the other. In that case, the lower melting point particle can melt at the sintering temperature a nd, b ecause of s urface t ension w et a nd s urround the p article that has not melted.

The phenomenon of wetting can be characterized by the contact angle, θ_c, formed between the liquid and solid surfaces (Figure 2-10). The angle is determined by the following equilibrium equation

$$\gamma_{SV} = \gamma_{SL} + \gamma_{LV}\cos\theta_c \qquad (2.9)$$

where γ_{SV}, γ_{SL}, and γ_{LV} are the forces along the solid-vapor, solid-liquid, and liquid-vapor interfaces respectively. Clearly, lower the value of θ_c, greater is the wetting effect.

Since a wetting liquid occupies the lowest energy position, it leaves the larger pores and flows to the smaller capillaries exhibiting the highest energy per unit volume. An example is cobalt that melts during sintering of tungsten carbide (WC) cutting tool inserts. When there is insufficient liquid to fill the pores, it pulls the particles together to minimize the energy. Thus, the solid particles are drawn together by the action of viscous flow under capillary pressure. This is called rearrangement densification (Figure 2-11). This is usually followed by solution precipitation owing to grain growth by migrating atoms followed by final densification. However, if wetting is inadequate, the parts can swell owing to possible exclusion of liquid from the surface pores. Presence of oxide contaminants can decrease wetting.

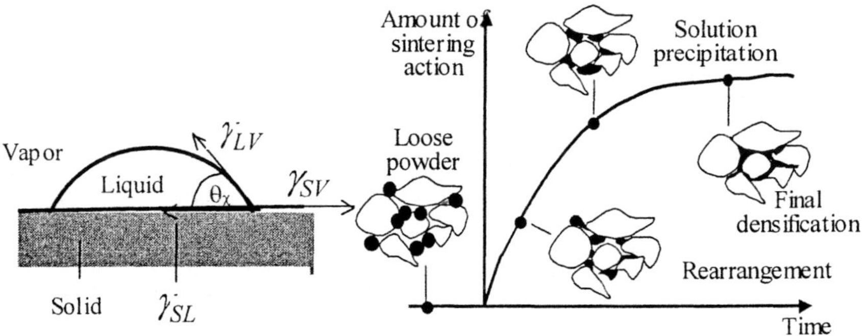

Figure 2-10. Wetting.   *Figure 2-11.* Stages in liquid phase sintering (Van der Schueren 1995).

Liquid phase sintering can be aided by coating the particles with materials (glass, silicates, etc.) that promote diffusion. A properly selected and controlled coating could melt and wet the higher melting point constituents. Subsequently, the liquid phase cools down to become a glassy (amorphous) or crystalline substance. The overall shrinkage (hence, overall increase in density) can be a few percentage points although part porosity remains essentially unaffected. Shrinkage during sintering is higher with more lightly compacted green parts.

2.5 COMPOSITES

Composites refer to structures made of two or more distinct starting materials so that the composite structure has significantly improved mechanical properties when compared to those of the individual starting materials. The oldest example of composites is perhaps the addition of straw to clay for making bricks for structural use, dating back to 4000BC.

An important class of composites are reinforced plastics that consist of fibers are dispersed in a discontinuous manner within a continuous matrix of polymer. Reinforced plastics with more than one type of fiber are said to be hybrid. The commonly used fiber materials include glass, graphite, aramids, boron nylon, silicon nitride, silicon carbide, aluminum oxide, sapphire, steel, tungsten, molybdenum, boron carbide, boron nitride, and tantalum carbide. Typically, fiber materials exhibit high specific strength (strength-to-weight ratio) and high specific modulus (stiffness-to-weight ratio). However, they are generally abrasive and lack toughness (they are brittle). In contrast, the plastic matrix is less strong and less stiff but much tougher. Thus reinforced plastics combine the advantageous properties of the fiber and plastic phases.

Fibers used in general reinforced plastics are usually less than $10\mu m$ thick so that their cross-sections are so small that the probability of defects being present is low. As a result, the tensile strength and stiffness values of the fiber are very high. For example, a glass fiber can exhibit a tensile strength value significantly higher than that of even steel in bulk. When the aspect ratio (length to diameter ratio) of a fiber is less than 60, it is said to be short. Many long fibers usually have aspect ratios between 200 and 500.

The matrix material in a reinforced plastic serves three purposes. Firstly, it supports and transfers the stresses to the fibers so that the latter carry much of the externally applied load. Secondly, it protects the fibers from physical damage c aused b y e xternal e nvironment. F inally, b y v irtue o f i ts d uctility and toughness, it discourages the propagation of cracks through the composite. Matrix materials used in general reinforced plastics include epoxy, polyester, phenolic, fluorocarbon, polyether-sulfone, and silicon.

Among the important factors influencing the properties of a reinforced plastic are the kind, shape orientation of the fiber; mean fiber length, and volume fraction of the fibers. The properties depend upon whether the fibers are long or short, continuous or discontinuous, uniformly oriented along the loading direction or in a different direction, and uniformly or randomly oriented. Short fibers yield inferior results because they are not as good as long fibers at transmitting loads through the matrix. Another problem with short fibers is that it is difficult to achieve high volume fractions, V_f, with them. Weak bonding between the fiber and matrix can lead to fiber pullout and delamination. However, bonding can be improved by means of special surface treatments such as coatings. Fibers oriented in the direction of tensile stress are particularly effective. However, when all the fibers are unidirectional, the transverse strength can suffer.

Consider n ow the s trength a nd e lastic m odulus o f a reinforced p lastic with a single fiber phase laid out in the longitudinal direction. Clearly, the total load on the composite, P_c, will be shared by the fiber and the matrix so that

$$P_c = P_f + P_m \qquad (2.10)$$

where P_f and P_m are the loads on the fiber and matrix respectively.

We can rewrite the above equations in terms of the stresses (σ_c, σ_f, and σ_m) and areas (A_c, A_f, and A_m) corresponding to the composite, f iber, a nd matrix respectively as

$$\sigma_c A_c = \sigma_f A_f + \sigma_m A_m \qquad (2.11)$$

Now, if V_f and V_m are the volume fractions of the fiber and matrix respectively,

$$\sigma_c = V_f \sigma_f + V_m \sigma_m \qquad (2.12)$$

Assuming perfect bonding between the fiber and matrix, the longitudinal strains experienced by the fiber and matrix can be taken to be equal, i.e., $e_f = e_m = e$, so that

$$e = \frac{P_f}{A_f E_f} = \frac{P_m}{A_m E_m} = \frac{P_c}{A_c E_c} \qquad (2.13)$$

where E_f, E_m, and E_c are the elastic moduli of the fiber, matrix and composite respectively.

The above equation enables us to calculate the loads on the fiber and matrix. Likewise, we can arrive at the following 'rule of mixtures' for calculating the elastic modulus of the composite in terms of those of the fiber and matrix:

$$E_c = V_f E_f + V_m E_m \qquad (2.14)$$

This simple model has been found to be capable of predicting the variation of the longitudinal modulus quite well provided that the fibers are long. When the fibers are short but aligned in the load direction, equation 2.14 may be modified as follows

$$E_c = \chi_1 \chi_2 V_f E_f + V_m E_m \qquad (2.15)$$

where χ_1, is the fiber orientation efficiency factor (Hull 1981), and χ_2 is the fiber length correction factor. According to (Hull 1981), χ_1 is equal to 3/8 for in-plane uniformly distributed fiber orientations and 1/5 for three-dimensional uniformly distributed orientations. The magnitude of χ_2 may be estimated on the basis of Cox's model described below.

Cox's model (also known as the shear lag model) assumes complete elastic transfer of load between the short fiber and matrix and considers an ideal c ase o f s hort f ibers aligned in the s tress d irection a nd a rranged in a particular pattern, e.g., a square or hexagonal array. T he model yields the following expressions for χ_2 and the tensile stress, σ_f, at a distance x from the center of the fiber, and the composite elastic modulus, E_c, in the alignment direction (Cox 1952):

$$\chi_2 = \left\{ 1 - \frac{\tanh(n_c l_f / d_f)}{n_c l_f / d_f} \right\} \qquad (2.16)$$

$$\sigma_f(x) = E_f e_c \left\{ 1 - \frac{\cosh(2n_c x / d_f)}{\cosh(n_c l_f / d_f)} \right\} \qquad (2.17)$$

where e_c is the strain of the composite in the loading direction, l_f and d_f are the length and diameter of the fiber respectively, and

$$n_c = \sqrt{\frac{E_m}{(1 + v_m) E_f \sqrt{\pi / (4V_f)}}} \qquad (2.18)$$

where v_m is the matrix Poisson ratio.

Modeling the strength of a composite is complicated because the failure of a c omposite usually occurs via onset of fracture r ather t han a yielding mechanism. Composite strength models are usually presented in the following form:

$$\sigma_c = \chi_3 \chi_4 V_f \sigma_{fu} + V_m \sigma_m^* \qquad (2.19)$$

where σ_{fu} is the ultimate tensile strength of the fiber, σ_m^* is the tensile stress in the matrix at the composite failure strain, χ_3 is the fiber length correction factor, and χ_4 is the fiber orientation efficiency factor. Formulas for estimating χ_3 and χ_4 are available in (Piggott 1994) and (Kelly 1973) respectively.

Chapter 3

LASERS FOR RP

Laser, a device that controls the way energized atoms release photons, is a key element in many RP technologies. In fact, many RP technologies would not have been possible without the availability of appropriate lasers. Hence it is essential for RP practitioners and researchers to have a basic understanding of laser technology.

There are many books describing laser principles (Hecht 1994, Elliott 1995, Silfyast 1996, Svelto 1998, Hitz *et al.* 2001 and Parker 2003), laser interactions with materials (Duley 1996), and laser applications in materials processing (Duley 1996, Migliore 1996 and Steen 1997) and manufacturing (Crafer and Oakley 1992 and Migliore and Walker 1994). While advanced readers might prefer to peruse one or more of such books, less experienced readers should find it useful to have at hand a summary of the characteristics of lasers that are of importance to RP. This chapter aims to provide such a summary.

3.1 THE PRINCIPLE OF LASER

3.1.1 The Nature of Light

'Laser' is an abbreviation for Light Amplification by Stimulated Emission of Radiation. To appreciate the principle of laser, we need to understand each of the highlighted terms.

The first term, 'Light', refers to any electromagnetic radiation. In the early 19th century, light was described in terms of waves, but experiments later showed that it also exhibits properties of particles with momentum. A 'particle' made up of light energy is called a photon.

All types of electromagnetic radiation travel at the same speed in vacuum. However they can vary in frequency and wavelength. As a result, they interact with matter differently. Further, light is the basis for the sensation of sight, and for the perception of color. The eye distinguishes the color of an object as the 'color' (wavelength or frequency) of light that the object reflects or transmits.

From the viewpoint of wave theory, monochromatic light (light of single color) may be viewed as a sinusoidal function of time at a fixed point in space. The main characteristics of the wave in the time domain are the amplitude, A, and period, T (or, frequency $f = 1/T$). Alternatively, it may be viewed as a function of space at constant time. In this view, light wave is characterized by its amplitude and wavelength, $\lambda = vT = v/f = (c/n)T = (c/n)/f$.

By convention, the full spectrum of light is divided into certain ranges. Gamma rays have a wavelength of the order of 10^{-3}nm whereas x-rays are around 10^{-2}nm. The visible spectrum is between 0.4 and 0.7nm. As we proceed to higher wavelengths within this range, we encounter the sequence of colors violet, indigo, blue, green, yellow, orange and red. The spectrum between x-rays and the visible range is called the ultraviolet range (of particular interest to RP) whereas that immediately next to the visible range is called the infrared range. Proceeding beyond the infrared range, we have microwaves with wavelengths of the order of a centimeter and radio waves in the meter range.

Light travels through empty space at a speed, c, equal to about 3×10^8 m/s. When traveling though a medium (e.g., air or glass), light slows down to a lower velocity, v. The ratio, c/v, is called the refractive index, n, of the medium. When a wave enters a new medium at an angle of less than 90°, the change in speed occurs sooner on one side of the wave than on the other, causing the wave to bend, i.e., refract. The refractive index of most materials transparent in the visible spectrum is between 1.4 and 1.8, while those of materials transparent in the infrared range are in the range 2 to 4. The differential bending at different wavelengths is the reason for the dispersion of white light into a 'rainbow-type' visible spectrum when it enters a triangular glass prism with a suitable angle of incidence.

"Ordinary light" (such as that emitted by the sun or lamps) is composed of many different wavelengths, radiating in all directions, and there is no phase relation (no coherence) between the different waves out of the source. In contrast, as we will see, laser is monochromatic, highly directional (i.e., the cross-section of a laser beam does not expand significantly as it traverses space), and highly coherent (i.e., all the waves in beam are in phase). Much of the attraction of lasers lies in these three properties.

As noted earlier, light can also be modeled as packets of energy called photons. The relation between the amount of energy, *E*, carried by the photon, and its frequency, *f*, is determined by the formula (first given by Einstein) $E = hf$ where h is Planck's constant (= 6.626×10^{-34} Js). This implies that a photon of higher energy corresponds to a light wave of higher frequency (smaller wavelength). Thus, ultraviolet light is more energetic than visible light.

3.1.2 Emission Radiation

It was noted in Chapter 2 that, among the electrons in an atom in the ground state, those in the outermost shell are the most energetic. However, there exist several permissible but unoccupied levels with energies larger than that of these 'outermost ground state' electrons. The energy differences — $\Delta W_i = W_i - W_0$, $i = 1, 2, 3$, — are of particular interest in the present discussion.

Suppose that an atom in the ground state receives a photon with energy exactly equal to the energy difference, ΔW_i. As a result of such *absorption* of light energy, one of the outer electrons will be raised to energy level W_i. An atom with its energy content so raised is said to be excited. In accordance with the principle of minimum energy, after a certain period of time (typically, in the 10^{-8} second range), the excited electrons return to the normal lower energy state by emitting a photon of energy equal to ΔW_i. This process occurs without any external interaction. Hence it is called spontaneous emission of light. Note that the emitted photon will have exactly the same energy (hence the same wavelength) and phase as those of the previously absorbed photon. However, the emission occurs in a random direction.

So far, we have considered spontaneous emission radiation from a single excited atom. However, a medium consists of a vast number of atoms, i.e., it has a large population of atoms made up of collections at different energy levels. We need to consider how atoms at different excitation levels can remain in thermodynamic equilibrium within a medium. This equilibrium state is determined by the Boltzmann equation. This equation can be rearranged to yield the following relationship for the normal relative populations of atoms at different energy levels:

$$N_i / N_0 = \exp\left(\frac{\Delta W_i}{k\theta}\right) \tag{3.1}$$

where N_i is the average number of atoms (population number) at energy level W_i ($i=1,2,,$), k is Boltzmann constant (=1.38×10^{23} J/^{0}K), and θ is the absolute temperature in ^{0}K. In the present discussion, we are interested only in the relative magnitudes of N_i. Note that population numbers increase with increasing temperature. Further, the higher the energy level, the lower the population number, i.e., $N_0 > N_1 > N_2$ and so on in a normal population (Figure 3-1a).

Every energy level has a characteristic average lifetime (typically of the order of 10^{-8} seconds), which is the time after which only $1/e$ (about 37%) of the excited atoms still remain in the excited state. According to the quantum theory, the probability of transition from a higher energy level to a lower one is inversely proportional to the lifetime of the higher energy level. Further, the probability of different transitions is a characteristic of each transition that behaves in accordance with certain selection rules (these are not of significance to our present discussion). However, it is important to note that if the transition probability is low for a specific transition, the lifetime of this energy level is longer (about 10^{-3} sec). Atoms in such a state are said to be meta-stable. Most importantly, only a medium with sufficient atoms in the meta-stable state is a candidate for the lasing process.

3.1.3 Light Amplification by Stimulated Emission Radiation

Although Einstein had arrived at the principle of stimulated emission radiation as early as in 1917, it was not until the late 1940s that engineers began to utilize it for practical purposes. At that stage, the focus of many engineers was on microwave amplification via stimulated emission of radiation (MASER) so as to enable microwave communication. However, Townes and Prokhorov independently tried to create powerful beams of light using higher frequency energy. (The pioneering work of these two scientists led to them receiving Nobel prizes.) However, it was Maiman who solved the practical problems to invent the first ever laser in 1960. This laser used ruby as the lasing medium (optical cavity) stimulated by high-energy flashes of intense light.

Suppose that a medium has been previously excited (temporarily) such that N_u atoms are at an upper energy level W_u. Suppose further that a lower energy level, W_l, is available, to which de-excitation could occur. Suppose now that this excited medium receives light input with photon energy equal to $(W_u - W_l)$. Einstein discovered that, under such conditions, the incoming optical signal (photons) causes the atoms at the upper energy level, W_u, to oscillate (resonate), thus triggering a transition to the lower energy level, W_l. Upon interaction with an excited atom, the incident photon will pass unchanged. However, in the process, the resonance effect stimulates the atom to drop back to its lower ground state. As a result a second photon with energy (hence wavelength), direction, and phase identical to the incident photon would be emitted. We had supplied two photons, one for exciting the atom and the other to stimulate the atom, and we have been returned the two photons. Hence, at this stage, we have no amplification of light.

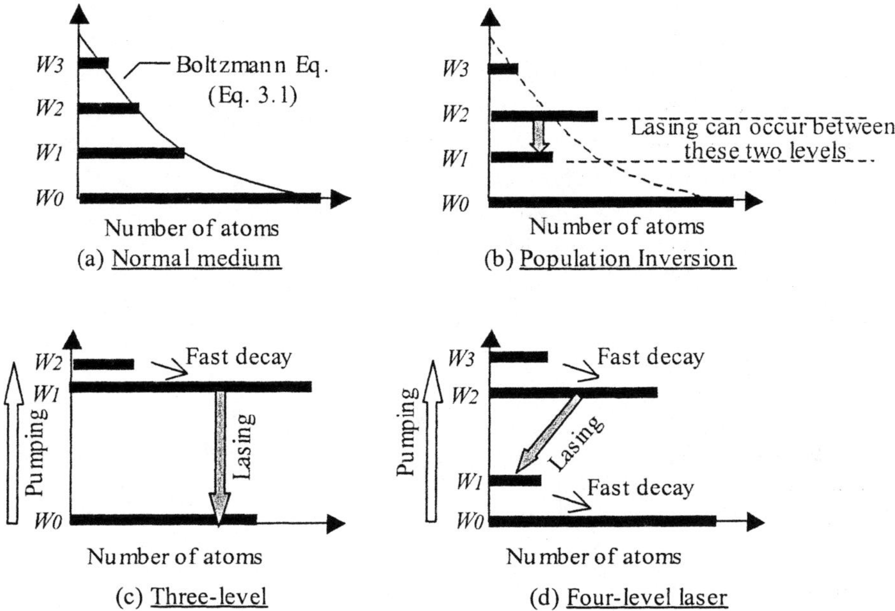

Figure 3-1. Numbers of atoms at different energy levels in a medium.

When conditions are right, the photon resulting from stimulated emission can resonate with another excited atom in its path, thus leading to a chain reaction. The photon density (number of photons per unit volume of medium), $\bar{N}(t)$, at time t is determined by the balance between the absorption and stimulation emission rates of photons and is given by the following relationship:

$$d\,\bar{N}(t)/dt = -K[N_u(t) - N_l(t)]\bar{N}(t) \tag{3.2}$$

where K is the relative strength of the response of the atom to the incoming radiation in the specific transition.

Thus, when the excitation state is normal (following equation 3.1), $N_u(t) > N_l(t)$, so photon density decays with increasing time. Thus, if a material is in a thermal equilibrium, only net absorption but no amplification can occur. However, if we can find a way of inverting the population, i.e., arrive at the condition $N_u(t) < N_l(t)$, the term $\{N_u(t)-N_l(t)\}$ becomes positive and we can attain light amplification. This is called population inversion where at least one of the higher energy levels has more atoms than a lower energy level (3-1b). To produce amplification, the material must be in a population inversion in which more atoms are pumped to an excited state as compared to a lower state.

One method of achieving population inversion is to use the three-level scheme shown in Figure 3-1c. In this scheme, lasing occurs between the first upper energy level, W_1, and the (lower) ground state, W_0. However, atoms are pumped (i.e., the required excitation energy is supplied) not to level 1 directly but to a higher level 2. The atoms at level 2 stay there for a very short average time, so they decay in a non-radiative manner to the meta-stable energy level 1. Since the lifetime of the meta-stable energy level (E_1) is relatively long (of the order of 10^{-3} seconds), many atoms remain at this level for sometime. When pumping is strong enough, there could be more than 50% of the atoms at energy level E_1. Thus, a transient state of population inversion is created where there are more atoms at level 1 than at the ground level so that light amplification by stimulated emission radiation (the lasing process) can occur. However, the need for high pumping limits the operation of a three level laser to pulsed operation.

One way of achieving continuous lasing is to adopt the four-level mechanism shown in Figure 3-1d. Here the upper lasing level is shifted upwards to W_3, so that an extra energy level, W_1, higher than the ground state, W_0, and with a very short lifetime can be introduced as the lower lasing level. Owing to the short lifetime, the atoms at level 1 get drained to the ground state quickly so that a strong population inversion exists between levels 2 and 1. This phenomenon facilitates continuous laser operation.

Light amplification follows Lambert's law,

$$\frac{I}{I_0} = \exp(-\alpha L) \tag{3.3}$$

where I is the output light intensity, I_0 is the input light intensity, α is the gain coefficient, and L is the length of lasing medium. It follows that light amplification increases with increasing length of lasing medium.

So far we have idealized the upper and lower energy levels inducing lasing as being discrete. However, in practice, transitions occur between bands consisting of closely spaced discrete energy levels. As a result, the plot of lasing frequency versus output light power will be a bell-shaped curve. Further wavelength broadening of output energy may arise due to the Heisenberg uncertainty principle, Doppler broadening and pressure (collisions) broadening. (Discussion of these broadening effects is beyond the scope of this book.) Similar effects are found with respect to fluorescence exhibited by the medium. Hence, the bell-shaped curve is called the fluorescence line. The fluorescence (hence, lasing) line width is the width of the fluorescent line at half its maximum height.

3.2 LASER SYSTEM

A laser system comprises four components: the active medium, the excitation mechanism, the feedback mechanism, and the output coupler (Figure 3-2). The active medium can be a solid, liquid or gas that can be made to experience population inversion and stimulated emission. Depending upon the type of laser, it can be a collection of atoms, molecules, ions, or semiconductors. The specific energy transitions occurring in the medium determine the possible wavelengths of the laser.

The excitation mechanism provides the energy needed to attain population inversion in the active medium. The excitation energy is usually supplied by means of flash lamps or another laser so that the electromagnetic radiation (photons) is absorbed in the active medium. When the medium is a gas, excitation may be achieved through an electrical discharge within the medium. Alternatively, the energy may be supplied through collisions among atoms of the gaseous medium. This is the standard excitation mechanism in commercial gas lasers, e.g., Helium-Neon laser and Carbon dioxide laser. The method depends on at least two gasses being placed inside the laser tube. One gas receives the energy from collisions with the accelerated free electrons. The second gas receives energy from collisions with the excited molecules of the first gas.

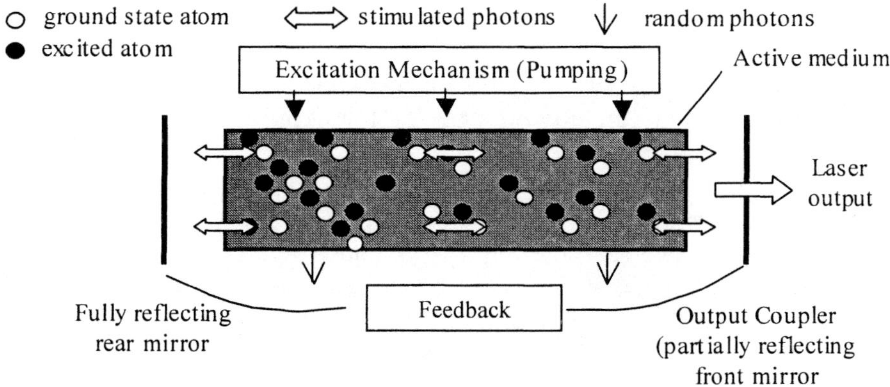

Figure 3-2. A typical laser system.

The feedback mechanism mainly consists of two mirrors each placed at one end of the length of the active medium. The intention is to let the light produced (following excitation, population inversion and stimulated emission) to bounce back and forth between the two mirrors. This process dramatically increases the effective length (see L in equation 3.3) traversed by the light beam within the medium, thus producing a corresponding increase in light amplification. The active medium confined between the two mirrors is called the optical cavity. The cavity essentially acts as what is known as Fabry-Perot Resonator. The line connecting the centers of the two mirrors is called the optical axis of the laser.

Three effects arise from the back and forth bouncing of the stimulated photons along the cavity. Firstly, as we have already noted, more stimulated emission occurs.

Secondly, any photons that do not travel along the optical axis are lost from the sides of the cavity. This feature establishes the directionality of the laser.

Finally, any photons that are not of the correct frequency are lost. The correct wavelength is determined by the principle that the optical length of the medium divided by light beam wavelength must be an integer. Note that the initial light beam has a wavelength spread (as determined by its fluorescent line width). Hence it will contain 'rays' of different wavelengths within the line width. While some of these might satisfy the principle of integral wavelength, others might not. When a 'ray' satisfying the principle hits a mirror, the incident and reflected rays will be in phase, thus promoting the conditions for the appearance of a standing wave. A ray that does not satisfy the principle would experience progressive phase shift upon each reflection. If the phase reaches $180°$, it will get annihilated. If it reaches $360°$, it will become part of the standing wave. In practice, there are many standing waves arising within the optical cavity, since there could be several wavelength satisfying the condition that the length of the optical cavity is exactly divisible by the wavelength. Of these, only those exhibiting optical gains greater than unity will survive. Each of the surviving standing waves is called a longitudinal mode of the laser.

Figure 3-2 shows an optical cavity bounded by parallel plane mirrors. The advantages of parallel plane cavity are that the active medium is fully utilized and there is no focal point within the medium causing electrical breakdown. However, owing to its high sensitivity to misalignment, the system is very difficult to operate. Hence, other mirror systems consisting of different combinations of plane, concave, or convex mirrors have been considered.

For instance, in a confocal cavity, both the end mirrors are spherical-concave with curvature radii larger than the length of the active medium.

With such an arrangement, the optical cavity is thinner in the middle than at the ends. The system is less prone to misalignment, exhibits fair utilization of the active medium, and exhibits low diffraction losses.

In most Helium-Neon lasers, the front mirror is plane and the rear mirror is spherical-concave with curvature radius equal to the length of the optical cavity. The result is an optical cavity that tapers to a point from the rear to the front. The plane mirror costs less and easily adjusted to correct misalignment.

The optical coupler enables amplified light to be drawn out from the optical cavity for external use. This is achieved by equipping the output-side end-mirror with a partial reflective coating, so there is optical feedback within the cavity while enabling some light to escape as the laser beam. The reflectivity ranges from 10% for pulsed lasers (more energy comes out in short bursts) to 90% for continuous lasers.

Figure 3-3 shows the output coupler for a semi circular optical cavity resulting from a plane rear mirror and a concave front mirror. The shape of the beam on the front is determined by the curvature of the mirror-side of the coupler. This shape is retained as the beam leaves the output coupler. Hence, The curvature of the mirror-side of the coupler is of particular importance. However, the output-side o f the end-mirror may be shaped l ike a convex lens so as to achieve a focused beam at the desired location.

In a continuous laser, laser power can be kept constant over time. In contrast, during each pulse produced by a pulsed laser in general, laser power exhibits a bell-shaped curve over time. Pulse duration (Δt) can vary from 10^{-14} second to a fraction of a second. The pulse can be single or repeated. It is possible to achieve pulse rates of over 10^9 pulses per second (PPS). In many material-processing applications, laser power is a very important parameter affecting the process. Hence, stability of power is important for maintaining quality.

3.3 LASER BEAM CHARACTERISTICS

Although lasers are expected to produce monochromatic light, there is a wavelength spread in practice owing to several effects including the effect of multiple longitudinal modes each producing a different standing wave within the optical cavity. In many applications, it is necessary to restrict the emitted spectrum to certain wavelengths. One way of achieving this is to pass the laser beam through a triangular prism so that spatially separated beams of constant wavelength emerge out of which we can select the desired one. Some lasers use a reflective grating to achieve a similar effect.

Each longitudinal mode is associated with a certain wavelength. Higher order modes exhibit smaller wavelengths. The gain (amplification) is a function of wavelength. Nearer the longitudinal mode to the center of the gain curve, greater is its amplification. Therefore, it is possible to choose a specific longitudinal mode by introducing an optical element that causes high l osses f or t he o ther longitudinal m odes i nside t he c avity. E talon is a device that exploits this principle.

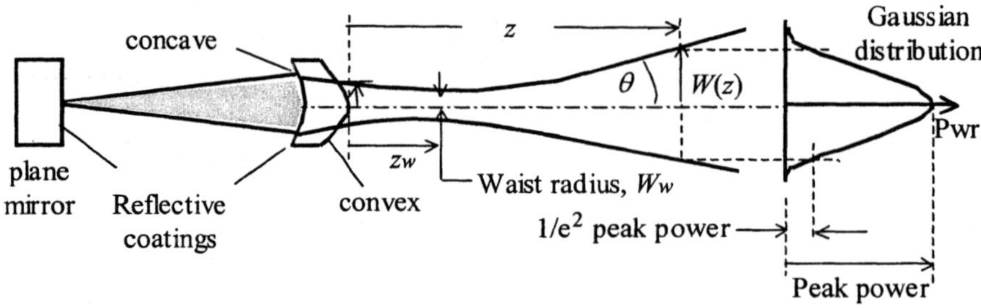

Figure 3-3. Output coupler, beam waist, divergence, and Gaussian power distribution.

Etalon consists of an additional optical cavity inserted into the main optical cavity. The additional cavity consists of two mutually parallel surfaces. The result is that the standing waves have to simultaneously satisfy the constraints imposed by the two distinct cavity lengths. By choosing the distance between the Etalon's two mirrors appropriately, one can control the frequency of the emitted laser beam.

In addition to the longitudinal modes, lasers usually exhibit several transverse (diagonal) electromagnetic modes (TEM). Further, if the end-mirrors are misaligned, the path lengths of different 'rays' inside the optical cavity would be different. While millions of TEMs can exist in theory, only those exhibiting significant power intensity need be considered in practice. Each TEM is associated with its own characteristic cross-sectional shape.

Figure 3-4 shows the beam shapes associated with a selection of transverse modes. Note that some parts o- the beam are un-illuminated (shown in white). By convention a transverse electromagnetic mode is specified a s TEM_{mn} where m a nd n a re t he n umbers o f z eros r espectively along the orthogonal axes (x and y axes) specified over the beam cross-section at a point along the beam propagation direction, z. Note that different beam shapes appear when different transverse modes are superposed (note the bagel shape obtained by combining TEM_{01} and TEM_{10}). The advantages of multi-modal beams are that they yield higher overall power and are less sensitive to misalignments of end-mirrors. Hence, a multimodal beam may be desirable in many RP applications.

It is possible to limit the number of modes in a laser beam by introducing a pinhole inside the optical cavity. As the pinhole diameter is decreased, the number of higher order TEM included can be reduced. When a limiting pinhole size is obtained, only TEM_{00} is included.

The overall laser beam power increases as the number of transverse modes utilized is increased. In general, laser power is distributed non-uniformly across the cross-section of the laser beam. This distribution is the combined result of the transverse electromagnetic modes comprising the laser beam. However, an approximately Gaussian power distribution can be achieved by isolating TEM_{00}. In view of the smoothing effect resulting from superposition of multiple transverse modes, even a multi-modal beam can be approximated for convenience as a Gaussian beam.

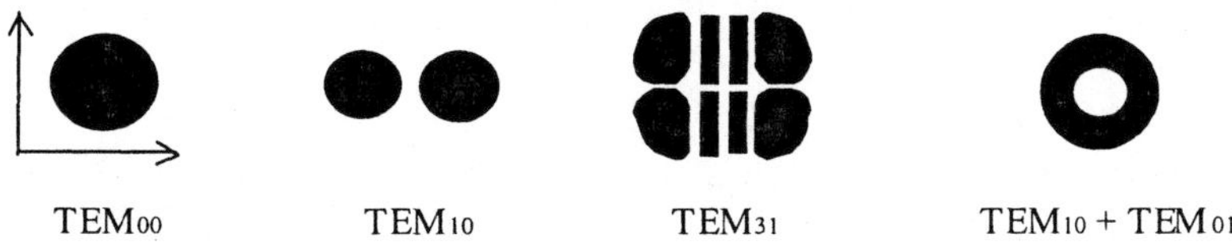

Figure 3-4. Examples of transverse electromagnetic modes (TEM).

The Gaussian power intensity distribution of a laser beam is illustrated on the right side of Figure 3-3. The mathematical form of the distribution is

$$H_r = H_0 \exp\left(\frac{-2r^2}{w^2}\right) \tag{3.4}$$

where H_r is the power intensity (W/m^2) at radius, r, along the beam cross-section, H_0 is the power intensity for $r = 0$ (peak intensity), and w is the magnitude of r at which the power intensity is a fraction $1/e^2$ (= 0.13534) of H_0. Note that the distribution is symmetric about axis z with a single peak located at $r = 0$.

Total radiant power of the laser beam over an elemental annular strip of radial width, dr, at radius r is equal to $H_r(2\pi r dr)$. By integrating this expression over the range of r from 0 to ∞, it can be shown (Jacobs 1992a) that the total power, P_L, delivered by a Gaussian laser beam is given by

$$P_L = (\pi/2)w^2 H_0 \tag{3.5}$$

Combining equations 3.4 and 3.5,

$$H_r = \frac{2P_L}{\pi w^2} \exp\left(\frac{-2r^2}{w^2} \right)$$
(3.6)

Since the power intensity asymptotically approaches zero as radius, r, increases, it is difficult to detect the edge of the beam. Hence, the beam size is specified by convention. One convention is to take the beam radius as being equal to w, i.e., equal to the magnitude of r at which the power intensity is equal to H_0/e^2. This value is called 'the $1/e^2$ Gaussian half width' of the beam.

Sometimes, the size of a laser beam is specified in terms of its 'zone of influence' defined as the radius of the beam, R_L, over which 99.99% of the total power, P_L, is delivered. By performing integration of irradiance function over the appropriate beam radius, it can be shown (Jacobs 1992a) that

$$R_L = 2.146w$$
(3.7)

Figure 3-3 shows the longitudinal shape of the laser beam after emerging from the output coupler. Note that the beam size converges initially to a minimum. The location of minimum beam radius is called the beam waist. Beyond the waist, the beam diverges.

For a given laser beam, beam size (diameter) can be expressed as a function of beam propagation distance, z, from the output coupler. The beam propagation distance from the beam waist over which the beam's cross sectional area doubles, i.e., the distance over which the beam radius increases $\sqrt{2}$ times is called the Rayleigh Range, z_R. It can be shown that

$$z_R = \pi W_w^2 n / \lambda$$
(3.8)

where n is the refractive index of the beam's medium. Interestingly, the magnitude of the waist radius is about the same irrespective of whether the beam consists of single or multiple transverse modes.

Many applications require the waist radius w_w, to be smaller than that produced by the laser unit. A common method is to use the arrangement of lenses typical of Galileo's telescope where two convex lenses (one of them is the lens embedded in the output coupler) are placed at a distance equal to the sum of focal lengths of the two lenses. However, one must be aware that beam divergence increases when the beam size is reduced.

The magnitude of beam propagation distance, z_w, between the output coupler and beam waist depends on the coupler lens. The range, $z \ll z_w$ is called the 'near field' whereas range $z \gg z_w$ is called the 'far field'. Beam radius, $w(z)$, varies in a nonlinear manner in the near and intermediate fields. However, in the far field range, $w(z)$ is well defined and increases in a linear fashion, i.e., the beam is conical. It can be shown that

$$w(z) = w_w \sqrt{1 + \frac{z^2}{z_R^2}} \tag{3.9}$$

Further, divergence angle, θ, in the far field is given by the following approximate relationship:

$$\theta \approx \arctan\left(\frac{\lambda}{\pi W_w n}\right) \tag{3.10}$$

Usually, the divergence of a laser beam is of the order of 0.001 radian.

As noted earlier, a multi-modal beam comprises several transverse modes starting from TEM_{00}. Of these, TEM_{00} is closest to being Gaussian and the smallest in size. Hence, the closeness of a multi-modal beam to being Gaussian can be characterized by the ratio of the overall diameter, d_{multi}, of the multi-modal beam to the diameter, d_{00}, of TEM_{00}. This ratio is called the 'mode purity number', M. A smaller M indicates greater mode purity. Further, it can be shown that M is equal to $0.5\pi\theta w_w/\lambda$. Note that, other parameters remaining the same, a higher value of M is associated with a larger divergence angle or a smaller wavelength.

3.4 LASER BEAM CONTROL

Light is a radiation resulting from oscillating electric and magnetic vectors that, while being orthogonal to each other, lie in the plane normal to the direction of beam propagation. The electric vector of ordinary light is directed along all possible directions within the plane normal to the propagation direction. However, it is possible to produce light with its electric vector oriented in a particular (single) direction within the transverse plane by passing the light through an optical device called a 'polarizer'. If the light so polarized is passed through a second polarizer with its 'polarization axis' perpendicular to that of the first polarizer, the light will be annihilated. The plane containing the direction of polarization and the direction of propagation of the light beam is called the 'plane of polarization'.

When a ray of light strikes the surface of a transparent medium, the atoms on the surface to oscillate. However, because of the presence of the surface, this oscillation can only occur in the plane of the surface. As a result, the incident ray is resolved into the reflected and refracted components that are partially polarized n directions perpendicular and normal to the beam plane respectively. In 1812, Brewster discovered that maximum polarization is achieved when the reflected ray is perpendicular to the refracted ray. Hence, the angle of incidence for maximum polarization of the reflected ray is equal to $\tan^{-1}(n_2/n_1)$ n_1 where n_2 and are the refractive indices of the 'incident' and refractive media respectively. This angle is called the Brewster Angle (θ_B). The angle is approximately equal to $57°$ for refraction from air into glass.

A Brewster Window is a device used in gas lasers for enhancing the performance of the optical cavity by taking advantage of the above observations. If we place the output-side plane end- mirror at the Brewster angle, all the light that is polarized perpendicular to the beam plane escapes out through the mirror at an early stage. The rest of the light (polarized parallel to the beam plane) is reflected back into the optical cavity. The advantage is that there will be no reflection losses, so the efficiency of the laser is improved.

The optical cavity stores electromagnetic energy through the formation of standing waves. The capability for such storage is indicated by the ratio (Q-factor) of the energy stored inside to that wasted or emitted during each round trip of the standing wave between the end-mirrors. As the Q-factor increases, more energy is stored instead of being emitted. The magnitude of Q-factor is proportional to the length of the optic cavity and inversely proportional to the wavelength of the laser and absorbance (1- reflectance) of the end-mirrors.

A Q-switch is frequently used to control the pulse rate of the laser beam. Several Q-switching methods are in use. When the desired pulse rate is small, a rotating mirror or an axial disc with a hole may be used to periodically interrupt the laser beam. Higher pulse rates may be obtained by means of electro-optic or acousto-optic transducers. These devices rely on the fact that light amplification through certain media can be changed by the application of a transverse voltage, or by applying an acoustic signal. When such a transducer is placed between the active medium and one of the end-mirrors, one can switch the laser beam on or off at the desired rate by controlling the electrical or optical signal applied to the transducer. Some lasers rely on a dye solution that saturates and becomes opaque when the laser intensity reaches certain magnitude.

3.5 TYPES OF LASERS USED IN RP

A wide variety of laser systems is currently available. However, only a few types are used in rapid prototyping machines. The basic principles of the five types of lasers commonly used in RP equipment are outlined below.

In an Nd-YAG laser, the active medium is a crystal of yttrium-aluminum-garnate ($Y_3Al_5O_{12}$) doped with 1 to 4% by weight of neodymium (Nd^{3+}) ions. Such a medium can be manufactured in a shape of disk or rod, with diameters of 2 to 15mm and lengths of 2 to 30cm. YAG laser rods can be expensive since growing crystals is a slow and complicated process. These lasers can be designed to operate in pulsed or continuous modes.

Excitation in Nd-YAG lasers utilize four energy levels. Nd ions have two absorption bands. Pulsed lasers use flash lamps to achieve optical pumping. Continuous lasers use arc lamps. Nd ions are transferred from these excitation levels by a nonradiative transition to upper levels. Stimulated emission at 1.064µm (infrared) occurs from the upper energy levels to the lower energy levels. The total efficiency is low (1 to 2%).

Many RP processes such as stereolithography seek to utilize ultraviolet light rather than infrared light. Different mechanisms are used to achieve this transformation. For instance, the neodymium-yitrium vanadate ($Nd-YVO_4$) crystal based lasers used in models SLA 3500/5000/7000 of 3D Systems also output infrared light (3D Systems 2001). Pumping is done by means of diode bar consisting of many diode lasers emitting infrared light at 0.808 µm. Two nonlinear crystals called 'doubler' and 'tripler' are then used to convert the infrared beam into an ultraviolet (0.355µm) beam. The optical cavity is operated in the Q-switched mode by means of an acousto-optic coupler that exploits the fact that the refractive index of an optical material is changed by acoustic waves. The device enables light pulses at the rate of tens of thousands per second. Switching the laser beam on and off is achieved by means of a separate acousto-optic coupler.

In an Argon-ion laser, the active medium is plasma consisting of Argon ions and free electrons. An electrical discharge through a tube containing flowing (recirculating) Ar gas, ionizes (positive, Ar^+) heats (to about 3000°K), and excites it. Production of the plasma requires high current densities of the order of 1 A/mm^2. High current density is achieved by surrounding the discharge tube with a magnetic field. As a result, Ar^+ ions move along a helical path parallel to the axis of the tube thus preventing loss of ions via the walls of the tube. Overheating is prevented by choosing low thermal conductivity beryllium oxide (BeO), graphite, or a metal-ceramic tube construction. Further, the tube is surrounded by a water-cooled chamber. In addition, metal discs may be inserted into the tube so as to act as heat exchangers. Owing to the cooling requirements, these lasers tend to be large in size.

The ground level of Ar^+ ion is higher than Ar atom and has a shorter lifetime than that for higher levels. This helps drain the immediate set of upper levels and creates population inversion. Lasing occurs through transitions between two energy bands consisting of several closely spaced energy levels. The output typically consists of blue (0.488μm), green (0.5145μm), and ultraviolet (0.3511μm) light.

The active medium in a He-Cd laser active laser medium is a mixture of Helium gas (at 3-7 millitorr pressure) and vaporized cadmium (a few millitorr) obtained by heating to 250 to 300°C. The need for vaporizing Cd results in large start up times (about 10 minutes). Since Cd liquefies before vaporizing, the laser cannot b tilted or used upside down.

The active laser beam is ionized by a process called Penning Ionization. Helium is a noble gas with much higher excitation energy (24.46 eV) than that (8.96 eV) of Cd vapor. The mixture is enclosed in discharge tube sealed at each end by a mirror. Pumping is done through a glow discharge initiated by a high voltage applied to two electrodes each placed near one end of the tube. The moving electrons in the discharge collide with the lighter He atoms and get excited. The excited He atoms transfer energy to the heavier Cd atoms through collisions. This leads to population inversion. The transitions lie between the energy levels of singly ionized Cadmium atoms. About twelve levels are available. The output is in the violet, ultraviolet, and shorter wavelength region.

Since Cd atoms are electrically positive they move towards the cathode and get deposited there. Hence designers of He-Cd laser tubes have to spend much effort towards minimizing the loss of Cd. He-Cd lasers mainly output blue light (0.4416μm wavelength)) and ultraviolet light (0.325μm). The UV power output is about a third of that of blue light. The efficiency of this type of laser is usually under 1%.

The CO_2 laser was first demonstrated by C. Patel in 1964. It is the preferred laser for many materials processing applications including cutting, welding, and annealing. The active medium of standard carbon-dioxide laser consists of CO_2, N_2, and He gases typically mixed in the ratio 1:1:8. In high power applications, the gas is made to flow continuously through the tube to discharge into the atmosphere (carbon-dioxide is non-poisonous). Lower power lasers use a sealed tube. However, to prevent disassociation of CO_2 into CO and O, a catalyzing agent is added.

Excitation is achieved by an electrical discharge through the laser tube. The discharge causes the lighter He molecules to accelerate and then transfer energy to and molecules through collisions. CO_2 is a linear molecule so that it can vibrate in three modes: symmetric stretch, bending, and asymmetric stretch. Since the first energy level of nitrogen is close to the energy level of the asymmetric stretch mode of CO_2, energy pumped to the energy level of nitrogen is easily transferred to CO_2. Helium helps through its lower energy level with short lifetime so that population inversion can be maintained. Lasing occurs when transition occurs from the higher energy asymmetric stretch mode to the other two lower energy modes. The transition to bending results in 9.6μm emission whereas that to symmetric stretch emits at 10.6μm. However, since each of these three modes actually consists of several closely spaced energy levels there is a frequency spread in the output.

Semiconductor lasers were first demonstrated independently in three different laboratories in the USA in 1962. These lasers are also called junction lasers or injection lasers. The energy bands in semiconductors can be classified into two types. In a valence band electrons are tied to the atoms in the semiconductor. In a conduction band they are free to move around. There exists a wide energy gap between the two types of energy bands. In a normal state, no electrons are found between the two energy levels. However, upon receiving enough energy, and electron in the valence band can jump up to the conduction band. In an insulator, the energy gap is large so electrons cannot jump. In a conductor (metal), the bands overlap so movement of electrons is easy. In a semiconductor, the energy gap is finite but very small so a small energy input is sufficient to make electrons in the valence band to jump to the conduction band. When an electron (negatively charged) executes such a jump, a 'hole' (positively charged) remains in the valence band. Both electrons and holes participate during current flow. In a pure s emiconductor, t he m aterial d etermines t he e nergy l evels (hence, t he energy gap). However, it is possible to add additional energy levels within the energy gap by introducing some impurities into the pure semiconductor material (doping). For instance, silicon and germanium are among the most commonly used semiconductor materials. Each atom in the crystalline lattices of these materials shares electrons with four neighbors. It is possible replace some of these atoms with donor atoms of materials (such as phosphorous or arsenic) that have five bonding electrons so that extra electrons become available. The maximum energy level of the electrons is called the Fermi level. A material thus doped is said to be of n-type (negative-type). The presence of n-type donor adds extra energy levels close to the conduction band of the semiconductor material.

When doping is done with a material such as gallium that contains lesser electrons, the material is said to be of the *p-type* (positive-type). Such a material will have extra energy levels close to the valence band. A p-n junction is formed when a p-type layer contacts an n-type layer. Figure 3-5a illustrates the energy configuration in such a junction. Note the additional energy bands on the conduction and valence sides. If we produce a forward bias voltage by connecting the positive terminal of a battery to the p-side and the negative terminal to the n-side, a current will flow. If we reverse the battery connections (negative bias voltage), no current will flow. In effect, we have created a diode (device with a preferred current flow direction).

Forward bias results in extra charge carriers being injected through the junction. The region through which the current passes can be controlled by shaping the metal electrodes on the upper and lower sides of the laser. When the current exceeds a threshold, there occurs selective pumping that creates population inversion in the region through which the current passes. As a result, electrons start from the n-side to meet and recombine with holes on the p-side. When an electron from the conduction band in the "n" side is injected through the junction to an empty "hole" in the valence band on the "p" side, a process of recombination of (electron + hole) associated with energy release takes place. In a laser diode, this energy appears as a thin and slightly diverging beam of electromagnetic radiation (light).

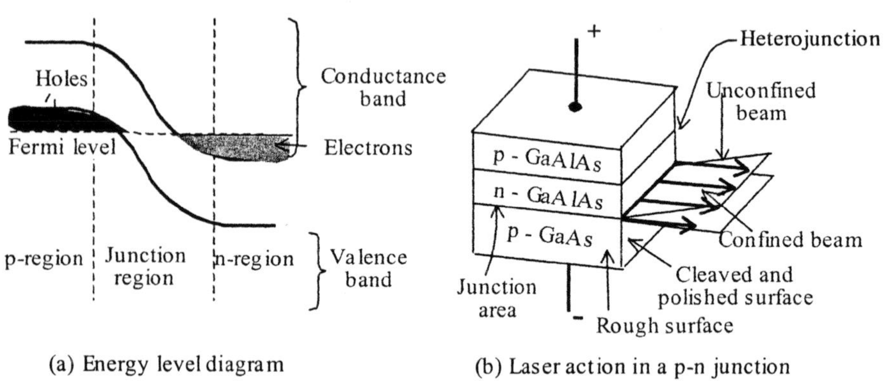

(a) Energy level diagram   (b) Laser action in a p-n junction

Figure 3-5. The principle of semiconductor diode laser.

Different methods are in use for confining (i.e., preventing the spreading of) the light. In one method, a layer of material with a different energy gap is placed on one side of the active layer (see Figure 3-5b). This heterojunction changes the distribution of refractive index across the junction and creates a waveguide structure capable of confining the light beam. Other methods rely on a gain-guided stripe, or double or buried heterojunction, etc. Discussion of these confinement methods is beyond our present scope.

The advantages of diode lasers are that they are small in size, highly reliable, inexpensive, efficient (>20%), last up to 100 years of continuous operation, and exhibit a narrow spectrum band (down to a few kHz). One can also directly modulate the output up to the GHz level.

Chapter 4

REVERSE ENGINEERING AND CAD MODELING

4.1 BASIC CONCEPT OF REVERSE ENGINEERING

In mechanical engineering, the development of a product from conceptual design to final production encompasses initial product description, mock ups for evaluation of alternative designs, product modeling, prototyping and tooling. This is an iterative process involving many refinements. The critical path of the product development cycle has many potential bottlenecks, which can slow down the whole process. One such bottleneck is a step called reverse engineering (RE). RE continues to be widely used in the development of products having freeform geometry. Since it involves creating a CAD model from a physical model, it is the inverse of the normal process of producing a part from a CAD model. The major applications of reverse engineering for product design can be classified into four cases shown in Figure 4-1:

- *New product design:* In many application areas, the design of a new product is often expressed as a physical model first. Typical applications can be found in automobile, aircraft and ship building industries, mold and die making industries, and industries for many household products. In the automobile industry, for example, the initial conceptual and aesthetic design of a car body is often done by stylists who visualize their ideas by making clay or wooden models, the so called mockups. A complete CAD model has to be created from the mockups for further design and downstream applications.
- *Existing product redesign:* When an existing product needs to be redesigned or the original design information is partially or

completely unavailable, one can also use reverse engineering techniques to create a CAD model from existing physical parts. The CAD model thus created can be used for further design modification or for reproduction.

- *Custom product design:* Another application area likely to become very important in the future is the generation of a custom CAD model for producing just one or a few product models. Orthopedic footwear, for example, can be designed and manufactured by measuring the foot shape of the customer and using the data to prepare a CAD model that fits the customer's foot closely.

- *Physical model-based shape modification:* During some other stages of the product development cycle, physical models are frequently used for shape verification, performance evaluation, safety and reliability analysis, manufacturability and assemblability testing. As a result, the geometrical shapes of the parts are often modified on the shop floor by directly adjusting the manufactured prototypes. The CAD model must be modified/updated according to the changes made to the physical models.

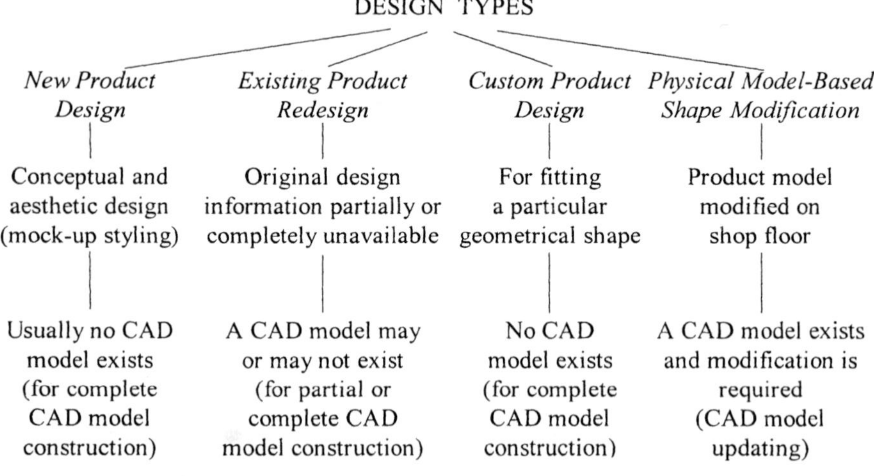

Figure 4-1. Typical applications of reverse engineering for product design.

In general, the entire reverse engineering process for product design can be subdivided into three major steps: data capturing, pre-processing and CAD modeling and/or model updating. Model updating can only be applied if a CAD model already exists. In this case, it is not necessary to re-compute the entire surface model. The major activities needing attention in each stage are as follows:

- *Data capturing:* The first step in reverse engineering is to acquire the geometrical shape of the physical models using some measuring devices. The captured information is usually represented as a set of three-dimensional (3D) data points, 2D slices or images. A variety of equipment can be used for data capturing.
- *Data pre-processing:* The captured data are first pre-processed to facilitate subsequent surface modeling. Typical data pre-processing steps include registration, data cleaning, data reduction and segmentation. Registration transforms the measured results obtained from different setups into a single coordinate system. Data cleaning wipes out all points not on the part surface. Data reduction reduces the number of points if they are too many to handle. Segmentation is the most important step during data pre-processing. It subdivides the combined data set into individual regions, each representing a single geometric feature that can be mathematically represented by a simple surface.
- *CAD modeling:* The segmented data are further transformed into individual surfaces. Several mathematical schemes for representing geometrical shapes are available. One can also use interpolation or fitting techniques for identifying the parameters of the geometrical shape. After fitting individual surfaces to measured points, the surfaces must be further processed to form the complete CAD model.
- *CAD model updating:* When a product design is directly modified on the shop floor through adjustments to the physical prototypes, the CAD model must be further updated according to the changes made to the physical part.

The process of capturing part shape and transforming the measured data into a complete CAD model is a complex task especially for parts having freeform geometry. Many encouraging results have been obtained during the past decade. To mention just a few, a general overview can be found in Chivate and Jablokow (1995) and Vàrady *et al.* (1997). One can also find some recent topics in (Benkó and Várady 2001, Pottmann and Leopoldseder 2003, Langbein *et al.* 2003) on modeling and model beautification. Approaches based on subdivision surface fitting can also be found in (Hoppe *et al.* 1994, Suzuki *et al.* 1999, Ma and Zhao 2000, and Ma et al. 2003a).

This chapter introduces some commonly used reverse engineering methodologies. Section 4.2 summarizes techniques for data capturing. Major approaches for surface reconstruction in reverse engineering are described in Sections 4.3-4.6. Issues concerning CAD model updating and surface local updating are addressed in Section 4.7. Some selected examples are presented in Section 4.8.

4.2 DIGITIZING TECHNIQUES FOR REVERSE ENGINEERING

During the past two decades, various 3D measuring techniques have been developed for industrial applications. Typical application fields include industrial metrology, robot navigation, machine vision, medical imaging, geomorphology and micro-geometric topography. The development of 3D measuring techniques in these fields has provided much support to reverse engineering. Thus, many of the 3D measuring techniques are directly applicable in reverse engineering applications.

The commonly used 3D measuring techniques used for reverse engineering include mechanical contact and various non-contact measuring techniques. Non-contact measuring techniques are highly diversified and have a rich family of measuring methods. In terms of energy source, non-contact measuring techniques can be classified as optical, ultrasonic, electrical, electromagnetic, radioactive, or a combination of some of these sensing strategies. In terms of signal processing principles, non-contact measuring techniques can be classified as triangulation, moiré, photogrammetry, focus, imaging radar (time-of-flight), and CT methods.

Regardless of the type of measuring technique utilized, these measuring systems usually comprise four main units as shown in Figure 4-2, i.e., probing unit, detecting unit, scanning unit and processing/control unit. The probing unit provides the action(s) necessary between the measuring equipment and an area of the part surface being digitized. The action(s) could be either mechanical contact, or non-contact operations. The detecting unit, such as a position sensitive device (PSD) or a charge coupled device (CCD), generates related signals corresponding to the range/depth information from the measuring probe to the probing area. The scanning unit enables one to change the probing areas in the required directions so that the entire surface of the object could be measured. The processing/control unit takes care of the overall control of the measuring system and transforms the acquired range information to 3D data. The following components are included, for example, in the four units of a typical coordinate measuring machine (CMM) equipped with a laser triangulation probe (a single laser beam plus a PSD detector):

- *Probing unit*: a laser diode, projecting lens, and a laser beam.
- *Detecting unit*: receiving lens, a PSD detector, and related circuitry.
- *Scanning unit*: the x, y, z guiding mechanism and their scaling units, the mechanisms of the two rotational axes of the probe head, the servomotors, etc.
- *Processing/control unit*: interfacing hardware, control box, processing computer and software.

The probing unit and the detecting unit are usually integrated into a single sensor head. Certain measuring equipment, however, may not include all of the four units mentioned above. For example, some passive 3-D acquisition systems using ambient light t o capture 3D information do not have a p robing u nit. I n s uch c ases, o ne c an suppose t hat the e nvironment itself is acting as the probing unit.

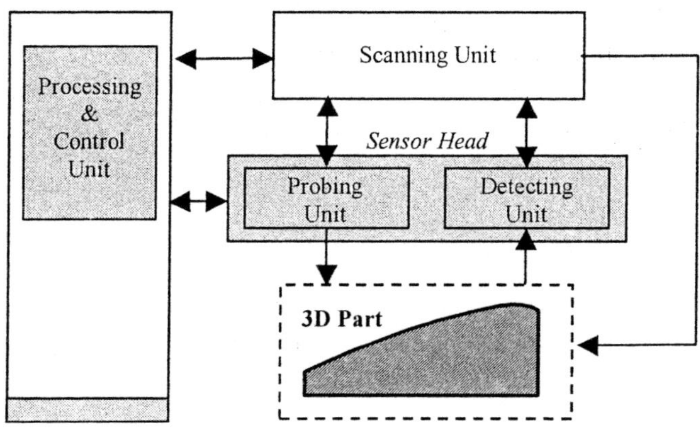

Figure 4-2. The four basic construction units of a digitizing setup.

The rest of this section summarizes the principles of some important measuring techniques for reverse engineering applications, such as mechanical contact digitization, optical non-contact measuring techniques and X-ray CT-scanning. One can also find some commercially available digitizing systems based on several of these measuring principles in (Wohlers 1995, Muthukrishnan 1994).

4.2.1 Mechanical Contact Digitizing

Mechanical contact measurement is commonly used for industrial metrology. A typical example of this approach is a CMM equipped with a touch trigger probe or mechanical scanning probe. The measuring principle of mechanical contact measurement is quite simple. For instance, a CMM usually has three or more axes. Each axis has its own measuring scale for position control and feedback. When the probe tip touches a part surface, a tiny deflection of the probe is detected that triggers a process of reading off the current position of the contact point along the three axes. The readings can easily be converted into three-dimensional coordinates of the center of the probe in accordance with the configuration of the CMM. Thus, a single point on the part surface is measured and the process is then repeated.

CMMs have various configurations in terms of the relative movements of the axes. The commonly used configurations include bridge type, column type, horizontal arm type and gantry type (Farago and Curtis 1994). A CMM can usually be operated in manual, teach-in and direct computer control (DCC) modes. CMMs are high accuracy measuring machines with a typical resolution of about 1μm. The linear accuracy can be up to 10μm in a volume of one cubic meter. Since measurement using a touch trigger probe is point-by-point, the measuring efficiency is very low when a large number of points need to be measured, as on complex surfaces.

An effective approach to improving the measuring efficiency of a CMM is the use of a mechanical scanning probe. The linear accuracy of this device may be slightly inferior, a few tens of micrometers, but it is often sufficient for reverse engineering applications. In the case of a mechanical contact scanning system, the position of the scanning probe is controlled such that the probe tip is in continuous contact with the part surface while following a pre-defined profile. The coordinates of the probe center are computed at the sampling frequency of the scanning process. A CMM can also be equipped with other non-contact scanning probes as detailed in the following subsections.

The use of a CMM for reverse engineering has the following advantages:

- very high accuracy, usually a few micrometers to several tens of micrometers;
- very large measurement range depending on the size of the CMM;
- machine configuration often with 3 linear axes plus 2 rotational axes for probing; and
- universal and flexible measuring software support.

The main limitations associated with the use of a CMM for reverse engineering applications are:

- low measuring speed especially with a touch trigger probe,
- low degree of automation while measuring unknown features,
- non direct surface measurement and probe radius compensation required,
- not suitable for measuring soft and fragile objects with touch trigger probes, and
- high cost of CMM equipment.

One way of avoiding the high investment cost of CMMs is to use a multi-axis CNC machine or jig borer as the scanning platform. Instead of a cutting tool, one may mount a probe or sensor head on the spindle head. These machines can then digitize 3D parts with acceptable accuracy. A typical example is the Retroscan system developed by Renishaw (www.renishaw.com) that converts CNC milling machines into 3D digitizers. Other cost effective digitizing systems use inexpensive guiding and scaling elements in the scanning unit, or inexpensive electric deflection type p robes for the s ensor h ead although a t t he c ost o f reduced a ccuracy. Another typical inexpensive scanning unit is based on mechanisms driven by stepping motors under open-loop control. This configuration can be used for some low accuracy reverse engineering applications.

Some digitizing systems seek to maximize the flexibility and portability of coordinate measuring machines by using 3D articulated arms similar to robotic manipulators as the scanning units. The articulated arms usually have up to 7 degrees of freedom. Typical examples include desktop digitizing systems called MicroScribe made by Immersion Corporation and 3-D digitizers called FaroArm made by Faro Technologies, Inc. (Wohlers 1994, www.immersion.com, and www.faro.com). The stylus is manually driven to move over the part surface, so a representative set of surface points is obtained. The accuracy of such digitizers is usually of the order of several tenths of a millimeter while the measuring range can be up to several meters. These digitizers can be used for RE applications requiring low accuracy.

4.2.2 Optical Non-Contact Measurement

Optical non-contact approaches are commonly used in industrial applications. They can be further subdivided into two categories: passive approaches, and active approaches. Passive approaches use ambient light source as the energy source for the probing action while active approaches use the energy supplied by the 3-D sensor itself. Active light sources include coherent and incoherent sources. Coherent sources based on lasing are superior in terms of monochromaticity and directionality. Hence they are frequently used in the probing units of 3-D sensors. The commonly used lasers are of the He-Ne or semiconductor diode type. A major drawback of coherent light sources is that they generate undue speckle noise. Further, the use of high-energy laser may compromise the safety of both the operator and the part being measured. So, many probing units use an incoherent source produced by a tungsten-filament lamp, a tungsten-halogen lamp, or an arc lamp. The advantages of using incoherent light sources are the lack of speckle noise and higher stability compared with laser coherent sources.

In terms of the direction of the light beam used in an active probing unit, two types of optical beams are popular: collimated beam or focused beam. A collimated beam can travel a long distance with little spread. Further, measuring equipment using a collimated beam usually have a larger measuring range or depth of field than one using a focused beam. Equipment using focused beams, however, is superior to those based on collimated beams in terms of the ability to measure very fine object details. In terms of the shape of the light beam, there are single spot beam, double beam, grid beam, and other structured or patterned beams. In terms of the intensity of the light beam, there are binary beams and multi-gray level beams.

The greatest advantage of optical non-contact measuring techniques is their high measuring speed. Soft and fragile objects can also be measured. A general limitation of optical approaches, however, is their inability for penetration. This limitation makes it impossible to use optical approaches for measuring internal geometric features.

4.2.2.1 Optical triangulation method

The optical triangulation method is based on the principle of triangulation. As shown in Figure 4-3, a triangle has six parameters in total: three lengths (a, b, c) and three angles (α, β, γ). If one has prior knowledge of three of them such that at least one of the three is a length value, the other three p arameters c an then be e stimated e asily. B ased o n t his t riangulation principle, there have been many types of 3D sensor heads developed for CMMs, CNC machines and dedicated laser scanners. Some portable systems based on the laser triangulation principle are also commercially available.

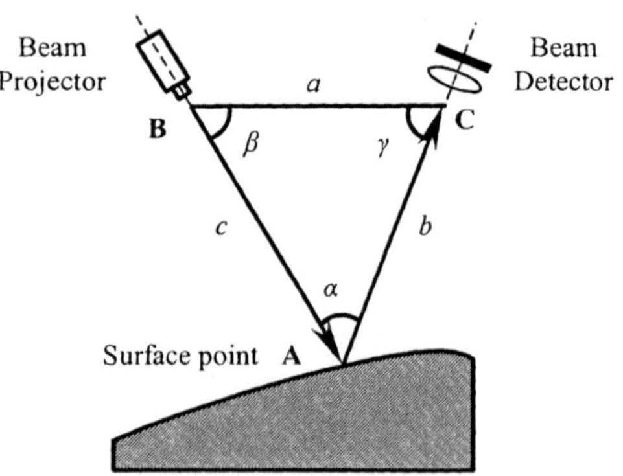

Figure 4-3. The triangulation principle.

Optical triangulation can be applied to measure both soft and hard objects with very high measurement efficiency at a reasonable cost. Hence it has become one of the most commonly used techniques in reverse engineering. In t erms o f t he c onfigurations o f t he p robing u nit for o ptical triangulation sensors, one can find point-to-point (single beam), line-by-line and area/volume measurement.

Table 4-1. Configurations based on single beam triangulation and key properties.

Case	Key elements	Resolution	Accuracy	Meas. range	Response speed
(a)	Laser diode, collimated beam, lenses, diffuse alignment, 1D PSDs detector	low	low	large	fast
(b)	Laser diode, focused beam, lenses, diffuse alignment, 1D PSDs detector	high	medium	small	fast
(c)	Laser diode, collimated beam, lenses, diffuse alignment, tilted 1D PSDs detector	low	low	very large	fast
(d)	Laser diode, focused beam lenses, diffuse alignment, 2D CCD detector	high	high	small	low
(e)	Incoherence light source, focused beam lenses, diffuse alignment, 1D CCD detector	high	high	small	medium
(f)	Incoherence light source, focused beam, lenses, specular alignment, 1D CCD detector	high	high	very small	medium

4.2.2.1.1 Single beam configuration

A commonly used configuration uses a single beam and determines the depth information from one side of the triangle and its two adjacent angles. In this case, the surface point to be measured and the two center points of the projecting and receiving lenses form a triangle. The side between the two center points of the projecting and receiving lenses is called the base-line of the triangle (a in Figure 4-3). One keeps constant the baseline (a in Figure 4-3) and one of its adjacent angles (β in Figure 4-3) between the projecting beam and the baseline itself. The other adjacent angle of the baseline (γ in Figure 4-3) is varied to track the surface point and the angle is detected through a detector placed behind the received lens. As illustrated in Figure 4-2, the triangle can be 'solved' and the absolute depth of probing beam (AB in Figure 4-3) can be determined with reference to the local coordinate system (see Figure 4-3 for an illustration). Table 4-1 and Figure 4-4 illustrate some configurations of 3D sensors based on the single beam triangulation principle.

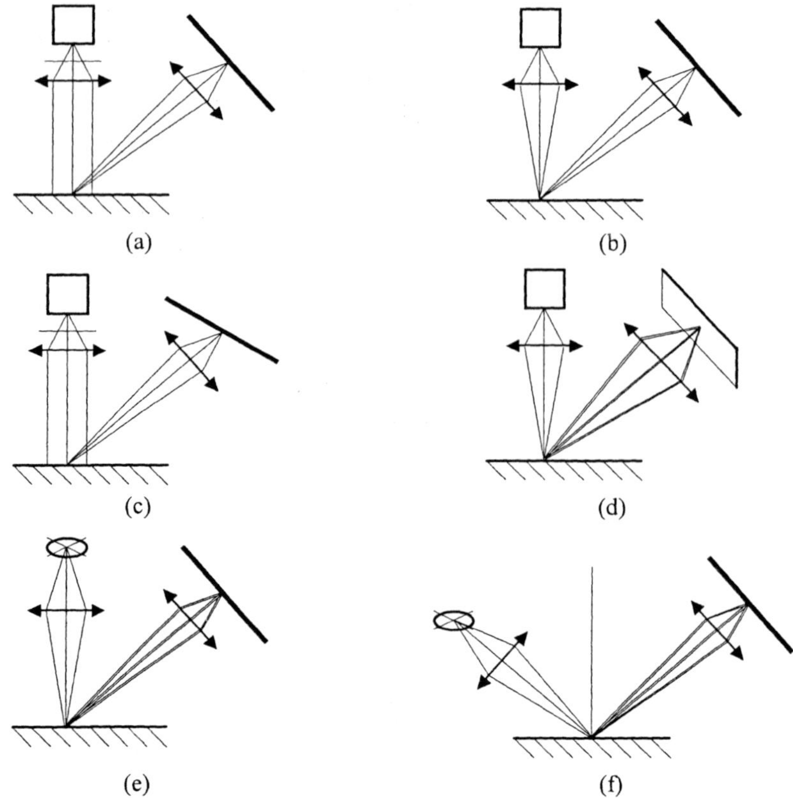

Figure 4-4. Illustrations of key configurations based on single beam triangulation.

Resolution and detection range are mainly affected by the sensor configuration and the alignment of the optical system. Resolution is mostly affected by the shape of the projecting beam, numerical aperture of the receiving lens, magnification value of the receiving lens, and lateral resolution of the detector. One can obtain a good system resolution by employing a focused beam, adequate numerical aperture, high optical magnification by the receiving lens, and high resolution of the detector. The detection range is usually affected by the depths of focus of the projecting and receiving lenses, and the effective area of the detector. One can obtain a large detection range by adopting a collimated projection beam, low magnification o f t he r eceiving l ens, a nd a t ilted a lignment o f t he d etector (Wang 1995). Usually, the ratio of measuring range over resolution (MR/R) of a triangulation-based 3D sensor head remains constant due to the limited ratio between the sensitive element and the sensitive area on the detector. In other words, the measuring range usually becomes smaller when a better resolution is sought. One may overcome this problem by using a low magnification of lenses with a low resolution for rough digitizing followed by a high magnification o f l enses w ith h igh r esolution f or final d igitizing (Godin *et al.* 1994). In spite of the modifications to the configuration or changes in the magnification of the optical lenses, the MR/R ratio remains limited for the optical triangulation method, so an eclectic approach, e.g., a normal measuring range with a reasonable resolution, has to be applied to achieve a practical configuration.

Regardless of the scanning unit and the processing/control unit, one can arrive at a variety of sensor head configurations based on a single beam by only varying the probing and detecting units. Considering the factors such as the light sources, the direction of the projecting beam, the alignment of the optical system and its type, dimension and orientation of the detector, the number of possible configurations can reach up to one hundred.

Some key factors must be considered when developing or selecting triangulation based 3-D sensors for reverse engineering. These include resolution, detection range, accuracy, surface features to be measured, surface orientation, and shading effect.

Many factors affect the accuracy of an optical 3D sensor. These may be classified into the following groups.

- *Factors related to the optical sensor itself*: With respect to the sensor itself, the quality of the lenses, speckle noise, the background light and the secondary reflections are some of the important factors affecting t he accuracy. T he u se o f g ood q uality l enses, i ncoherent light sources, and CCDs as detectors can improve the accuracy of an optical sensor. For optical systems with PSDs as detectors one can also use polarization filters for improving the accuracy.

- *Material properties and the measuring environment*: The accuracy of an optical 3D sensor head is affected not only by the optical system itself but also the geometric properties of the part surface being measured and the external environment. The reflectivity, color, roughness and inclination of the surface being measured are some of the important factors affecting accuracy (Fan *et al.* 1993b, Kakunai *et al.* 1999). Shining objects and severely inclined surfaces may not be properly measured with an optical measuring system. In addition, one also needs to keep in mind the shadowing effect associated with any triangulation based 3D sensor.

- *The scanning unit of the measuring system*: The overall accuracy of an optical 3D measuring system is also affected by the accuracy of the scanning unit, which is the same as mechanical contact systems discussed in Section 4.2.1.

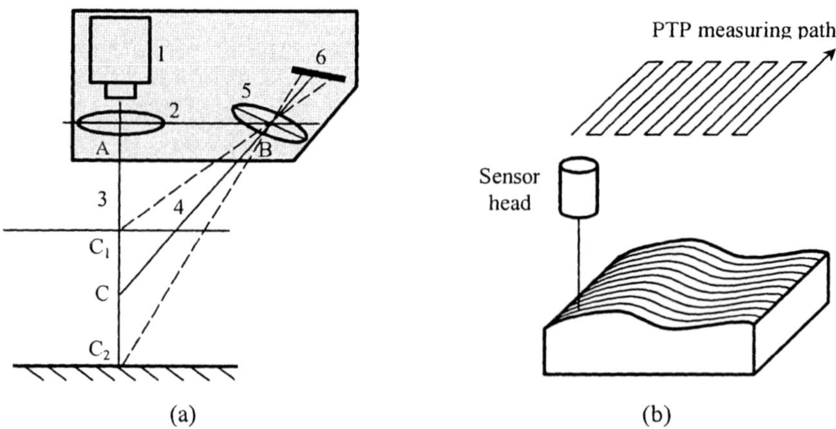

Figure 4-5. Optical triangulation systems: (a) configuration of an optical scanner using a single probing beam; and (b) illustration of the corresponding measuring path. In (a), individual elements/components are 1. laser diode; 2. projecting lens; 3. probing beam; 4. reflecting beam; 5. receiving lens; 6. detector; A. center of projecting lens; B. center of receiving lens; and C1, C and C2 are probing points.

Surface inclination is an important factor affecting reverse engineering applications. The maximum inclination angle between the sensor axis and the surface normal is usually within ±45° for diffuse alignment and ±8° for diffuse/specular compound alignment. Severely inclined surface may lead to a deterioration of the detection range and result in measurement failure. The use of double detectors aligned with the two sides of the sensor axis (Zhuang *et al.* 1994 and Mashimo *et al.* 1995) or of a sophisticated scanning unit with 5-axis freedom (Massen *et al.* 1995) are effective in reducing the surface inclination. Painting also helps in decreasing the effects of the reflectivity, surface color and surface roughness.

Shadowing or occlusion is also a common phenomenon associated with a triangulation based 3D sensor. Shadowing occurs when the projecting light beam is prevented from reaching some area of surface or the reflecting light is prevented from reaching the detector. To avoid the shadowing effect, the base line of triangulation should be kept as small as possible (Rioux 1984). One can also use an auxiliary detector installed on the other side of the projecting optical axis, so it can become active when the shadowing effect occurs. A smart scanning unit can also be used to rotate the sensor head around the sensor axis.

4.2.2.1.2   Other configurations

Besides the single beam configuration, one may also use other light beams to enhance the measuring efficiency. Among the commonly used light beams are parallel-collimated multi-beams, single slit beam, multi-slit beams and intersectional multi-slit beams—all still based on the triangulation principle. For example, in configurations using a parallel-collimated multi-beam, three or more separate collimated beams are projected simultaneously on to the surface being measured. The detector in the detecting unit receives the multiple spots and determines their centroidal positions on the detection plane. The triangulation principle is then used to calculate the depth information of these probing points. Some skillful algorithms such as the perspective transformation principle (Theodoractos 1994) can be used to accelerate the processing peed.

A sensor head with a single slit beam measures an area section-by-section and thus has plane scan ability. Such a configuration is illustrated in Figure 4-6a. It is also called light sectioning or strip method. The single slit beam is usually created by focusing a slit mask or by using a collimated laser beam through a cylindrical lens (Moss *et al.* 1989). It can also be created with a scanning axicon diffraction pattern (Hausler and Heckel 1988) or by means of Fresnel diffraction of an edge (Lewandowski *et al.* 1993) for enhancing the depth of focus. In order to improve the vertical resolution, a cylindrical lens can be inserted in front of the received lens (Lewandowski and Desjardins 1995).

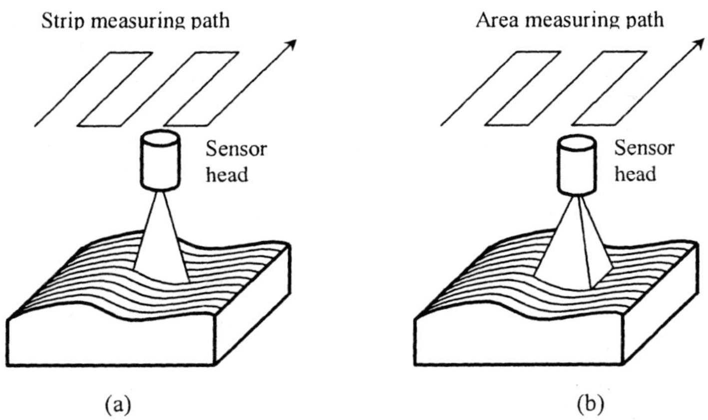

Figure 4-6. Section-by-section scanning.

A sensor head with multi-slit beams can also have an area/volume scanning ability. Figure 4-6b illustrates such a configuration. However, a problem is the ambiguity of slits when a set of dense parallel slits is projected. An effective way of dealing with this problem is to use a grating with a spare multi-slit. By mechanically orienting the grating perpendicular to the optical axis, the spare multi-slit image can be moved over the object's surface, so the detector is able to sample a series of images (Ko *et al.* 1994). A LCD (Liquid Crystal Display) unit can also act as a grating for generating the spare multi-slit. The electrical phase shift technique is usually used to move the grating automatically, so there is no need for any mechanical movement. Another problem is that the quality of the multi-slit image received by the detector can be affected by an abrupt reflection from the surface. Consequently, the results of binary image processing can be far from being perfect. This problem can, however, be minimized by projecting multi-slit beams with multi-gray-level or adapted color in combination with a multi-gray-level image processing technique (Massen *et al.* 1995, Kakunai *et al.* 1999).

In addition to the multi-slit configurations summarized above, one can also find approaches using an intersection multi-slit or a slit pattern (Guisser *et al.* 1992), and a sequence series of coding beams (Muller 1995, Osawa *et al.* 1995, Takamasu *et al.* 1993, McDonald *et al.* 1994) so as to ensure high speed measurement and less ambiguity in practical applications. One can also find portable 3D optical systems based on the triangulation principle for the reverse engineering of certain types of parts (Hageniers 1994).

To summarize, the major advantages of triangulation optical measurement are that

- both soft and hard objects can be measured;
- very high measuring speeds, usually of the order of several tens of points per second to millions of points in a few seconds, can be achieved;
- the surface points are directly measured, so there is no need for probe radius compensation; and
- moderate measuring accuracy can be achieved at moderate cost.

However, sensors based on optical triangulation have the following limitations:

- limited ratio of measuring range over resolution (MR/R);
- sensitivity to sharply varying surfaces due to problems related surface inclination;
- inability to measure black, shinning or transparent objects; and
- inability to measure internal geometric features.

4.2.2.2 Other optical measuring techniques

4.2.2.2.1 Moiré method

Moiré effect was first reported by Lord Rayleigh in 1874 and was first applied industrially in 1948 for measuring strain and displacement. In combination with advanced digital image processing and computer technology, Moiré technique has also been applied to 3D digitizing for reverse engineering (Shellabear *et al.* 1992). Moiré-based 3D digitizing heads have volume scan ability and are usually mounted on CMMs or CNC milling machines for high-speed measurement of free-form surfaces for reverse engineering.

The basic principle of the Moiré technique is as follows. When two sets of multi-slit beams are superimposed spatially at a certain angle, a beat pattern or Moiré pattern is created and a set of optical parallel planes called Moiré planes are generated. These Moiré planes are perpendicular to the depth direction and are distributed either equidistantly or according to a variable depth spacing between the planes depending on the distribution of the projected multi-slit beams. If a 3D surface is placed in the field of view, the Moiré planes will cut it into slices and produce the Moiré contour lines, the so-called equivalent height contour lines. Instead of viewing the multi-slit beams as discussed in Section 4.2.2.1, a 2D detector detects the Moiré contour lines and an image of the contours can be recorded. The 3D information is then analyzed to produce surface points in 3D coordinates. If the multi-slit beams are parallel, the resulting Moiré planes are equidistant, otherwise they are at variable depth spacing. Usually, the two sets of multi-slit beams are produced by separate gratings, namely, the deformed specimen grating and the reference grating. A 3D digitizer head usually uses the projection Moiré method to generate the deformed specimen grating.

Critical issues related to projection Moiré based scanners for reverse engineering are how to obtain the absolute coordinates and how to increase the depth resolution. The triangulation technique can be combined with Moiré technique to enable absolute coordinate measurement (Steinbichler 1994), while the phase shift technique or interpolation technique can be used to enhance the depth resolution (Gasvik 1989).

4.2.2.2.2 Photogrammetry

Photogrammetry is also called stereovision. In literature, there are also some other names such as photogrammetric method, stereo photogrammetry, stereo disparity, stereoscopy and binocular image system. The measuring principle of photogrammetry is also based on the triangulation principle. Conventional photogrammetry has been applied for air surveying and is based on photo-film and manual static processing. Industrial 3D digitizers based on photogrammetry usually use CCD cameras and dynamic processing techniques. A typical configuration of an industrial photogrammetric system is shown in Figure 4-5a. Note that, for the purpose of generating the probing beam, laser diode 1 is replaced by a detector and a specific optical system. The known parameters are again the base line between two centers of the lenses and its two adjacent angles. Here, both of the angles are variables and are detected by two 2D detectors via perspective projection.

Industrial photogrammetry systems usually have a maximum resolution of 1:30,000 (a ratio of resolution over measuring range, R/MR). The correspondence between the two detected images is a critical technical issue associated with such a photogrammetry system. However, photogrammetry has the advantages of high measuring efficiency, low cost, and area/volume scanning. It has also the special ability to measure very large objects. Several RE applications of the technique have been explored in (Clarke *et al.* 1993, Mass 1991).

4.2.2.2.3   Imaging radar

The imaging radar approach is also called a time-of-flight (TOF) method. The principle is to measure the time interval between the projected energy signal and the received energy signal with a TOF detector. The 3D distance is then computed from the measured time interval and the known traveling speed of the signals. Different energy sources, such as ultrasonic, electromagnetic, infrared light, laser and Gamma-ray energies, can be used in the probing unit. The corresponding digitizers are often called ultrasonic radar, electromagnetic radar, infrared light radar, laser radar and Gamma-ray radar respectively. Commonly used TOF detectors include pulse detectors, phase detectors and coherent heterodyne detectors (Makynen *et al.* 1995, Marinos 1993, Slocum 1992, Grabowski *et al.* 1989).

The laser radar approach is more suitable for reverse engineering since it is easily generated and controlled for focusing and scanning. In laser radar, the transmitter in the probing unit and the time interval detector in the detecting unit are usually coaxial. The shadowing effects caused by the triangulation principle are thus eliminated. After transmitting the probing pulse, the reflecting pulse is received, and the time interval is detected by a TOF detector. The depth information is then calculated as a half of the product of the time interval and the pulse traveling speed. An optical scanning unit using a beam scanning mirror or an oscillating mirror is usually adopted for a completed measurement.

The resolution of laser radar depends on the TOF detection technique adopted. Laser radar using pulsed TOF technique typically has a resolution of one to several millimeters that is independent of the measuring range. Laser radar using amplitude modulation (AM) technique or frequency modulation (FM) TOF technique can reach a resolution of the order of 0.1 millimeter depending on the measuring range. The advantages of imaging radar include portable configuration, high measuring speed, long work distance, large detection range or volume, freedom from the shadowing effect, and good resolution even when measuring a large part. An imaging radar can be used in reverse engineering applications involving the measurement of very large parts, such as a ship-body or aircraft wings. Applications of the technique in reverse engineering can be found in (Marinos 1993, Moring and Ailisto 1989).

4.2.2.2.4　Focus method

The focus method is also called an optical stylus method. Other names include focus detection method, shape-from-focus or range-from-focus. It is a kind of non-contact optical measuring technique. A typical configuration of focus based 3D digitizers can be found in (Stout *et al.* 1994). A collimated or diverging beam is projected on to the part surface through an objective lens. The reflected or scattered beam from the part's surface goes through the objective lens again in the reverse direction and is received by the receiving optics. The projecting optics and the receiving optics are coaxial so that shadowing effects are eliminated. With the help of some focus principles, such as the confocal principle, the displacement between the focal point of the probing beam and the probed point can be detected and transformed into a defocusing signal by a detector placed in the receiving optics. After appropriate amplification and pre-processing, the defocusing signal can be used to drive the movement of the objective lens and to freeze the Z-axis scale of the objective lens. In some cases, the objective lens is oscillated, so a defocusing signal to drive is not needed. A variety of focus principles c an b e u sed in the r eceiving o ptics, s uch as intensity d etection, confocal, differential detection, critical angle and astigmatic methods (Stout et al. 1994, Cielo 1988, Loughlin 1993).

The major advantages of focus based 3D digitizers include high lateral and high vertical resolutions, freedom from shadowing, and applicability to the m easurement o f p recise p arts. I ts l ateral r esolution i s usually w ithin a range of one to several microns and the vertical resolution is of the order of a fraction of one micron, which is at least one order higher than that reached by optical triangulation based techniques. Therefore, focus based 3D digitizers can be used in reverse engineering applications for the measurement of precise and small parts, such as precise metal models, coin-like parts with relief and some precision plastic parts.

4.2.3 CT Scanning Method

CT stands for computed-, computerized- or computer-tomography commonly used for medical applications. It is a non-contact measuring technique with X-ray as its energy source. X-ray CT scanning distinguishes itself by virtue of its ability to measure both external and internal part features for reverse engineering applications. X-ray CT technique is widely used in reverse engineering. When so used, it is sometimes labeled as CT-assisted reverse engineering (Yancey *et al.* 1994 and 1995, Stanley *et al.* 1995, Ma *et al.* 1997, and Liu and Ma 2000). Combined with rapid prototyping technologies, a variety of products or prototypes with complex internal and/or external structures, such as net-shape castings or models, complex valve-body components, rotor or propeller components with blades and turbo-pump prototypes can be directly duplicated in the virtual domain with this approach. One can even exactly reproduce artificial bone structures and the skull of a live person for various applications. In addition to providing both internal and external geometry structures, the measuring efficiency of a CT-scanner is independent of the internal and external geometric complexity. It requires no special positioning of the object to be measured. There is no programming need either, so one can directly apply available image processing techniques to the measured results/images for enhancement, compression, retrieval and feature/contour extraction. It can be used to measure a variety of parts, such as metallic, non-metallic, solid, fibrous or transparent objects, and parts with coarse, irregular, or low reflectivity surfaces.

CT-scanners measure an object section by section, thus producing a series of adjacent thin slices along the z-direction. Each slice or layer is a 2D cross-sectional image, often called a CT-image, of the object being measured. The image corresponds to a map of the material property of the section, i.e., the linear attenuation coefficient $\mu(x, y)$ of the sectioned thin slice on the x-y plane (Kalender 2000). The numerical value, or the so-called CT number, of each small element, i.e., a pixel of the CT-image, is proportional to its relative value of linear attenuation coefficient. Because an abrupt change of adjacent CT numbers corresponds to the location of material contours, an X-ray CT image can be regarded as a geometric cross-section image. Hence, the contours of the slices can be derived directly with the aid of appropriate edge detection algorithms. A stack of contours of the adjacent slices is the 3D map of the part being measured.

To create a CT image, i.e., to measure a single slice, the operations of 1D multi-projection and 2D reconstruction are incorporated. An X-ray beam from an X-ray source is collimated into a single pencil beam or a thin slice of f an b eam a nd t ransmitted t hrough t he p art. The X-ray b eam s hould b e energetic enough to be able to penetrate the part. The required energy level is a function of the density and atomic number of the part material. A high contrast i mage c an b e o btained b y h aving e nough p hoton f lux. A s patial resolution can be achieved by having a small spot size. A single detector or a detector array is often used to record the transmitted X-ray attenuation. The attenuation is governed by the principles of Compton scattering, photoelectric absorption, and pair production. By translating/rotating the part or the X-ray source and its detector simultaneously, a number of 1D projection profiles of the slice are obtained at different angle ranging between 0° and 180°. Reconstruction algorithms can be further applied to produce a 2D CT image of the sectioning slice. In literature, one can find a number of reconstruction algorithms, such as filtered back projection (FBP), iterative algorithms and Fourier algorithms. Figure 4-7 (Liu 2000) illustrates the principle of the FBP algorithm for CT image reconstruction. In addition to reconstruction algorithms, edge and feature detection algorithms are needed to further convert/process the CT-images into cross-sectional surface contours for RE applications.

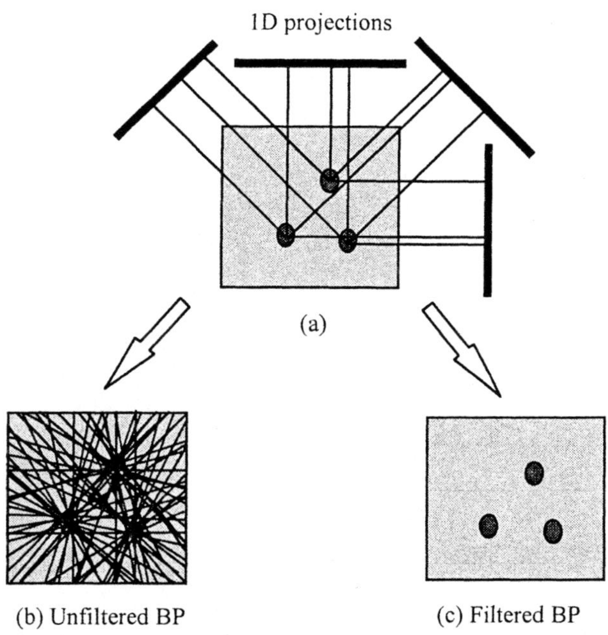

Figure 4-7. Principles for CT-image reconstruction through Filtered BP.

Similar to other 3D measuring equipment, a CT-scanner has four basic units, namely a probing unit with an X-ray tube, a detecting unit with an X-ray detector, a scanning unit mounted on a translational or rotational gantry, and a processing and control unit with a computer and mathematical algorithms. The instrumentation of CT-scanners is a challenging area involving various technical fields ranging from computer hardware, mathematical algorithms, control technology, mechanical engineering to techniques for X-ray generation and detection. In terms of applications, CT-scanners can be classified as medical or industrial CT-scanners. The architectures of medical CT-scanners have been well developed and, hence, utilized in industrial CT-scanners.

Medical CT-scanners are designed especially for the measurement of human bodies. The X-ray energy is usually less than hundreds of keV (kilo-electron-Volt). The scanning time is kept as short as possible so as to reduce the motion artifacts and the exposure to X-ray radiation by human bodies. The configurations used for medical CT-scanning can be traced back to as early as the 1970s. Figure 4-8 illustrates four generations of CT-scanners (Kalender 2000 and Hsieh 2003). Hounsfield made a prototype CT-scanning device with a single pencil beam and a single sodium iodide detector. The prototype device took 9 days for data acquisition and additional 2 days for 2D reconstruction. The device was soon improved and put into market by EMI Ltd., a British firm where the first scanners were developed, as EMI scanners for brain detection (Kalender 2000). A typical configuration is illustrated in Figure 4-8a. The X-ray tube and the detector can be regarded as a scan head or digitizer head. Since only line scanning is provided, the scan head has to translate for a complete view and rotate for multiple views. The incremental step for rotation is usually 1°. Several minutes are required for scanning one slice. Such configurations are usually called the first generation of CT-scanners. By using a fan-beam X-ray and multiple detectors, one obtains the configuration of the second generation CT-scanners shown in Figure 4-8b. Only a few tens of seconds are required for scanning one slice. Rotation and translation are still required for single slice scanning due to the limited fan plan scanning provided by the scan head. In the third generation of CT-scanners, the fan beam angle is widened and a larger number of detectors are adopted, as shown in Figure 4-8c. The scan head only requires a rotational movement. Several seconds are required for single slice scanning. In the fourth generation of CT-scanners (see Figure 4-8d), the stationary detectors are arranged along a fixed circle. Only the X-ray tube rotates around the object. One or two seconds are required for one slice scanning. In addition to the four generations of CT-scanners, more advanced ultra-fast CT-scanners have been developed to enable scanning of moving organs. Such CT-scanners can scan several tens to one hundred slices within one second (Boyd and Lipton 1983). Some CT-scanners also adopt continuous spiral scanning instead of the step movement of the object through the scan head (Kalender *et al.* 1990).

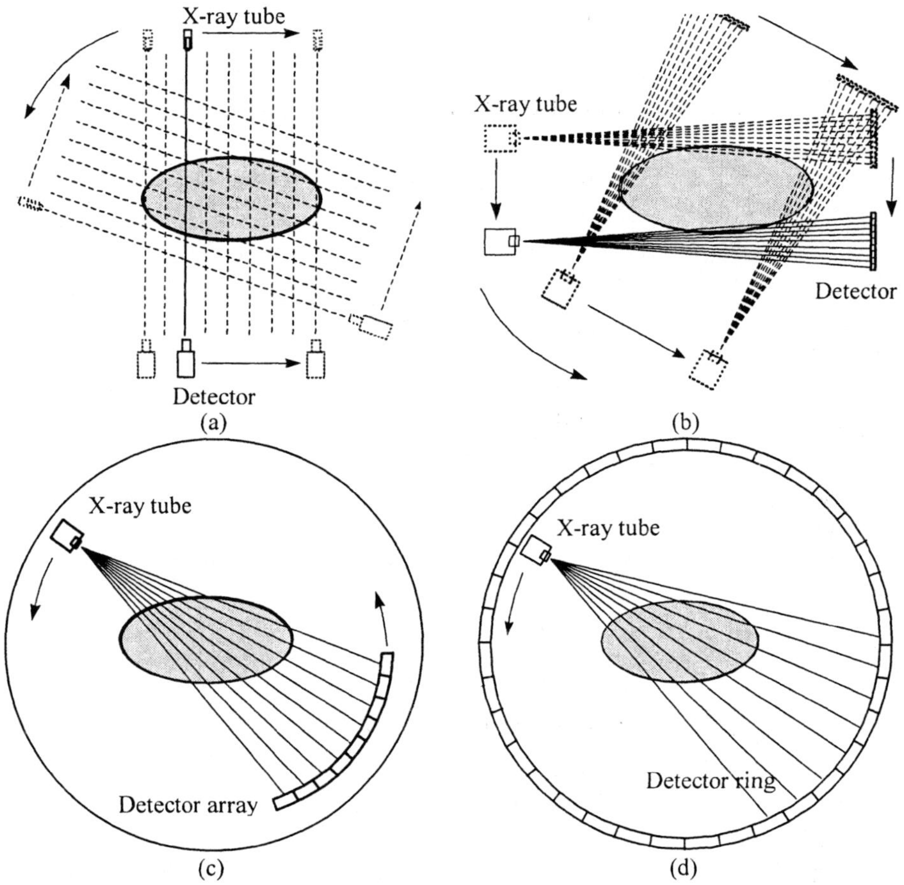

Figure 4-8. Illustrations of CT-scanners of four generations.

Industrial CT-scanners are often designed for the measurement of industrial parts with material densities ranging from 8×10^{-3} to 5g/cm^3 and of sizes ranging from less than 10mm to several meters. There are no limitations such as allowable X-ray radiation dose. The construction of industrial CT-scanners began in the late 1970s. Two of the earliest industrial CT systems were the X-ray Inspection Module (XIM) from General Electric and the Air Force Advanced CT-System (AFACTS) from ARACOR (Advance Research and Applications Corp., www.aracor.com). XIM was used for the inspection of small turbine blades and AFACTS for the inspection of rocket motors (Copley *et al*. 1994). The configurations of industrial CT scanners are usually similar to those of the second or the third generation medical CT scanners. Some portable CT-scanners also make use of the structure of the first generation. The energy levels of the X-rays used in industrial CT-scanners range from 5keV to over 15MeV. This range is wider than that used in medical CT-scanners. These X-rays mainly come from two kinds of sources, namely electric X-ray sources and isotope sources. Electric X-ray sources include industrial X-ray tubes, microfocus X-ray tubes, and linear accelerators corresponding to energy ranges of 10-450kVp (Kilovolts Peaks), 10-200kVp and 1-15MeV respectively. The main drawback of electric sources arises from beam-hardening artifacts resulting from the polychromatic spectrum of the X-ray beam. Typical isotope sources include those using Ir-192 (iridium), Co-60 (cobalt), Ce-137 (cerium), Thu-170 (thulium) and Na-24 (sodium). The beams from isotope sources are also called gamma-ray, with energy ranging from several hundreds keV to a few MeV. Due to their monochromatic nature, the isotope sources can eliminate the beam-hardening artifacts. Typical beam shapes range from a single, small diameter beam (so called pencil beam) to a fan beam or cone beam. Two types of radiation detectors, namely scintillation detectors and gas ionization detectors are commonly used. Scintillation detectors usually have high detection efficiency and a low energy resolution. In contrast, gas ionization detectors have low detection efficiency and high spatial resolution. Among the various reconstruction algorithms available, the FBP methods are particularly popular in industry.

4.2.4 Data Pre-processing for Surface Reconstruction

The geometrical shape of a physical part is often composed of various surfaces ranging from simple planes to complex freeform surfaces. The measured points are usually divided into several subsets for surface modeling through a process called data segmentation (Vàrady *et al.* 1997, Liu and Ma 1999, Benkó and Várady 2003, and references therein). Each of these subsets represents a single geometric feature and is devoid of any abrupt changes in shape. The major segmentation techniques can be classified as being either edge-based or face-based (Wani and Batchelor 1994, Vàrady *et al.* 1997).

- *Edge-based approaches*: Edge-based techniques identify the boundaries or edges between surfaces from a dense set of points. In case of a sharp edge, the direction of the surface normal is estimated from abrupt point data changes across the boundaries. In case of a smooth edge, surface curvatures or higher order derivatives might have discontinuities across the boundaries. Otherwise the boundaries are inferred from the topological information already available.
- *Face-based approaches*: Face-based techniques connect regions of points with similar properties. Each identified area represents a surface. Edges can be derived from surfaces through intersection

and/or other algorithms. Face-based approaches can be bottom-up and top-down. The former start from seed points. A small area of points that consistently belongs to a single surface is constructed and grown. Area-growing stops when there are no more consistent points in the vicinity of the current region. In contrast, the top-down method starts from all measured points that are considered to be on a single surface. The hypothesis is then tested following certain rules. If the properties of the points are in agreement with each other, one proceeds with segmentation. Otherwise the points are subdivided into two or more new sets and the single-surface hypothesis is applied recursively to these subsets. The process continues until all generated subsets satisfy the hypothesis.

Certain measuring techniques require pre-processing in order to obtain 3D coordinates of the surface points. Typical examples are CT-scanners and some passive non-contact optical measuring techniques. In the case of CT-scanning, one obtains slices containing cross-sectional density images with known distances of separation between two adjacent slices. Boundary detection techniques are then applied to identify the coordinates of points on the part surface on that slice. Similarly, one also needs to transform the measured individual images into 3D data points on the part surface for other passive image-based measuring techniques.

While segmentation partitions the entire set of points, often called the point cloud, into individual surface areas or individual point sets, surface reconstruction transforms each of the segmented areas/point sets into a compact mathematical surface model. Depending on the type of surfaces/features to be identified and the number and distribution of the acquired points, one may use either interpolation or fitting techniques to transform the points into a surface model. For interpolation, the final constructed surface goes through all the data points. In other words, the measured points are exactly on the constructed surface. However the final interpolated surface may not be smooth in case of defects on the part surface or in the presence of undue measurement errors. There might be oscillations on surface areas when there are too few measured points. For surface fitting, the final constructed surface may not go through all the data points, although still approximating the data points on the basis of certain criteria. The most commonly used fitting criterion is the least squares fitting technique. This method is mainly suited for transforming data points with errors or noise into a smooth surface.

4.3 MODEL REPRESENTATION

After segmentation, one also needs to determine a mathematical model for representing the geometrical feature before surface construction. In the area of computer aided geometric design, a variety of mathematical surfaces are available. The simplest surfaces are algebraic surfaces, such as a plane, a cylinder, a cone or a sphere. Commonly used surfaces for free-form geometry are parametric surfaces such as Bézier, B-spline and Non-Uniform Rational B-Spline (NURBS) surfaces (Farin 2002, Piegl and Tiller 1995). Subdivision surfaces are also becoming popular for CAD applications, especially for modeling shapes with arbitrary topology, for animation, data compression and texture mapping (Zorin and Schröder 2000).

4.3.1 Basic Geometric Features

Basic geometric features refer to regular or analytic geometric elements, such as lines, circles, planes, cylinders, cones and spheres. These features cover more than 80% of the engineering features found in our daily lives. Each of such features is often defined by a few parameters. A line can be defined by four parameters, a unit vector of the line segment (two parameters) and the intersection point of the line segment with one of the coordinate planes (two parameters). A plane can be defined by three parameters with a unit vector (two parameters) for defining its normal and an intersection point of the plane with one of the coordinate axis (one parameter). A cylinder, on the other hand, needs five parameters, four for defining the cylinder axis (same parameters for defining a line) and one for defining the cylinder radius.

4.3.2 General Algebraic Surfaces

General algebraic surfaces, an important class of implicit surfaces (Bloomenthal *et al.* 1997), can be defined by a polynomial equation of the type $f(x, y, z) = 0$. A typical example is a quadric surface defined by the following equation:

$$f(x, y, z) =$$
$$\alpha x^2 + \beta y^2 + \gamma z^2 + 2\mu xy + 2\nu xz + 2\omega yz + 2\delta x + 2\sigma y + 2\eta z + \rho = 0 \tag{4.1}$$

where $\alpha, \beta, \gamma, \mu, \nu, \omega, \delta, \sigma, \eta$ and ρ are free-coefficients. Equation (4.1) can be used effectively to represent shapes such as planes, spheres, cones, cylinders and ellipsoids. Algebraic surfaces and implicit surfaces in general are capable of representing many other engineering types.

4.3.3 Parametric Surfaces

Parametric surfaces are widely used in geometric design. Commonly used parametric surfaces are Bézier, B-spline and Non-Uniform Rational B-Spline (NURBS) surfaces. They are finite surfaces defined using certain basis functions and control points. A NURBS surface is a generalization of Bézier and B-spline surfaces. It can also be used to represent most of the afore-mentioned algebraic surfaces, such as a cylinder, cone or sphere. NURBS is currently the de-facto standard in the CAD industry, so most of today's CAD systems support NURBS. We will now briefly describe B-splines and NURBS to support the subsequent discussion of surface reconstruction.

4.3.3.1 B-spline curves and surfaces

A B-spline curve is defined by an order k (k-1 being the so-called degree), a set of knots $\mathbf{t} := \{t_i\}_{i=1}^{n+k}$ and a set of n control points $\{\mathbf{v}_i = [x_i, y_i, z_i]^T\}_{i=1}^{n}$. A B-spline curve is defined by the following equation

$$\mathbf{p}(t) = \sum_{i=1}^{n} B_i(t) \cdot \mathbf{v}_i \qquad (4.2)$$

where t is the location parameter of the curve that uniquely locates a point on the curve, $\{B_i(t)\}_{i=1}^{n}$ are normalized B-splines that can be uniquely defined by their order k and the knots $\mathbf{t} := \{t_i\}_{i=1}^{n+k}$. For an untrimmed B-spline curve, the entire curve is defined within $t_k \leq t \leq t_{n+1}$.

A B-spline surface, on the other hand, is defined by its orders k_u and k_v, two set of knots $\xi := \{\xi_i\}_{i=1}^{n_u+k_u}$ and $\varsigma := \{\varsigma_j\}_{j=1}^{n_v+k_v}$, and a set of $n = n_u \times n_v$ control points $\{\{\mathbf{v}_{ij} = [x_{ij}, y_{ij}, z_{ij}]^T\}_{j=1}^{n_v}\}_{i=1}^{n_u}$. A B-spline surface is defined by the following equation

$$\mathbf{p}(u,v) = \sum_{i=1}^{n_u} \sum_{j=1}^{n_v} B_{ui}(u) B_{vj}(v) \cdot \mathbf{v}_{ij} \qquad (4.3)$$

where u and v are location parameters uniquely locating a point on the surface, and $\{B_{ui}(u)\}_{i=1}^{n_u}$ and $\{B_{vj}(v)\}_{j=1}^{n_v}$ are normalized B-splines,. The basis functions, $\{B_{ui}(u)\}_{i=1}^{n_u}$, can be uniquely defined by its order k_u and the set of knots $\xi := \{\xi_i\}_{i=1}^{n_u+k_u}$ for the u-direction, whereas the basis functions $\{B_{vj}(v)\}_{j=1}^{n_v}$ can be uniquely defined by its order k_v and the set of knots $\varsigma := \{\varsigma_j\}_{j=1}^{n_v+k_v}$ for the v-direction. For an untrimmed B-spline surface, the entire surface is defined within $\xi_{k_u} \leq u \leq \xi_{n_u+1}$ and $\zeta_{k_v} \leq v \leq \zeta_{n_v+1}$.

B-spline curves and surfaces are piecewise polynomial curves and surfaces, respectively, and are generalizations of Béziers (Farin 2002) for defining a single curve segment or surface patch. If the order k of the a B-spline curve equals its number of control points n and if a special set of knots expressed as

$$\mathbf{t} := \{t_i\}_{i=1}^{n+k} = \left\{ \underbrace{0,0,\cdots 0}_{k}, \underbrace{1,1,\cdots 1}_{k} \right\} \tag{4.4}$$

is used, the resulting B-spline curve becomes a Bézier curve. Similarly, if the order is the same as the number of control points for both the u- and v-directions, respectively, and the following sets of special knots are used

$$\xi := \{\xi_i\}_{i=1}^{n_u+k_u} = \left\{ \underbrace{0,0,\cdots 0}_{k_u}, \underbrace{1,1,\cdots 1}_{k_u} \right\} \tag{4.5a}$$

$$\zeta := \{\zeta_i\}_{i=1}^{n_v+k_v} = \left\{ \underbrace{0,0,\cdots 0}_{k_v}, \underbrace{1,1,\cdots 1}_{k_v} \right\}, \tag{4.5b}$$

the resulting B-spline surface becomes a Bézier surface. For both B-spline curves and surfaces as well as the NURBS curves and surfaces to be discussed later, the normalized basis functions can be stably evaluated by the recursive Cox-de Boor algorithm (de Boor 1978)

In the context of reverse engineering, one needs to fit or interpolate a B-spline curve or surface through the measured sample points. For B-spline curve fitting or interpolation, the objective usually is to find a set of control points $\{\mathbf{v}_i\}_1^n$ with fixed number of control points n, order k, and a set of knots $\mathbf{t}$. Similarly, for B-spline surface, a set of control points $\{\{\mathbf{v}_{ij}\}_{j=1}^{n_v}\}_{i=1}^{n_u}$ need to be defined with fixed number of control points n_u and n_v, orders k_u and k_v, and two set of knots ξ and ζ.

4.3.3.2 NURBS curves and surfaces

NURBS stands for Non-Uniform Rational B-Spline. A NURBS curve is defined by an order k, a set of knots $\mathbf{t} := \{t_i\}_{i=1}^{n+k}$, a set of n control points

$\{\mathbf{v}_i = [x_i, y_i, z_i]^T\}_{i=1}^n$, and the corresponding weights $\{w_i\}_{i=1}^n$. A NURBS curve is defined by the following equation

$$\mathbf{p}(t) = \frac{\sum_{i=1}^n B_i(t) \cdot w_i \mathbf{v}_i}{\sum_{i=1}^n B_i(t) \cdot w_i} \tag{4.6}$$

where t is again the location parameter of the curve, $\{B_i(t)\}_{i=1}^n$ are normalized B-splines that can be uniquely defined by its order k and the knots $\mathbf{t} := \{t_i\}_{i=1}^{n+k}$. For an untrimmed B-spline curve, the entire curve is also defined within $t_k \le t \le t_{n+1}$.

A NURBS surface is defined by its orders k_u and k_v, two sets of knots $\xi := \{\xi_i\}_{i=1}^{n_u+k_u}$ and $\varsigma := \{\varsigma_j\}_{j=1}^{n_v+k_v}$, a set of $n = n_u \times n_v$ control points $\{\{\mathbf{v}_{ij} = [x_{ij}, y_{ij}, z_{ij}]^T\}_{j=1}^{n_v}\}_{i=1}^{n_u}$ and the corresponding weights $\{\{w_{ij}\}_{j=1}^{n_v}\}_{i=1}^{n_u}$. A NURBS surface is defined as follows

$$\mathbf{p}(u,v) = \frac{\sum_{i=1}^{n_u} \sum_{j=1}^{n_v} w_{ij} \mathbf{v}_{ij} B_{ui}(u) B_{vj}(v)}{\sum_{i=1}^{n_u} \sum_{j=1}^{n_v} w_{ij} B_{ui}(u) B_{vj}(v)} \tag{4.7}$$

where u and v are the location parameters, $\{B_{ui}(u)\}_{i=1}^{n_u}$ are normalized B-splines that are uniquely defined by its u-order k_u and the u-knots $\xi := \{\xi_i\}_{i=1}^{n_u+k_u}$, $\{B_{vj}(v)\}_{j=1}^{n_v}$ are normalized B-splines that are uniquely defined by its v-order k_v and the v-knots $\varsigma := \{\varsigma_j\}_{j=1}^{n_v+k_v}$. For an untrimmed NURBS surface, the entire surface is also defined within $\xi_{k_u} \le u \le \xi_{n_u+1}$ and $\varsigma_{k_v} \le v \le \xi_{n_v+1}$.

NURBS curves and surfaces are further generalizations of Béziers and B-splines. When all the weights of a NURBS curve are equal and constant, i.e., when $w_i = c$ for all i's, a NURBS curve becomes a B-spline curve. When all the weights of a NURBS surface are equal and constant, i.e., when $w_{ij} = c$ for all i's and all j's, a NURBS surface becomes a B-spline surface.

4.3.4 Subdivision Surfaces

In addition to B-splines and NURBS, one can also use subdivision surfaces for modeling objects having arbitrary topology. The basic idea of subdivision is to define a smooth shape from an initial control polygon (or polygonal mesh) by repeatedly and infinitely adding new vertices and edges according to certain subdivision rules. Subdivision techniques were first proposed for surface modeling by Catmull and Clark (1978) and Doo and Sabin (1978) and were initially conceived as an extension of traditional splines. Both spline and subdivision surfaces are defined by an initial control mesh. Different from splines, a subdivision surface can be defined by a control mesh of arbitrary topology, but not spline surfaces. Similar to spline surfaces, stationary subdivision surfaces with fixed subdivision rules can also be defined analytically by the initial control points and a set of basis functions for surface evaluation. Today, many subdivision schemes are available developed for modeling and animation (Zorin and Schröder 2000).

4.3.5 Other Approaches and Recommendations

Apart from the surface modeling approaches mentioned above, there are also other approaches used for reverse engineering applications. Typical examples are direct solid modeling from measured points (Chivate and Jablokow 1993), deformable object modeling (Terzides *et al.* 1993) and a neural network model for reverse engineering (Gu and Yan 1995). For product design, the most widely used geometric representations are algebraic surfaces and parametric surfaces in rectangular patch. Subdivision surfaces are becoming popular in CAD modeling due to their capability of modeling objects with arbitrary topology.

As a thumb of rule for reverse engineering, one should always use the simplest surface if it can satisfy the tolerance requirement. One should also try to capture the original intention of the designers/stylists who made the physical model. If the original intention of an area is for a plane, one should use a planar surface. Similarly, one should use a rotational surface if functionally it should be. A freeform surface is used only if no other alternative is available. Even in this case, the surface must be smooth and one should use as few parameters as possible. A large number of surface parameters could result in defects appearing on the surfaces of the physical model and may degrade the smoothness conditions of the surface in the presence of measuring errors. Complex surfaces often require long processing times and are difficult to approximate. On the other hand, one should use as large surface patches as possible to avoid extra work while smoothly connecting the final fitted surfaces.

4.4 B-SPLINE BASED MODEL RECONSTRUCTION

In this and the following subsections, we assume that the data has been properly segmented (see Section 4.2.4) such that each data set will only represent one unique geometric feature, whether a regular feature or a freeform feature. The construction of a B-spline curve or surface from a set of segmented data points is a typical problem of B-spline curve or surface fitting or interpolation. One needs to determine the order(s), the required set(s) of knots and the corresponding control points. The entire fitting process can be divided into three major steps:
- Parametrization of the measured points for determining approximate location parameter(s) for each of the measured points.
- Knots-selection for determining the order, the number of control points and one or two sets of knots for curve or surface fitting, respectively.
- Least squares fitting for computing the control points.

4.4.1 Parametrization of Measured Points

Let $\overline{\mathbf{p}}_i = [\overline{x}_i, \overline{y}_i, \overline{z}_i]^{\mathrm{T}}$, for i=1, 2, ..., m, be a set of measured points, $\{u_i | i = 1,2,\cdots m\}$ be the corresponding location parameters in case of curve fitting and $\{u_i, v_i | i = 1,2,\cdots m\}$ be the corresponding location parameters in case of surface fitting. In literature, one can find two types of parametrization methods capable of assigning location parameters to the measured data points during curve or surface fitting. Methods of the first type are good for parametrizing a chain of selected curve points or a regular grid of surface points (Lee 1989). In addition, there is also a base surface approach (Ma and Kruth 1995a) for parametrizing randomly located or distributed points for curve and surface fitting or interpolation. Further discussions on parametrization for surface fitting can also be found in Piegl and Tiller (2001) and Azariadis (2003). The parametrization methods discussed in this section can be used for both B-spline and NURBS surface fitting. They can also be used for parametrization-based subdivision surface fitting (Ma and Zhao 2000, 2002).

Uniform parametrization is the simplest method of assigning parameters to digitized data. In this case,

$$u_i = \frac{i-1}{m-1} \qquad 1 \le i \le m \tag{4.8}$$

for m curve points, and, analogously,

$$\begin{cases} u_i = \dfrac{i-1}{m_u - 1} & 1 \le i \le m_u \\[2ex] v_j = \dfrac{j-1}{m_v - 1} & 1 \le j \le m_v \end{cases} \tag{4.9}$$

with $(u_i,\ v_j)$, for a set of $m = m_u \times m_v$ regular grid surface points $\overline{\mathbf{P}}_{ij} = [\overline{x}_{ij}, \overline{y}_{ij}, \overline{z}_{ij}]^T$, for i=1, 2, ..., m_u, j=1, 2, ..., m_v. However, this method is sometimes unsatisfactory sometimes owing to the fact that the distribution of the measured points is not taken into account. When the measured points are unevenly spaced, a better choice is the cumulative chord length parametrization defined by

$$u_i = u_{i-1} + \frac{\left\| \overline{\mathbf{P}}_i - \overline{\mathbf{P}}_{i-1} \right\|}{\sum_{j=1}^{m-1} \left\| \overline{\mathbf{P}}_{j+1} - \overline{\mathbf{P}}_j \right\|} \qquad 2 \le i \le m \tag{4.10}$$

with $u_1 = 0.0$ for curve points. A set of $m_u \times m_v$ grid points for surface fitting can also be parametrized analogously using the following formulas

$$u_{ij} = \frac{\sum_{l=1}^{i-1} \left\| \overline{\mathbf{P}}_{l+1,j} - \overline{\mathbf{P}}_{l,j} \right\|}{\sum_{l=1}^{m_u -1} \left\| \overline{\mathbf{P}}_{l+1,j} - \overline{\mathbf{P}}_{l,j} \right\|} \qquad \text{for} \quad 2 \le i \le m_u, \quad 1 \le j \le m_v \tag{4.11a}$$

$$v_{ij} = \frac{\sum_{l=1}^{j-1} \left\| \overline{\mathbf{P}}_{i,l+1} - \overline{\mathbf{P}}_{i,l} \right\|}{\sum_{l=1}^{m_v -1} \left\| \overline{\mathbf{P}}_{i,l+1} - \overline{\mathbf{P}}_{i,l} \right\|} \qquad \text{for} \quad 1 \le i \le m_u, \quad 2 \le j \le m_v \tag{4.11b}$$

with $(u_{ij},\ v_{ij})$, for a set of $m = m_u \times m_v$ regular grid surface points, and with $u_{1j} = 0.0$ for j=1,2, ..., m_v and $v_{i1} = 0.0$ for $i = 1,2,...,m_u$. In case of surface fitting, the arc length refers to that of the isometric curve of interest on the surface. If the data points are more or less evenly spaced, equation (4.10) approximates to equation (4.8) of uniform parametrization and similarly equation (4.11) approximates to equation (4.9). One can also use a centripetal model (Lee 1989) for data parametrization , where

$$u_i = u_{i-1} + \frac{\left\| \overline{\mathbf{P}}_i - \overline{\mathbf{P}}_{i-1} \right\|^{\frac{1}{2}}}{\sum_{j=1}^{m-1} \left\| \overline{\mathbf{P}}_{j+1} - \overline{\mathbf{P}}_j \right\|^{\frac{1}{2}}} \qquad 2 \le i \le m \tag{4.12}$$

with $u_1 = 0$. This method observes the changing nature of the curvature for curve fitting. A generalization of the above three methods leads to the following exponent method (Lee 1989)

$$u_i = u_{i-1} + \frac{\left\| \overline{\mathbf{P}}_i - \overline{\mathbf{P}}_{i-1} \right\|^e}{\sum_{j=1}^{m-1} \left\| \overline{\mathbf{P}}_{j+1} - \overline{\mathbf{P}}_j \right\|^e} \qquad 2 \leq i \leq m \qquad (4.13)$$

with $u_1 = 0.0$ and $0.0 \leq e \leq 1.0$. When $e=0.0$, equation (4.13) reduces to the uniform parametrization and when $e=1.0$, equation (4.13) reduces to the cumulative chord length parametrization. The centripetal model is a method between the uniform and chord length parametrization with $e=0.5$. Both equation (4.12) and equation (4.13) for u_i's can be extended to a grid of surface points for u_{ij}'s similar to equation (4.11).

Methods discussed in the previous paragraphs apply to a set of chain points for curve fitting or a set of regular grid points for surface fitting. In case of randomly distributed points, however, all these methods fail. The base surface parametrization reported by Ma and Kruth (1995a), on the other hand, provides a stable solution and is a convenient approach for the parametrization of randomly distributed points. It can be applied to parametrize both regular and randomly distributed points. With the base surface approach, one also obtains a final fitted surface with very smooth parameter lines.

A base surface is actually a first approximation of the final fitted surface and is used to allocate parameters to the captured points for surface fitting. It can be constructed solely from the boundary information or boundary curves plus some other internal cross-section curves. The allocation of the parameters is realized by a projection process. Each of the measured points is first projected onto the base surface. The parameters of the projected point are then used as the parameters of the measured point. The projection is realized by minimizing the distance between the corresponding measured point and a surface point. Base surface parametrization can be applied to parametrize both regularly distributed and randomly distributed points.

4.4.2 Knots Allocation

A complete set of knots includes the order k, the number of control points n and a set of $n+k$ knots for curve fitting. For surface fitting, one needs to determine an order, the number of control points and a set of knots for each of the u- and v-directions, respectively.

The determination of the number of control points depends on the number of measured points available and how accurate the final fitted curve or surface is. For the case of curve fitting, the number of control points should be selected such that $2 \leq n \leq m$. The smaller the number of control points, the smoother the final fitted curve or surface. The larger the number of control points, On the other hand, the better the approximation of the final fitted curve or surface to the measured points. In the extreme case, one obtains a straight line segment if only two control points are used for curve fitting with $n=2$. If the number of control points equals the number of measured points with $n=m$, one obtains an interpolation B-spline curve, i.e., the fitted curve will exactly pass through all the measured points with zero residuals. The number of control points is often determined by controlling the sum of residuals of the final fitted curve or surface to be more or less equal to a pre-specified tolerance.

The order for curve fitting should be selected such that $2 \leq k \leq n$ and the orders for surface fitting should be selected such that $2 \leq k_u \leq n_u$ and $2 \leq k_v \leq n_v$. The higher the order, the smoother the final fitted curve or surface. However the lower the order, the better the controllability for shape local modification. In case $k=2$, the fitted curve will be a piecewise linear curve and, in case of $k_u = 2$ and $k_v = 2$, the fitted surface will be a piecewise bilinear surface. When the order equals the number of control points, i.e., with $k=n$ for curve fitting and with $k_u = n_u$ and $k_v = n_v$ for surface fitting, the final fitted curve or surface will be a Bézier curve or surface. For freeform shapes, one usually uses fourth order B-splines and the final fitted B-spline curve or surface is a cubic B-spline curve or surface.

Once the order and the number of control points have been determined, one can use either uniform knots or average knots for curve and surface fitting (Ma and Kruth 1995a). For uniform parametrization, the knots for curve fitting $\mathbf{t} := \{t_i\}_{i=1}^{n+k}$ are defined by

$$t_i = \begin{cases} 0 & i \leq k \\ \dfrac{i-k}{n-k+1} & k+1 \leq i \leq n \\ 1 & i \geq n+1 \end{cases} \tag{4.14}$$

In case of surface fitting, the knots are defined as

$$\xi_i = \begin{cases} 0 & i \leq k_u \\ \dfrac{i-k_u}{n_u-k_u+1} & k_u+1 \leq i \leq n_u \\ 1 & i \geq n_u+1 \end{cases} \tag{4.15a}$$

$$\zeta_j = \begin{cases} 0 & j \le k_v \\ \dfrac{j - k_v}{n_v - k_v + 1} & k_v + 1 \le j \le n_v \\ 1 & j \ge n_v + 1 \end{cases} \qquad (4.15b)$$

A better approach for setting the knots is to use average knots. In the case of curve interpolation with $n=m$, the average knots can be defined as

$$t_i = \begin{cases} u_1 & i \le k \\ \dfrac{\sum_{j=i-k}^{i-2} u_j}{k-1} & k+1 \le i \le n \;, \\ u_m & i \ge n+1 \end{cases} \qquad (4.16)$$

i.e., each of the knots is defined as the average of k-1 consecutive parameters of the measured points. In a similar way, one can also define average knots for the case of surface interpolation. Compared with uniform knots, average knots can greatly improve the fitting conditions (Ma and Kruth 1995a). Average knots for the general case of B-spline curve and surface fitting can also be found in (Ma and Kruth 1995a).

4.4.3 Least Squares Fitting

After parametrization, the remaining parameters requiring identification are the coordinates of the control points, which is a linear least squares fitting problem. To proceed further, we rewrite equations (4.2) and (4.3) in matrix form and switch the left and right sides to yield a uniform representation,

$$\begin{cases} \mathbf{b}^T(\cdot) \cdot \mathbf{x} = x(\cdot) \\ \mathbf{b}^T(\cdot) \cdot \mathbf{y} = y(\cdot) \\ \mathbf{b}^T(\cdot) \cdot \mathbf{z} = z(\cdot), \end{cases} \qquad (4.17)$$

where $(\cdot) = (u)$ for B-spline curves and $(\cdot) = (u,v)$ for B-spline surfaces representing the location parameters, $x(\cdot)$, $y(\cdot)$ and $z(\cdot)$ represent a point $\mathbf{p}(\cdot) = [x(\cdot), y(\cdot), z(\cdot)]^T$ on the curve or surface, $\mathbf{x}$, $\mathbf{y}$ and $\mathbf{z}$ represent the collection of x, y and z coordinates of the control points, and $\mathbf{b}(\cdot)$ stands for a collection of the basis functions. In case of a B-spline curve, we have,

$$\begin{cases} \mathbf{x} = [x_1, x_2, \cdots, x_n]^T \\ \mathbf{y} = [y_1, y_2, \cdots, y_n]^T \\ \mathbf{z} = [z_1, z_2, \cdots, z_n]^T, \end{cases} \qquad (4.18)$$

$$\mathbf{b}(u) = [B_1(u), B_2(u), \cdots, B_n(u)]^T \qquad (4.19)$$

and for B-spline surface, we have

$$\begin{cases} \mathbf{x} = [x_1, x_2, \cdots, x_n]^T = [x_{11}, x_{12}, \cdots, x_{1n_v}, x_{21}, \cdots, x_{n_u n_v}]^T \\ \mathbf{y} = [y_1, y_2, \cdots, y_n]^T = [y_{11}, y_{12}, \cdots, y_{1n_v}, y_{21}, \cdots, y_{n_u n_v}]^T \\ \mathbf{z} = [z_1, z_2, \cdots, z_n]^T = [z_{11}, z_{12}, \cdots, z_{1n_v}, z_{21}, \cdots, z_{n_u n_v}]^T, \end{cases} \qquad (4.20)$$

$$\begin{aligned} \mathbf{b}(u,v) &= [B_1(u,v), B_2(u,v), \cdots, B_n(u,v)]^T \\ &= [B_{u1}(u)B_{v1}(v), B_{u1}(u)B_{v2}(v), \ldots, B_{u1}(u)B_{vn_v}(v) \\ &\quad B_{u2}(u)B_{v1}(v), \ldots, B_{un_u}(u)B_{vn_v}(v)]^T \end{aligned} \qquad (4.21)$$

with $n = n_u \times n_v$ being the total number of control points.

Let $\mathbf{m} := \{\overline{\mathbf{p}}_i = [\overline{x}_i, \overline{y}_i, \overline{z}_i]^T\}_{i=1}^m$ be a set of m measured points, $\mathbf{u} := \{u_i\}_{i=1}^m$ for curve points and $\mathbf{u} := \{u_i\}_{i=1}^m$ and $\mathbf{v} := \{v_i\}_{i=1}^m$ for surface points be the *location parameters* allocated to $\mathbf{m}$ as discussed in the previous paragraphs. By introducing the measured points together with their *location parameters* into equation (4.17), we obtain the following observation equations,

$$\begin{cases} \mathbf{b}^T(\cdot_i) \cdot \mathbf{x} = \overline{x}_i \\ \mathbf{b}^T(\cdot_i) \cdot \mathbf{y} = \overline{y}_i \qquad \text{for} \quad i = 1, 2, \ldots, m. \\ \mathbf{b}^T(\cdot_i) \cdot \mathbf{z} = \overline{z}_i \end{cases} \qquad (4.22)$$

Or, in a compact matrix form,

$$\begin{cases} \mathbf{B} \cdot \mathbf{x} = \overline{\mathbf{x}} \\ \mathbf{B} \cdot \mathbf{y} = \overline{\mathbf{y}} \\ \mathbf{B} \cdot \mathbf{z} = \overline{\mathbf{z}}, \end{cases} \qquad (4.23)$$

in which

$$\begin{cases} \bar{\mathbf{x}} = [\bar{x}_1, \bar{x}_2, ..., \bar{x}_m]^T \\ \bar{\mathbf{y}} = [\bar{y}_1, \bar{y}_2, ..., \bar{y}_m]^T \\ \bar{\mathbf{z}} = [\bar{z}_1, \bar{z}_2, ..., \bar{z}_m]^T \end{cases} \tag{4.24}$$

represent the collection of x, y and z coordinates, respectively, of the measured points,

$$\mathbf{B} = \begin{bmatrix} B_1(\cdot_1) & B_2(\cdot_1) & \cdots & B_n(\cdot_1) \\ B_1(\cdot_2) & B_2(\cdot_2) & \cdots & B_n(\cdot_2) \\ \vdots & \vdots & & \vdots \\ B_1(\cdot_m) & B_2(\cdot_m) & \cdots & B_n(\cdot_m) \end{bmatrix}_{m \times n} \tag{4.25}$$

denotes the observation matrix. In these equations $(\cdot_i) = (u_i)$ for B-spline curves and $(\cdot_i) = (u_i, v_i)$ for B-spline surfaces, for i=1, 2, ..., m. The observation e quation (4.23) i s a linear s ystem. If n =m, o ne c an a chieve a solution that interpolates through all the observation data:

$$\begin{cases} \mathbf{x} &= \mathbf{B}^{-1} \cdot \bar{\mathbf{x}} \\ \mathbf{y} &= \mathbf{B}^{-1} \cdot \bar{\mathbf{y}} \\ \mathbf{z} &= \mathbf{B}^{-1} \cdot \bar{\mathbf{z}} \end{cases} \tag{4.26a}$$

In case m>n, a least squares solution can be found. In terms of the normal equation of the observation system, we obtain:

$$\begin{cases} \mathbf{x} &= (\mathbf{B}^T\mathbf{B})^{-1}\mathbf{B}^T \cdot \bar{\mathbf{x}} \\ \mathbf{y} &= (\mathbf{B}^T\mathbf{B})^{-1}\mathbf{B}^T \cdot \bar{\mathbf{y}} \\ \mathbf{z} &= (\mathbf{B}^T\mathbf{B})^{-1}\mathbf{B}^T \cdot \bar{\mathbf{z}} \end{cases} \tag{4.26b}$$

Several very stable algorithms capable of numerically solving the above equation are available. The most commonly used is Gaussian elimination without pivoting. Since matrix $\mathbf{B}$ is a banded matrix after reordering the observation points, it can be solved efficiently with low memory requirements using Cholesky factorization for positive banded matrix and a least squares fitting solution can be achieved when $m > n$. To solve equation (4.23) for x-coordinate, e.g., $\mathbf{x}$ can be found by minimizing

$$\min_{\mathbf{x}} D(\mathbf{x}) = (\mathbf{B} \cdot \mathbf{x} - \overline{\mathbf{x}})^T \cdot (\mathbf{B} \cdot \mathbf{x} - \overline{\mathbf{x}}). \qquad (4.27)$$

Matrix $\mathbf{B}$ has the following properties (Ma and Kruth 1995a):

- Matrix $\mathbf{B}$ is of full column rank if and only if the Schoenberg-Whitney conditions (Ma and Kruth 1995a) are satisfied. For numerical c omputations, t he c ondition n umber o f $\mathbf{B}^T\mathbf{B}$ is a lso influenced by the actual positions of the available location parameters of the B-spline inside its definition domain. For example, if all the parameters available to $B_i(u)$ are located at positions where $B_i(u)$ becomes almost zero, $\mathbf{B}^T\mathbf{B}$ will then become an ill-conditioned matrix although the Schoenberg-Whitney conditions may be satisfied.
- If $n \gg k$ for curve fitting or $n_u \gg k_u$ and $n_v \gg k_v$ for surface fitting, $\mathbf{B}$ is a sparse matrix. After row permutations or reordering the points to be fitted, $\mathbf{B}$ will become a banded matrix. Under the same conditions, $\mathbf{B}^T\mathbf{B}$ is a banded, symmetric and positive definite matrix, i.e., $\forall \mathbf{x} \in R^n$ and $\|\mathbf{x}\| \neq 0$, $\mathbf{x}^T(\mathbf{B}^T\mathbf{B})\mathbf{x} > 0$. The bandwidth of $\mathbf{B}^T\mathbf{B}$ is $2k-1$ for curve fitting and $(2k_u - 3)n_v + 2k_v$ for surface fitting (depending on the ordering of the control points).

Solutions to the B-spline based fitting examples discussed in Section 4.8 were realized by a stable algorithm using Householder transformations with stepwise refinement in order to achieve an accurate solution (Wilkinson and Reinsch 1971). The numerical algorithm is available in a mathematical library NAG (Numerical Algorithms Grope Ltd.) (NAG 1991 and 2002). Other stable algorithms that observe the banded structure of matrix $\mathbf{B}$ can also be found in (de Boor 1978, Cox 1990).

4.5 NURBS BASED MODEL RECONSTRUCTION

The main task in NURBS based model reconstruction is again NURBS curve and surface fitting or interpolation. The general procedure for NURBS based model reconstruction is the same as that for the B-spline based approach, i.e., parametrization of the measured points, parametrization of the knots and the identification of the remaining parameters through least squares fitting. Both the parametrization of the measured points and the parametrization of the knots are also the same as that for B-spline curve and surface fitting. The main difference is at the third step for the identification of the remaining parameters and, in the case of a NURBS based approach, the remaining parameters to be identified include both control points and the corresponding weights.

In literature, there have been many approaches developed for NURBS curve and surface fitting. In terms of the weights to be identified, the existing approaches can be classified into the following groups:

- Most of the approaches use simple B-splines by setting all the weights of the control points to 1.0 as known parameters. With pre-defined location parameters of the measured points and the u- and v-knots, this results in a linear least squares fitting problem and the final fitted curve and surfaces are B-splines.

- With the same parametrization of the measured points and the knots, some authors suggested to interpolate or fit the weighted data points $\{\overline{w}_i\overline{\mathbf{p}}_i, \overline{w}_i\}_1^m$ in 4D homogeneous space (Piegl 1991, Farin 1992). This is however not applicable to practical applications since the weights are not available. Farin suggested to allocate weights according to the curvature of the underlying curve and allocate higher weights to regions where the interpolant is expected to curve sharply (Farin 2002).

- Some other researchers also investigated a non-linear approach for NURBS curve and surface approximation. An example can be found in (Hoschek and Schneider 1994). It identifies both the control points and the weights of a NURBS curve or surface simultaneously by minimizing the sum of squares of distances from the measured points to the corresponding fitted curve or surface points. The minimization is solved by local linearization involving an iteration process. To avoid negative weights, a special transformation is used. A similar approach is also reported in (Laurent-Gengoux and Mekhilef 1993) and constrained non-linear minimization techniques are used to identify the unknowns of the observation system. In addition to the weights and the control points, the knots are also considered as unknowns in this chapter.

- In addition, a two-step linear approach is also reported in (Ma and Kruth 1994, Ma and Kruth 1995b, Ma and Kruth 1998). During the first step, the weights are first identified from a homogeneous system through Symmetric EigenValue Decomposition (SEVD). The control points are solved in a way similar to B-spline curve and surface fitting and interpolation with the identified weights as known parameters.

In this section, a two-step linear approach for least squares fitting of NURBS curves and surfaces (Ma and Kruth 1995b, Ma and Kruth 1998) is introduced.

4.5.1 A Two-Step Linear Approach

To facilitate further processing, we introduce the homogeneous coordinates of a point $\mathbf{p}(\cdot) = [x(\cdot), y(\cdot), z(\cdot)]^T \in R^3$ with corresponding weight $w(\cdot)$ as $\mathbf{P}(\cdot) = [X(\cdot), Y(\cdot), Z(\cdot), w(\cdot)]^T = [x(\cdot)w(\cdot), y(\cdot)w(\cdot), z(\cdot)w(\cdot), w(\cdot)]^T \in R^4$. For a control point $v_i = [x_i, y_i, z_i]^T \in R^3$ with corresponding weight w_i as $V_i = [X_i, Y_i, Z_i, w_i]^T = [x_i w_i, y_i w_i, z_i w_i, w_i]^T \in R^4$. We also use the matrix form of a NURBS curve and surface. After rewriting equations (4.6) and (4.7) in matrix form, moving the denominators to the other sides of the equations, and switching the left and right side of the equations, both of the equations can be written in the following uniform representation

$$\begin{cases} \mathbf{b}^T(\cdot) \cdot \mathbf{X} = x(\cdot) \cdot \mathbf{b}^T(\cdot) \cdot \mathbf{w} \\ \mathbf{b}^T(\cdot) \cdot \mathbf{Y} = y(\cdot) \cdot \mathbf{b}^T(\cdot) \cdot \mathbf{w} \\ \mathbf{b}^T(\cdot) \cdot \mathbf{Z} = z(\cdot) \cdot \mathbf{b}^T(\cdot) \cdot \mathbf{w} \end{cases} \tag{4.28}$$

where $(\cdot) = (u)$ for NURBS curves and $(\cdot) = (u,v)$ for NURBS surfaces represent the location parameters, $x(\cdot)$, $y(\cdot)$ and $z(\cdot)$ represent a point $\mathbf{p}(\cdot) = [x(\cdot), y(\cdot), z(\cdot)]^T$ on the curve or surface, $\mathbf{b}(\cdot) = [B_1(\cdot), B_2(\cdot), \cdots, B_n(\cdot)]^T$ stands for a collection of the basis functions, $\mathbf{X}$, $\mathbf{Y}$, $\mathbf{Z}$ and $\mathbf{w}$ represent the collection of x, y, z and w coordinates of the control points in homogeneous space. For NURBS curves, the homogeneous coordinates are given by

$$\begin{cases} \mathbf{X} = [X_1, X_2, \cdots, X_n]^T = [w_1 x_1, w_2 x_2, \cdots, w_n x_n]^T \\ \mathbf{Y} = [Y_1, Y_2, \cdots, Y_n]^T = [w_1 y_1, w_2 y_2, \cdots, w_n y_n]^T \\ \mathbf{Z} = [Z_1, Z_2, \cdots, Z_n]^T = [w_1 z_1, w_2 z_2, \cdots, w_n z_n]^T \\ \mathbf{w} = [w_1, w_2, \cdots, w_n]^T \end{cases} \tag{4.29}$$

and for NURBS surfaces by

$$
\begin{cases}
\mathbf{X} &= \left[X_1, X_2, \cdots, X_n\right]^T \\
&= \left[w_{11}x_{11}, w_{12}x_{12}, \cdots, w_{1n_v}x_{1n_v}, \cdots, w_{n_u1}x_{n_u1}, w_{n_u2}x_{n_u2}, \cdots, w_{n_un_v}x_{n_un_v}\right]^T \\
\mathbf{Y} &= \left[Y_1, Y_2, \cdots, Y_n\right]^T \\
&= \left[w_{11}y_{11}, w_{12}y_{12}, \cdots, w_{1n_v}y_{1n_v}, \cdots, w_{n_u1}y_{n_u1}, w_{n_u2}y_{n_u2}, \cdots, w_{n_un_v}y_{n_un_v}\right]^T \\
\mathbf{Z} &= \left[Z_1, Z_2, \cdots, Z_n\right]^T \\
&= \left[w_{11}z_{11}, w_{12}z_{12}, \cdots, w_{1n_v}z_{1n_v}, \cdots, w_{n_u1}z_{n_u1}, w_{n_u2}z_{n_u2}, \cdots, w_{n_un_v}z_{n_un_v}\right]^T \\
\mathbf{w} &= \left[w_1, w_2, \cdots, w_n\right]^T
\end{cases}
$$

$$(4.30)$$

with $n = n_u \times n_v$ being the total number of control points.

With the same notations as used for B-splines, let $\left\{\overline{\mathbf{p}}_i = \left[\overline{x}_i, \overline{y}_i, \overline{z}_i\right]^T\right\}_{i=1}^m$ be a set of measured points, $\{u_i\}_{i=1}^m$ be the location parameters in case of curve fitting and $\{u_i, v_i\}_{i=1}^m$ be the location parameters in case of surface fitting. By introducing the measured points together with the location parameters into equation (4.28), we obtain the following observation equations,

$$
\begin{cases}
\mathbf{b}^T(\cdot_i) \cdot \mathbf{X} = \overline{x}_i \cdot \mathbf{b}^T(\cdot_i) \cdot \mathbf{w} \\
\mathbf{b}^T(\cdot_i) \cdot \mathbf{Y} = \overline{y}_i \cdot \mathbf{b}^T(\cdot_i) \cdot \mathbf{w} \qquad \text{for} \quad i = 1, 2, \ldots, m. \\
\mathbf{b}^T(\cdot_i) \cdot \mathbf{Z} = \overline{z}_i \cdot \mathbf{b}^T(\cdot_i) \cdot \mathbf{w}
\end{cases}
$$

$$(4.31)$$

where $(\cdot_i) = (u_i)$ for NURBS curves and $(\cdot_i) = (u_i, v_i)$ for NURBS surfaces for $i = 1, 2, \ldots, m$ are the location parameters. In a compact matrix form, we have

$$
\mathbf{A} \cdot \begin{bmatrix} \mathbf{X} \\ \mathbf{Y} \\ \mathbf{Z} \\ \mathbf{w} \end{bmatrix} = \begin{bmatrix} \mathbf{B} & \mathbf{0} & \mathbf{0} & -\overline{\mathbf{x}}\cdot\mathbf{B} \\ \mathbf{0} & \mathbf{B} & \mathbf{0} & -\overline{\mathbf{y}}\cdot\mathbf{B} \\ \mathbf{0} & \mathbf{0} & \mathbf{B} & -\overline{\mathbf{z}}\cdot\mathbf{B} \end{bmatrix}_{3m \times 4n} \cdot \begin{bmatrix} \mathbf{X} \\ \mathbf{Y} \\ \mathbf{Z} \\ \mathbf{w} \end{bmatrix} = [\mathbf{O}]_{4n \times 1} \qquad (4.32)
$$

where $\mathbf{B}$ is the usual observation matrix that is uniquely defined by the location parameters and the knots and is defined in the same way as equation (4.25) for B-splines, and $\overline{\mathbf{x}}$, $\overline{\mathbf{y}}$ and $\overline{\mathbf{z}}$ are diagonal matrices whose diagonal elements contain the x-, y- and z-coordinates of the measured points defined in the same way as equation (4.24) for B-splines. By manipulating the block matrix in equation (4.32), it is possible to divide the whole system into two subsystems. The weights and the control points are then solved in two subsequent steps. The separation is realized by first left-multiplying equation

(4.32) by the transpose of $\mathbf{A}$ and then eliminating the first three elements of the last row of the new left-side matrix. This yields the following equation

$$\begin{bmatrix} \mathbf{B}^T\mathbf{B} & \mathbf{0} & \mathbf{0} & -\mathbf{B}^T\overline{\mathbf{x}}\mathbf{B} \\ \mathbf{0} & \mathbf{B}^T\mathbf{B} & \mathbf{0} & -\mathbf{B}^T\overline{\mathbf{y}}\mathbf{B} \\ \mathbf{0} & \mathbf{0} & \mathbf{B}^T\mathbf{B} & -\mathbf{B}^T\overline{\mathbf{z}}\mathbf{B} \\ \mathbf{0} & \mathbf{0} & \mathbf{0} & \mathbf{M} \end{bmatrix}_{4n\times 4n} \cdot \begin{bmatrix} \mathbf{X} \\ \mathbf{Y} \\ \mathbf{Z} \\ \mathbf{w} \end{bmatrix} = [\mathbf{0}]_{4n\times 1} \tag{4.33}$$

where

$$\mathbf{M} = \mathbf{M}_x + \mathbf{M}_y + \mathbf{M}_z \tag{4.34}$$

is a $n\times n$ symmetric and non-negative matrix with

$$\begin{cases} \mathbf{M}_x = \mathbf{B}^T\overline{\mathbf{x}}^2\mathbf{B} - \left(\mathbf{B}^T\overline{\mathbf{x}}\mathbf{B}\right)\left(\mathbf{B}^T\mathbf{B}\right)^{-1}\left(\mathbf{B}^T\overline{\mathbf{x}}\mathbf{B}\right) \\ \mathbf{M}_y = \mathbf{B}^T\overline{\mathbf{y}}^2\mathbf{B} - \left(\mathbf{B}^T\overline{\mathbf{y}}\mathbf{B}\right)\left(\mathbf{B}^T\mathbf{B}\right)^{-1}\left(\mathbf{B}^T\overline{\mathbf{y}}\mathbf{B}\right) \\ \mathbf{M}_z = \mathbf{B}^T\overline{\mathbf{z}}^2\mathbf{B} - \left(\mathbf{B}^T\overline{\mathbf{z}}\mathbf{B}\right)\left(\mathbf{B}^T\mathbf{B}\right)^{-1}\left(\mathbf{B}^T\overline{\mathbf{z}}\mathbf{B}\right) \end{cases} \tag{4.35}$$

Extracting the fourth row from equation (4.33), we obtain the following subsystem for identifying the weights

$$\mathbf{M} \cdot \mathbf{w} = [\mathbf{0}]_{n\times 1}. \tag{4.36}$$

This leads to the two-step linear approach for NURBS fitting. During the first step, the weights of the control points are solved from the above-mentioned homogeneous system (4.36). Further details will be explained in Section 4.5.2. During the second step, the control points are solved from equation (4.31) by taking the identified weights as known parameters. In matrix form, we revert to the following equation for identifying the control points in homogeneous space

$$\begin{cases} \mathbf{B} \cdot \mathbf{X} = \overline{\mathbf{x}} \cdot \mathbf{B} \cdot \mathbf{w} \\ \mathbf{B} \cdot \mathbf{Y} = \overline{\mathbf{y}} \cdot \mathbf{B} \cdot \mathbf{w} \\ \mathbf{B} \cdot \mathbf{Z} = \overline{\mathbf{z}} \cdot \mathbf{B} \cdot \mathbf{w} \end{cases} \tag{4.37}$$

The coordinates of the control points $\mathbf{v}_i = [x_i, y_i, z_i]^{\mathrm{T}}$ in Euclidean space R^3 can be further computed by back projection from the homogeneous space to the Euclidean space, i.e.,

$$\begin{cases} x_i = X_i / w_i \\ y_i = Y_i / w_i \\ z_i = Z_i / w_i \end{cases} \qquad \text{for} \quad i = 1, 2, \cdots, n . \tag{4.38}$$

4.5.2 Numerical Algorithms for Weights Identification

Since equation (4.36) is homogeneous, non-zero solutions exist if and only if the rank of matrix $\mathbf{M}$ is less than n, i.e., $rank(\mathbf{M}) < n$. In addition, if $\mathbf{w} \in R^n$ is a solution of equation (4.36), for all $\alpha \in R$ and $\alpha \neq 0$, $\alpha\mathbf{w}$ is also a solution of (4.36). Due to the last property, one can simply consider the following criterion for weights identification

$$\min_{\|\mathbf{w}\|_2 = 1} \|\mathbf{M} \cdot \mathbf{w}\|_2^2 \tag{4.39}$$

i.e., to minimize $\|\mathbf{M} \cdot \mathbf{w}\|_2^2$ from all $\mathbf{w}$ located on the unit circle in the n-dimensional Euclidean space. It is easy to prove that the following equation is equivalent to equation (4.39) for $\mathbf{w}$,

$$\min_{\|\mathbf{w}\|_2 \neq 0} R(\mathbf{w}) = \frac{\mathbf{w}^T \mathbf{Q} \mathbf{w}}{\mathbf{w}^T \mathbf{w}} \tag{4.40}$$

where

$$\mathbf{Q} = \mathbf{M}^T \mathbf{M} . \tag{4.41}$$

$R(\mathbf{w})$ is called the *Rayleigh quotient* in linear algebra (Golub and Van Loan 1989). The properties of $R(\mathbf{w})$ are well understood. Among others, the relationship between the stationary points of $R(\mathbf{w})$ and the SEVD of $\mathbf{Q}$, and the minimax theorem are very important for the identification of NURBS weights. As an application, both the general solution and solution with positive weights of equation (4.36) can be represented as a linear combination of some eigenvectors of $\mathbf{Q}$ corresponding to smaller eigenvalues.

In view of the importance of the relationship between the Singular Value Decomposition (SVD) of $\mathbf{M}$ and the SEVD of $\mathbf{Q}$ (Golub and Van Loan

1989), one can alternatively use the SVD of $\mathbf{M}$ for NURBS identification. Moreover, as $\mathbf{M}$ is a symmetric and non-negative matrix, the SEVD of $\mathbf{M}$ exists. Following the definition of SVD and observing the symmetric and non-negative nature of $\mathbf{M}$, the SEVD of $\mathbf{M}$ is actually a SVD of $\mathbf{M}$. Thus we can simply use the SEVD of $\mathbf{M}$ instead of the SEVD of $\mathbf{Q}$ for NURBS identification. This fact can also be directly proved. Let

$$\mathbf{M} = \mathbf{P}\mathbf{D}\mathbf{P}^{\mathrm{T}},\tag{4.42}$$

be the SEVD of $\mathbf{M}$, where

$$\mathbf{D} = diag\{d_1,d_2,...,d_n\}\tag{4.43}$$

is a diagonal matrix whose diagonal elements are the eigenvalues of $\mathbf{M}$ in decreasing order with $d_i \geq d_{i+1} \geq 0.0$, and $\mathbf{P}$ is an orthogonal matrix whose columns $\mathbf{p}_i$ for $i=1, 2, ..., n$ are eigenvectors of $\mathbf{M}$ corresponding to d_i. The SEVD of $\mathbf{Q}$ is then given by

$$\mathbf{Q} = \mathbf{P}\mathbf{D}^2\mathbf{P}^{\mathrm{T}}.\tag{4.44}$$

Both the L_2-norm of the residuals $\gamma = \sqrt{R(\mathbf{w})}$ and $R(\mathbf{w})$ can be used to test whether the final fitted NURBS curve or surface with the identified weights is good or not.

Once the SEVD of $\mathbf{M}$ is available, we are ready to identify the weights. Let equation (4.42) be the Symmetric EigenValue Decomposition (SEVD) of $\mathbf{M}$, $E^i = span\{\mathbf{p}_n,\mathbf{p}_{n-1},\cdots,\mathbf{p}_{n-i+1}\}$ be an invariant subspace of R^n and $\mathbf{w}_1 = [1,1,\cdots,1]^T \subset R^n$ be the uniform weights that define B-splines. Let us further define R^{n+} as a subspace of R^n such that $\forall \mathbf{w} \in R^n$, all of its elements are positive, i.e., $w_i > 0.0$ for $i=1, 2, ..., n$. This leads to the following solutions for equation (4.36).

4.5.2.1 General solution

The general solution of the weights $\mathbf{w} \in R^n$ with $\|\mathbf{w}\|_2 \neq 0$ are given by

$$\mathbf{w} = \sum_{i=p+1}^{n} \alpha_i \mathbf{p}_i\tag{4.45}$$

where p is an index such that $d_p > d_{p+1} = \cdots = d_n$, $\{\mathbf{p}_i\}_{i=p+1}^{n}$ are eigenvectors corresponding to the smallest eigenvalues $\{d_i\}_{i=p+1}^{n}$ of $\mathbf{M}$, and $\{\alpha_i\}_{i=p+1}^{n}$ are arbitrary coefficients with $\sum_{i=p+1}^{n} \alpha_i^2 \neq 0$. The corresponding L_2-norm of the residuals of equation (4.36) is given by

$$\gamma = \sqrt{R(\mathbf{w})} = \|\mathbf{M} \cdot \mathbf{w}_0\|_2 = d_{p+1} = \cdots = d_n \qquad (4.46)$$

i.e., the smallest eigenvalues of $\mathbf{M}$.

Equation (4.45) also implies that, for a given set of α_i's, $\mathbf{w}$ is a vector in subspace E^{n-p}, i.e., $\mathbf{w} \in E^{n-p}$. E^{n-p} is therefore called the solution subspace that contains all the solutions of equation (4.36). It gives the interpolation or exact solutions of equation (4.36) if the rank r of $\mathbf{M}$ satisfies $r < n$, i.e., $d_i = 0$ for $i=p+1$, $p+2$, ..., n. It also provides the best fitting solutions if the rank r of M satisfies $r = n$, i.e., $d_n \neq 0$.

4.5.2.2 Solution for positive weights

Equation (4.45) provides a collection of all interpolation or best fitting solutions with possibly negative weights. As negative weights may introduce singularities and are not expected in practical applications, the following algorithms can be used for obtaining a set of positive weights in the eigenspace of $\mathbf{M}$.

We first check if the general solution contains positive alternatives. If there exists $\mathbf{w} \in E^{n-p}$ such that all the elements of $\mathbf{w}$ are positive, $\mathbf{w}$ is then a set of positive interpolation or best fitting weights. $\mathbf{w}$ can be computed from the following minimization algorithm,

$$\begin{cases} \min_\beta \|\mathbf{w} - \mathbf{w}_1\|_2^2 \\ \text{subject to} : w_l \leq w_i \leq w_u \end{cases} \qquad (4.47)$$

where $\mathbf{w} = \sum_{i=p+1}^{n} \beta_i \mathbf{p}_i$ and $w_u \geq w_l > 0.0$ are positive upper and lower bounds for the weights. The objective function of equation (4.47) guarantees a set of stable solutions, i.e., far from singularities controlled by attracting the weights towards the unit weights $\mathbf{w}_1$ and subjecting all the weights to the lower and upper bounds. It also results in smaller convex hulls for the same sum of fitting residuals. If no positive interpolation or best fitting solutions exist, one can still achieve a set of best fitting positive weights. The basic strategy is as the following. We first look for the *best* subspace of R^n that contains positive weights and a set of feasible solutions in this subspace. Starting from this feasible solution, we try to optimize the weights in this subspace in the sense of least value of $R(\mathbf{w})$. The meaning of the *best* is twofold. On the one hand, the maximum value of the L_2-norm of the residuals γ inside this subspace is the smallest compared with other subspaces containing positive weights. On the other hand, this subspace is the largest one compared with other subspaces having the same maximum value of the L_2-norm of the residuals.

To find the best subspace containing positive weights and a set of feasible solutions in this subspace, let q be an index such that $d_q > d_{q+1} = \cdots = d_p$. Furthermore, let $l=n-q$ and $\mathbf{w} = \mathbf{p}_n$. We try to move $\mathbf{w}$ into $E^l \cap E^{n+}$, i.e., to find a set of positive weights $\mathbf{w}$ in E^l. If it is successful, following the minimax theorem (Golub and Van Loan 1989), E^l is then the best subspace containing positive weights and $\mathbf{w}$ is a feasible solution in this subspace. Otherwise, l is incremented to include a new group of eigenvectors corresponding to the next larger eigenvalue, i.e., to include a new eigen subspace, and the searching process is continued till the objective is satisfied. The algorithm used to move $\mathbf{w}$ into $E^l \cap E^{n+}$ is quadratic programming with a null objective function subjecting to positive constraints, i.e., $w_l \leq w_i \leq w_u$ for $i=1, 2, ..., n$ with $\mathbf{w} = \sum_{i=n-l+1}^{n} \beta_i \mathbf{p}_i$.

When we have the best subspace E^l and a feasible solution in E^l, the following minimization problem can be solved to find the set of best fitting solutions in this subspace.

$$\begin{cases} R(\beta) = \min_\beta \dfrac{\sum_{i=n-l+1}^{n} \beta_i^2 d_i^2}{\sum_{i=n-l+1}^{n} \beta_i^2} \\ \text{subject to}: w_l \leq w_i \leq w_u \end{cases} \qquad (4.48)$$

where $\mathbf{w} = \sum_{i=n-l+1}^{n} \beta_i \mathbf{p}_i$. The objective function of equation (4.48) is derived by introducing $\mathbf{w}$ into equation (4.40). According to the maximal theorem of $\mathbf{w}$ (Golub and Van Loan 1989), the L_2-norm of the residuals γ of $\mathbf{w}$ is bounded by

$$\gamma = \|\mathbf{M} \cdot \mathbf{w}\|_2 = \sqrt{\frac{\sum_{i=n-l+1}^{n} \beta_i^2 d_i^2}{\sum_{i=n-l+1}^{n} \beta_i^2}} \leq d_{n-l+1}. \qquad (4.49)$$

One may use any constrained minimization algorithms to solve equations (4.47) and (4.48) for a set of positive weights. Solutions to the examples discussed in Section 4.8 were obtained using a general minimization function E04UCF found in the NAG (Numerical Algorithms Group Ltd.) mathematical library (NAG 1991 and 2002). The feasible solution of equation (4.48) was also obtained by calling a general quadratic programming function E04NCF of NAG with a null/zero objective function.

4.6 OTHER APPROACHES FOR MODEL RECONSTRUCTION

4.6.1 Basic Geometric Features

For basic geometric features, the parameters can be identified by solving the following least squares problem.

$$\min_{\mathbf{x}} D(\mathbf{x}) = \min_{\mathbf{x}} \sum_{i=1}^{m} d_i^2 \qquad (4.50)$$

where, $\mathbf{x} = [x_1, x_2 \cdots, x_n]^T \in R^n$ represents the parameters to be identified, d_i is the normal distance from the i-th measured point $\overline{\mathbf{p}}_i = [x_i, y_i, z_i]^T$ to the geometric feature. In (Kruth and Ma 1992), equation (4.50) was solved for six basic features, i.e. line, circle, plane, cylinder, cone and sphere, using general optimization techniques. Some of the basic geometric features can also be identified through algebraic surfaces (see following Section 4.6.2). One can also find other approaches for the identification of these basic geometric elements from sample data points in (Varady *et al.* 1998, Ahn *et al.* 2001, Piegl and Tiller 2002).

4.6.2 General Algebraic Surfaces

In equation (4.1), if $\alpha = \beta = \gamma = \mu = \nu = \omega = 0$, a plane is defined. If $\alpha = \beta = \gamma = -\rho = 1$ and the remaining coefficients are zero, a unit sphere is defined at the origin. One could therefore solve a subset of the problems using methods for regular feature fitting. Given a set of measured points whose underlying geometry is defined by equation (4.1) in a general setting. One can also use the following approach. We assume that all of the points approximately satisfy equation (4.1). Thus for any point $\overline{\mathbf{p}}_i = [x_i, y_i, z_i]^T$, we obtain $f(x_i, y_i, z_i) = \varepsilon_i$, where ε_i is a residual fitting error. All parameters $\mathbf{x} = [\alpha, \beta, \gamma, \mu, \nu, \omega, \delta, \sigma, \eta, \rho]^T$ can then be calculated using the least squares fitting technique from the following equation, so an algebraic surface is defined.

$$\min_{\mathbf{x}} \sum_{i=1}^{n} \varepsilon_i^2 = \min_{\mathbf{x}} \sum_{i=1}^{n} f^2(x_i, y_i, z_i) \qquad (4.51)$$

Some further discussions on algebraic surface fitting and interpolation can be found in (Pratt 1987 and Bloomenthal *et al.* 1997).

4.6.3 Subdivision Surface Fitting

In literature, many subdivision schemes have been developed for modeling and animation (Zorin and Schröder 2000). One can also find several approaches for fitting subdivision surfaces to range or scanning data. Here we briefly introduce approaches in this area and leave it to the readers to consult the original papers for further details.

Among others, Hoppe *et al.* (1994) present an approach for automatically fitting subdivision surfaces from a dense triangle mesh. The fitting criterion is defined such that the final fitted control mesh after two subdivisions will be closest to the initial dense mesh, which provides sufficient approximation for practical applications. Both smooth and sharp models can be handled. This approach produces high quality visual models. However, it may need extensive computing time due to the large amount of data to be processed and a procedure involved for mesh optimization.

The method of Suzuki *et al.* (1999) for subdivision surface fitting starts from an interactively defined initial control mesh. For each of the control vertices, a corresponding limit position is obtained from the initial dense mesh. The final positions of the control vertices are then inversely solved following the relationship between the control vertices and the corresponding limit positions. The fitted subdivision surface is checked against a pre-defined tolerance. In case of need, the topological structure of the control mesh is further subdivided and a refined mesh of limit positions is obtained. A new subdivision surface is then fitted from the corresponding refined limit positions. The procedure is repeated until the fitted subdivision surface meets the allowed tolerance bound. The fitting criterion is defined such that the resulting subdivision surface interpolates the corresponding limit positions of the current control mesh. Since only a subset of vertices of the initial dense mesh is involved in the fitting procedure, the computing speed is extremely high. However, local and sharp features are not observed with the reported implementation.

Ma and Zhao (2000 and 2002) also report a parametrization-based approach for fitting Catmull-Clark subdivision surfaces. A network of boundary curves is first interactively defined for topological modeling. A set of base surfaces is then defined from the topological model for sample data parametrization. A set of observation equations is obtained based on ordinary B-splines for regular surface patches and an evaluation scheme of Stam (1998) for extraordinary surface patches. A Catmull-Clark surface is finally obtained through linear least squares fitting. The approach makes use of all known sample points and the fitting criterion ensures best fitting condition between the input data and the final fitted subdivision surface.

Apart from the parametrization procedure, the fitting process alone is pretty fast.

In a recent paper, Ma *et al.* (2003a) propose a direct method for subdivision surface fitting from a known dense triangle mesh. A feature- and topology-preserving mesh simplification algorithm is first used for topological modeling and a direct fitting approach is developed for subdivision surface fitting. For setting up the fitting equations, both subdivision rules and limit position evaluation masks are used. The final subdivision surface is obtained through linear least squares fitting without any constraints. Sharp features are also preserved during the fitting procedure. While only a modified Loop subdivision scheme (Loop 1987) was discussed in the paper, the algorithms reported in (Ma *et al.* 2003a) can be applied to all stationary subdivision schemes. The computing time is also extremely small. On the other hand, the quality of the fitted subdivision surface is sensitive to the quality of the simplified mesh and a procedure for mesh optimization needs to be used. Edge swapping and a few other geometric rules were used for mesh optimization when producing the examples reported in the paper.

All the above fitting approaches are built upon schemes for limit surface query, either at discrete positions (Hope *et al.* 1994, Suzuki *et al.* 1999, Ma *et al.* 2003a) or at arbitrary parameters (Ma and Zhao 2000, Ma and Zhao 2002). In literature, one can also find schemes interpolating a set of surface limit positions with or without surface normal vectors instead of fitting conditions (Halstead *et al.* 1993, Nasri and Sabin 2002). In theory, all interpolation schemes can be extended as a fitting scheme if the number of known surface conditions, such as limit positions, normal vectors and other constraints, are more than those of the unknown control vertices of the subdivision surface. One can also find an approach for adaptively fitting a Catmull-Clark subdivision surface to a given shape through a fast local adaptation procedure (Litke *et al.* 2001). The fitting process starts from an initial approximate generic model defined as a subdivision surface of the same type.

4.7 SURFACE LOCAL UPDATING

After CAD modeling, modifications to the physical part might still be needed. One may repeat the whole reverse engineering process to obtain a new CAD model. However it might not be justified to re-compute the entire geometric model if only a small portion of the physical model is changed. In this case, one only needs to measure the modified area and update the CAD model locally according to the newly measured data.

4.7.1 Related Work and General Strategies

In the broad area of computer aided geometric design (CAGD), there are a variety of techniques for shape modification, either global alteration or local editing. The available techniques depend on the representation schemes of the geometric models to be modified or updated. For example, for NURBS surfaces, the designer can modify a local surface area by interactively moving a control point or changing its corresponding weight. With this general method, it is however very difficult to exactly predict the outcome of the modifications. To this end, (Piegl 1989a, 1989b) presented methods for modifying the shape of rational B-spline curves and surfaces by relocating points on the curves and surfaces. The modification is realized by repositioning one or more control points and/or modifying the corresponding weights based on geometric interpretations. (Piegl and Tiller 1995) discussed details concerning surface wrapping, flattening, bending, stretching, and twisting using the same approach. (Chen *et al.* 1997) introduced a method that allowed direct modification of a highlight line on a NURBS surface by adjusting the control points of the surface. The process results in a modified surface that produces the desired highlight line. In addition, (Au and Yuen 1995) investigated how the shape of a NURBS curve changes when the weights and location of control points are modified simultaneously. One can also find a technique in (Sánchez-Reyes 1997) for modifying the shape of NURBS curves and surfaces based on perspective functional transformation.

Instead of the above mentioned shape modification techniques, (Ma and He 1998) presented an approach for local updating of freeform surfaces based on locally measured and randomly distributed points suited for reverse engineering applications. The reported algorithms are based on the principles of B-splines and local least squares fitting techniques. The first step in surface local updating is to evaluate which surface and which surface area need to be updated. When the areas for updating are identified, the local surface areas can then be modified to fit the newly digitized points through a local fitting procedure. Depending on the changes in the complexity of the surface, knot insertion algorithms may be applied before updating. The following discussions are based on B-spline mathematics. The whole procedure for local updating (Ma and He 1998) is subdivided into the following steps:

- Initial surface identification and data segmentation.
- Identification of surface patches for each of the identified surfaces for local updating.
- Knot insertion for region isolation and identification of control points for local updating.

- Computing the updated control points through a fitting procedure.

4.7.2 Pre-Processing Steps for Surface Local Updating

At the very beginning, all measured points are segmented and all individual surfaces needing updating are identified. The segmentation and identification are realized by projecting all measured points onto the CAD model. All surfaces including projected points are then updated. At the same time, the measured points are also segmented and partitioned into regions, each of which uniquely belongs to an individual surface for local updating.

Once the data have been partitioned and segmented, all affected surfaces of the CAD model can be updated one by one. For each surface, the affected surface patches are further identified. The identification can also be based on the projection results of the previous step for data partition. At the same time, the measured points are also parametrized based on the projection results (Ma and Kruth 1995a).

During the next step, all identified regions are isolated through knot insertion and control points used for local updating are identified. Region isolation is based on the local support property, one of the most important properties of B-spline curves and surfaces. If one changes the position of a control point, the shape of the surface will only change locally and there will be in total $k_u \times k_v$ patches affected by modifying a single control point. To decide whether to apply knot insertion or not, there are two cases:

- If one wants to update a single surface patch without any changes to its neighboring patches, k_u-1 and k_v-1 knots must be inserted within the patch along u and v directions respectively. After knot insertion, there will be one control point whose changes would not affect any of its neighboring patches. If there is more than one patch needing updating, the number of knots to be inserted will be less. However, if the area to be updated is very complex, one might need to insert more number of knots than the minimum requirement.
- Alternatively, if minor deviation is allowed on the neighboring patches, one could introduce artificial points evaluated on the neighboring patches and combine the artificial points with the measured points such that a region with sufficient number of contingent patches are produced and sufficient number of control points could be identified whose changes would not affect patches beyond the new area boundary.

The numbers of knots to be inserted before surface local updating depend on whether the neighboring patches are allowed for alteration, the numbers of patches involved for updating and the local fitting conditions after updating. If more knots are inserted in the region to be updated, one achieves better local fitting conditions. The use of excess knots may however degrade the smoothness of the updated surface and the condition for c ontrol p oint b ased s hape l ocal c ontrol. I n a ddition, t he u se o f e xcess knots will result in redundant parameters for defining the surface. The number of knots for insertion should therefore be minimized. In order to obtain a uniform distribution of knots and enhance ease of implementation, it is better to insert the same number of knots, if any, for all patches falling within the region to be updated along the same parametric direction. After knot insertion, all control points are extracted and are registered for later local surface fitting.

4.7.3 Computing Updated Control Points

After completing the registration of the control points, we should have all the necessary information for surface local updating through a fitting procedure. Let $\{V\}$ be a collection of all the control points of the initial surface, i.e., the surface to be updated after knot insertion, $\{V_u\}$ be the collection of the extracted or registered control points for surface local updating, and $\{V_n\}$ be the collection of the remaining control points of the surface to be updated, i.e., $\{V_u\} = \{V_u\} \cup \{V_n\}$ and $\{V_u\} \cap \{V_n\} = \{0\}$ using set operations. Other known information includes the set of locally measured points plus artificial points, if any, their location parameters and the refined knot vectors of the surface after knot insertion. Following a similar derivation discussed in Section 4.4 with regard to least squares fitting of B-spline surfaces, we can obtain the same observation equation as equation (4.23). For surface local updating, only some of the control points, i.e., $\{V_u\}$, a re u nknowns a nd t he remaining c ontrol p oints in $\{V_n\}$ a re k nown parameters. By re-organizing the columns of the matrix $\mathbf{B}$, we can separate the control points to be updated from others and they can then be solved in the same way summarized in Section 4.4.3.

After reorganizing the columns of $\mathbf{B}$ in equation (4.23), we obtain the following equation

$$\mathbf{B} \cdot \mathbf{x} = \begin{bmatrix} \mathbf{B}_u & \mathbf{B}_n \end{bmatrix} \cdot \begin{bmatrix} \mathbf{x}_u \\ \mathbf{x}_n \end{bmatrix} = \overline{\mathbf{x}} \qquad (4.52)$$

where the elements of $\mathbf{x}_u$ correspond to the x-coordinates of the control points $\{V_u\}$ to be updated and are unknowns of equation (4.52), and the elements of $\mathbf{x}_n$ correspond to the x-coordinates of the control points $\{V_n\}$

and are known parameters of equation (4.52). Moving all known elements to the right side,

$$\mathbf{B}_u \cdot \mathbf{x}_u = \overline{\mathbf{x}} - \mathbf{B}_n \cdot \mathbf{x}_n = \overline{\mathbf{x}}' \qquad (4.53)$$

We can also similarly obtain the observation equations of the y- and z-components for surface local updating. Equation (4.53) can be solved using the same technique as used for solving equation (4.23).

Since the number of unknowns in equation (4.53) is usually much smaller compared with the number of all the control points, the computing requirement of equation (4.53) is much lower. When $\{\mathbf{V}_u\}$ is resolved or updated, the complete updated surface is defined by the control points $\{\mathbf{V}_u\} = \{\mathbf{V}_u\} \cup \{\mathbf{V}_n\}$, and the new knot vectors after knots insertion, if any.

With regard to equation (4.53), if there are any patches to be updated/altered where there are no points for updating, one should also generate some artificial points on these patches for updating. Otherwise, matrix $\mathbf{B}_u$ will not be of full rank.

4.8 EXAMPLES ON MODEL RECONSTRUCTION

In this section, we provide some selected examples on model reconstruction for reverse engineering applications. Figure 4-9 illustrates a flow chart of the whole reverse engineering process. It also indicates possible strategies for data acquisition and CAD modeling. Starting from a physical model, one may use any digitizing system, such as a laser scanner, a coordinate measuring machine, a CT scanner or other measuring instruments, to capture the shape information. The measured data can then be further segmented and portioned followed by surface modeling. Upon verification and in case of need, the CAD model, whether newly reconstructed or existing, can be updated to meet the accuracy requirements. A further normal CAD session can then be carried out to transform the constructed surfaces into a complete CAD model. After CAD further processing, one should have a compact CAD surface model for further design a nd o ther d ownstream a pplications s uch a s f inite element a nalysis, CNC programming and rapid prototyping.

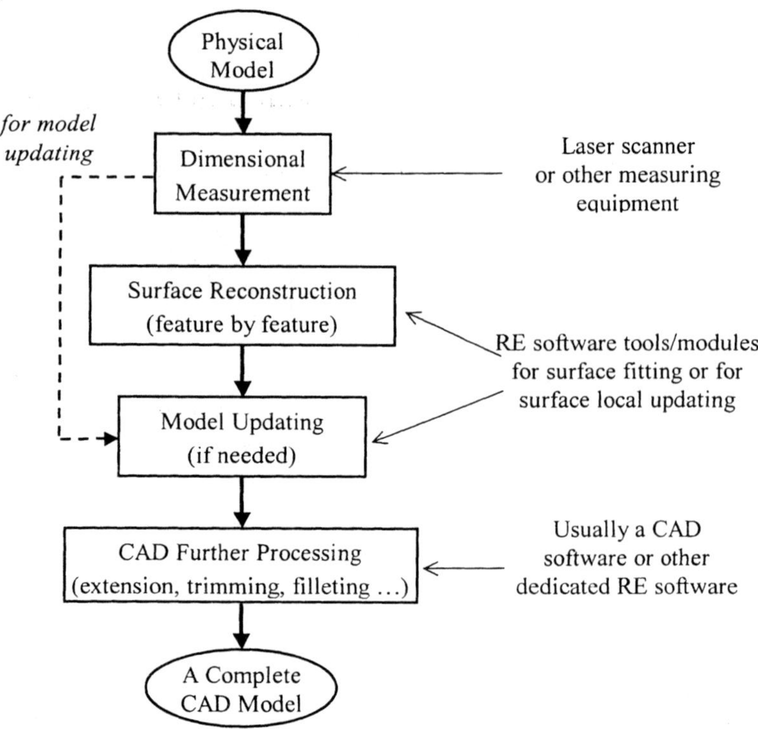

Figure 4-9. A general procedure for CAD modeling from a physical part.

4.8.1 Parametrization for Surface Reconstruction

As noted in Section 4.4.1, parametrization of the measured points is an important s tep in surface r econstruction. Figure 4-10 illustrates t he fitting results o f a c utter s urface with d ifferent p arametrization m ethods f rom a n irregular grid of pints measured on a CMM. Figure 4-10a illustrates the measured points. The parametrization methods used for fitting the surfaces shown in Figures 4-10b, 4-10c and 4-10d were uniform, centripetal and base surface, respectively. The base surfaces used for fitting the surfaces of Figures 4-10b, 4-10c and 4-10d were uniform, centripetal and base surface, respectively. The base surface knots (Ma and Kruth 1995a) with the same orders $k_u = k_v = 4$ and number of control points $n_u = 12$, $n_v = 8$ were used. The measured surface was a surface of a profile cutter measured on a coordinate measuring machine. Following the fitting results, it can be concluded that the fitted surface using base surface parametrization behaves very well for any arrangement of points, either regularly, irregularly or randomly distributed, while with other parametrization methods the fitted surface may be severely distorted if the measured points are not regularly distributed.

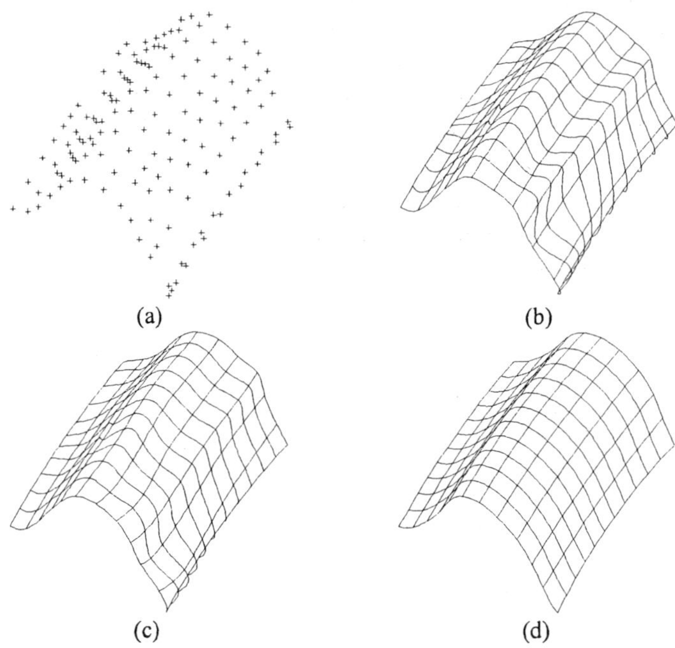

Figure 4-10. Parametrization for B-spline surface fitting (Ma and Kruth 1995): (a) 14 by 10 irregular grid points; (b) fitted surface with uniform parametrization; (c) fitted surface with centripetal parametrization; and (d) fitted surface with base surface parametrization.

4.8.2 B-Spline Surfaces

Figure 4-11 illustrates an example in fitting a set of B-spline surfaces from measured points for CAD modeling of a toy car model (Ma *et al.* 1993). This example also demonstrates the complete procedure for constructing a CAD model starting from a physical part. A CMM was used for digitizing. To avoid manual digitizing of the entire surface, a semiautomatic measuring approach was used for digitizing:

- *Initial preparation*: The physical model was first analyzed and the entire surface was partitioned into topological regions, each of which represents a B-spline surface in the final CAD model. The boundary curves were marked on the surfaces of the physical model.
- *Manual boundary digitizing*: All the boundary curves were then manually digitized on a coordinate measuring machine (CMM).

- *Rough CAD modeling*: All boundary curves were then created from the manually measured boundary points. A set of approximate surfaces was further created from the boundary curves, each of which for one of the topological regions defined during initial preparation. These surfaces served two purposes. They were used for defining a full measuring path for automatic digitizing of the entire surface through a DMIS (Dimensional Measuring Interface Specification) interface. They were also used as base surfaces for later surface fitting.

- *Automatic digitizing*: A DMIS module of a CAD software, Unigraphics, was then used to generate the automatic digitizing paths. The digitizing paths were exported in a DMIS file that could be read by a CMM. The final measured points are shown in Figures 4-11a and 4-11b. Since the model is symmetric, only half of the model was measured.

- *Base surfaces creation*: A base surface can be created from four boundary curves plus any internal section curves. The first approximate surfaces after rough CAD modeling can directly be used as base surfaces. Figure 4-11c shows the base surfaces finally selected for LSQ fitting of the car model.

- *Least-squares fitting of different features*: B-spline based least-squares fitting algorithms discussed in Section 4.4 were finally applied to the measured points. All the surfaces were fitted one by one. The final fitted surfaces are shown in Figure 4-11d.

Take the top surface of the car body as an example. There were 427 digitized points on this surface. Before applying least-squares fitting, the maximum deviation from the digitized points to the base surface was about 22mm and the standard deviation was 6.5mm. After least-squares fitting, the maximum deviation was reduced to about 1mm and the standard deviation was reduced to 0.4mm. In this example, the fitted surface has the same number of control points and same order as the base surface. After implementing the above procedure on the surface shown in Figure 4-11d, we have all the necessary surfaces to complete the CAD model. An ordinary CAD session can be further processed to complete the CAD model.

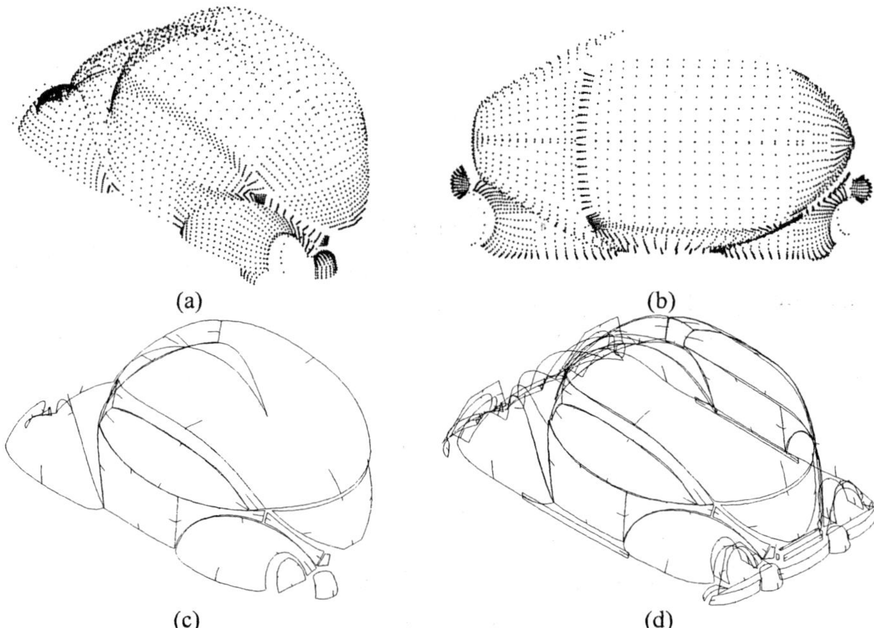

Figure 4-11. Construction of a car surface model through B-spline surface fitting (Ma *et al.* 1993): (a)-(b) all the automatically measured points in two views; (c) all the base surfaces used for later fitting; and (d) all the fitted surfaces (including a mirror of the fitted side surfaces from one side to the other side).

4.8.3 NURBS Surfaces

Figure 4-12 illustrates the construction of a CAD model of a car wheel in full scale using NURBS surfaces. The main feature of this example is shown in Figure 4-12b. The rotational NURBS surface was created by rotating three NURBS curve segments around an axis that was measured separately. The three curve segments are the internal area, the external ring and the intermediate fillet curve respectively. The NURBS curve segment for the internal area was fitted from 487 points with an order k=4 and number of control points n=6. Another feature of the car wheel is the insert surface in three duplicates as illustrated in Figures 4-12a and 4-12d. This is a general open free-form NURBS surface and it was fitted from 996 randomly distributed points measured on a coordinate measuring machine. The surface orders were chosen as $k_u = k_v = 4$ and the number of control points used for fitting the surface was $n = n_u \times n_v = 8 \times 8 = 64$. In this example, there is also a general rotational NURBS surface, i.e., the valve surface shown in Figure 4-12a, which was also fitted from measured points. In addition, there are also some other regular features fitted from the measured points for defining a complete CAD model.

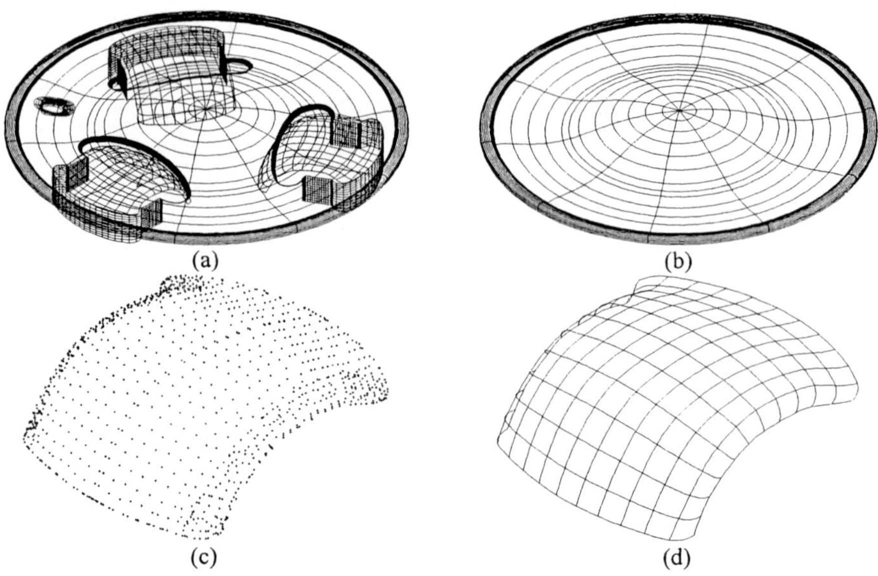

Figure 4-12. Construction of a car wheel model through NURBS fitting (Ma and Kruth 98): (a) all fitted NURBS surfaces in wire frame; (b) the main surface of the wheel created as a rotational NURBS surface; (c) the measured points for the insert surface (a general open NURBS surface); and (d) the fitted insert surface (a general open NURBS surface).

4.8.4 Subdivision Surfaces

Figure 4-13 shows an example of subdivision surface fitting. It is an example of fitting the entire surface of a vacuum cleaner model using a single Catmull-Clark surface using a parametric approach (Ma and Zhao 2000, Ma and Zhao 2002). The model was first subdivided into large topological domains, each of which was mainly represented as a cubic B-spline surface during the fitting process, except at extraordinary corner positions where the number of surface patches meeting at such corners was other than four. In this case, all extraordinary corner patches are defined as Catmull-Clark surface patches. Since Catmull-Clark surfaces are generalizations of cubic B-spline surfaces, the final fitted surfaces, a hybrid model of B-splines and subdivision surfaces, could be entirely exported as a single Catmull-Clark surface.

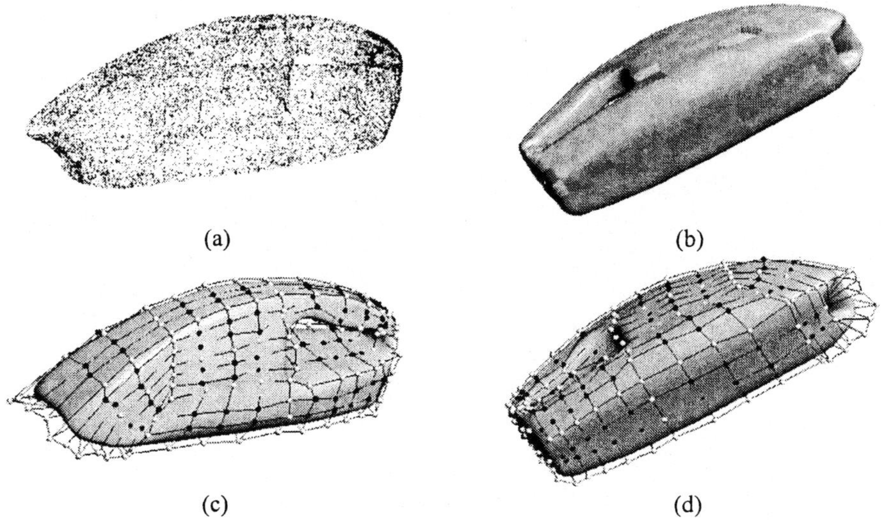

Figure 4-13. Subdivision surface fitting for CAD modeling of the entire surface of a vacuum cleaner model (Ma and Zhao 2002): (a) measured points for the entire surface; (b) final fitted Catmull-Clark subdivision surfaces in shaded image; (c)-(d) final fitted Catmull-Clark subdivision surfaces with control meshes in two views.

The original measured data of this example were acquired from a plastic body of a vacuum cleaner using a medical CT-scanner. The dimension of the physical object were about 300mm × 100mm × 105mm. A total number of 305 slices of cross-sectional images were obtained at an interval of 1mm. After preprocessing, a set of 3D CT-contours were obtained. The measured points, i.e., the contour points, are illustrated in Figure 4-13a. Figure 4-13b shows the final fitted Catmull-Clark surface in shaded image. Figures 3-13c and 4-13d illustrate the Catmull-Clark surface with control points.

For the model shown in Figure 4-13, the average, minimum, maximum and standard deviations of the fitting errors of all measured points were, 0.1864mm, 0.0022mm, 0.7987mm and 0.0005mm respectively. There were in total 13138 measured points involved in the final fitting procedure. The entire topological model in the hybrid form (B-splines and Catmull-Clark surface patches) contained a total number of 31 inter-connected large quadrilateral surface domains. There were in total 6 extraordinary corners of valence $N = 3$ and 8 extraordinary corners of valence $N = 5$. The total number of control points of all the 31 surfaces was 1248, of which 364 are shown in Figures 4-13c and 4-13d are independent control points for defining the entire surface as a Catmull-Clark surface. The entire topological model contained a total number of 345 individual surface patches as shown in Figure 4-13b.

This example also demonstrates the modeling power of subdivision surfaces. If one had used traditional B-splines, one would have had to maintain the continuity among individual surfaces, which is an extremely difficult task for today's CAD systems. Even if one can represent the entire model as a mosaic of B-spline surfaces with continuity conditions, it is not possible to modify the surface by moving a control points near domain boundaries, or one breaks all the continuity conditions. With Catmull-Clark surface, h owever, the continuity c onditions along a ll p atch b oundaries a re automatically maintained. One can modify the shape by moving any of its control points, just like manipulating the control points of a single B-spline surface.

4.8.5 Surface Local Updating

For surface local updating, one firstly needs to measure the modified areas as shown in the flow chart of Figure 4-14. The newly digitized points are then compared with the existing CAD model and all geometric features need to be updated are registered. For each of the surface features, all surface patches that need to be updated are also identified. Control points that only affect such surface patches are further identified and registered. In case there are insufficient control points for modifying a surface area, knot insertion needs to be applied. The insertion of new knots will increase the number of control points to be updated for a particular area. Surface local updating algorithms discussed in Section 4.7 can then be applied to update the affected control points. If the updated surface satisfies the tolerance requirement, the updating for this feature is completed. Otherwise, the algorithm will add some additional knots to the original B-spline surface to locations with maximum deviations. The process is then repeated until the required fitting tolerance is satisfied. Surface local updating is an important RE technique.

Figure 4-15 illustrates an example involving the addition of a local feature t o a n existing C AD s urface b y m odifying t he s urface l ocally w ith digitized points. Figure 4-15a shows the original surface and 975 measured points for surface local updating. The size of the original surface is about 140mm × 130mm. One may apply two strategies for updating the local area, i.e., allowing to insert some knots into the modified area or not; and allowing alteration to some neighboring patches or not. In this example, all internal patches were allowed for shape alteration. However, the neighboring patches were not allowed to be modified. The original surface shown in Figure 4-15a is a B-spline surface with orders 4×4 and the numbers of control points are 10×15 along u and v directions, respectively. For parametrization, the measured p oints w ere p rojected o nto the initial s urface a nd t he p rojection direction is chosen to be the surface normal. The identified region for updating was within the area of $(u,v) = \{\xi_6 \leq u < \xi_9; \; \zeta_5 \leq v < \zeta_8\}$. In this example, the neighboring patches changed whenever any of the registered control points are modified. Further, we had to apply knot insertion since the neighboring p atches w ere n ot a llowed for shape m odifications. F igures 4-15b, 4-15c and 4-15d show the results of local updating with one, two and three inserting knots, respectively, for each patch within the region for updating. The fitting errors of the updated local surfaces are summarized in Table 4-2.

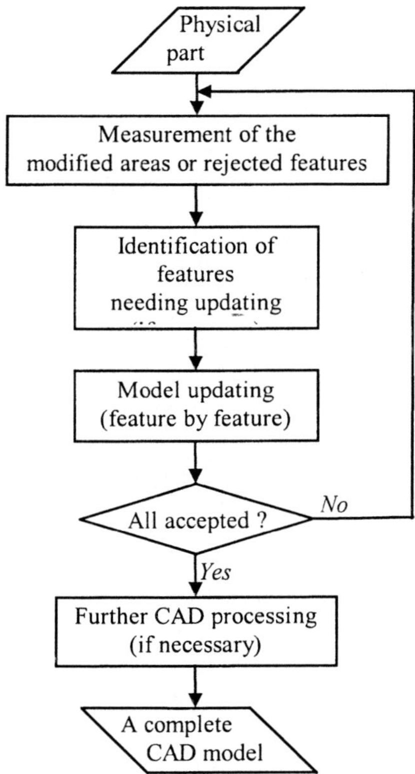

Figure 4-14. CAD model local updating based on measurement.

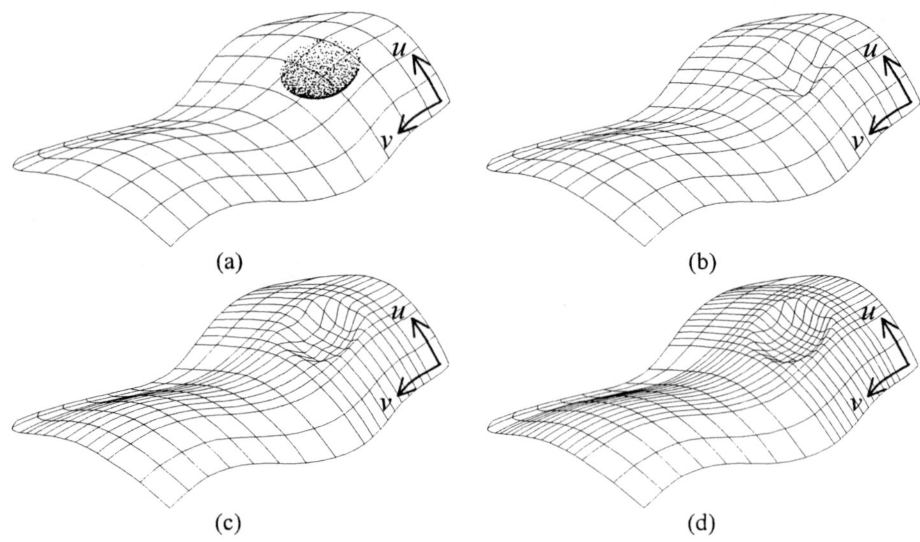

Figure 4-15. Surface local updating: (a) the underlying surface and the measured points; (b) the projection region of measured points; (c) the updated surface with one inserting knot per patch for both *u* and *v*; (d) the updated surface with three inserting knots per patch for both *u* and *v*.

As evident from Figure 4-15 and Table 4-2, the fitting results are gradually improved when more knots are inserted within the region for local updating. The maximum deviations on both sides of the surface occur along the boundary of the area being updated where there is no smooth transition between the underlying surface and the area being updated.

Table 4-2. Fitting errors of the updated local surface shown in Figure 4-15.

Ins. Knots	$r_u = r_v = 1$	$r_u = r_v = 2$	$r_u = r_v = 3$
Max. Dev.	2.083697	1.005867	0.469698
Min. Dev.	-2.135457	-1.040433	-1.025658
Average Dev.	0.093489	-0.002357	-0.002723
Std Dev.	0.699974	0.269104	.158274

It should be noted that excess knot insertion will result in parameter redundancy and difficulties in control point based modifications for future processing. While the flexibility of controlling the shape of the updated local surface is improved with additional knots, i.e., additional number of control points, the ability to control the shape in pushing or pulling a single control point is however decreased.

Chapter 5

DATA PROCESSING FOR RAPID PROTOTYPING

5.1 INTRODUCTION

In this chapter, we address various issues related to data processing for rapid prototyping. In contrast to traditional material removal processes, rapid prototyping techniques build a part by incrementally adding materials layer by layer (Kruth 1991, Barequet and Kaplan 1998), so they are also called layer-based manufacturing or layered manufacturing. The processes are fully automatic and offer a number of competitive advantages over traditional manufacturing processes. Hence, they are particularly useful for rapid product development. The entire process for model prototyping is illustrated in Figure 5-1 and can be summarized as follows:

- *CAD modeling*: The first step in rapid prototyping is to prepare a computer-aided design (CAD) model of the object to be fabricated using layer based manufacturing processes. For most of the available RP technologies, a solid model with complete topological and geometric information is required.
- *STL interfacing*: STL is the de-facto standard for the rapid prototyping industry. It is a file format for approximately defining an object using triangular facets. When a CAD model is available, the entire part geometry is converted into STL format based on a tolerance for accuracy control.
- *Part orientation*: Before processing for prototyping, a RP engineer needs to figure out the specific orientation in which the prototype model will be produced. Part accuracy, the amount of supporting

material required and ease of post-processing are important factors influencing part orientation determination.

- *Support generation*: Depending on the specific RP process to be used, one may need to further define support structures for supporting down-facing areas during part build-up. Support generation can be done on the basis of a STL model or the original CAD model.

- *Model slicing and tool path generation*: In contrast to material removal manufacturing technologies, rapid prototyping technologies refer to a class of layer-based material increase manufacturing processes. The digital model of the object and related support structures need to be sliced layer by layer. For each layer, a set of surface c ontours is obtained and a t ool path is d efined within t he material area for model production in a layer-wise fashion.

- *Model production on a RP machine*: The produced tool path is sent to a RP machine for building up the prototype model, including support, layer by layer.

- *Post-processing*: Depending on the RP process involved, a post-processing step might be needed for post-curing in the case of stereolithography, for infiltration and furnace sintering in the case of SLS, and/or for removing the support structures and surface polishing in the cases of most other RP processes.

In addition to active research areas such as new processes and new materials development, data processing is one of the most important areas that largely affect the efficiency of prototyping and part surface quality. Data processing includes initial CAD modeling, model conversion for data interfacing through an STL file specification, part orientation determination, support generation, and various approaches for model slicing and tool path generation. Several of these issues are of critical importance while producing accurate prototype models in a short time.

Several issues related to CAD modeling have already been addressed in Chapter 4. With current technologies, it is possible to directly interface CT-scanning for rapid prototyping. It is also possible to directly convert laser scanning data into STL format for model prototyping. For the sake of further product development and automation, including the use of rapid prototyping technologies, a valid solid model is a must. All surfaces created using reverse engineering techniques discussed in Chapter 4 have to be further transformed into a valid solid model before embarking on model prototyping, or production. On other occasions, one directly works with a CAD solid model for product design or prototyping. Some further issues concerning solid model representation will be briefly discussed in the following section.

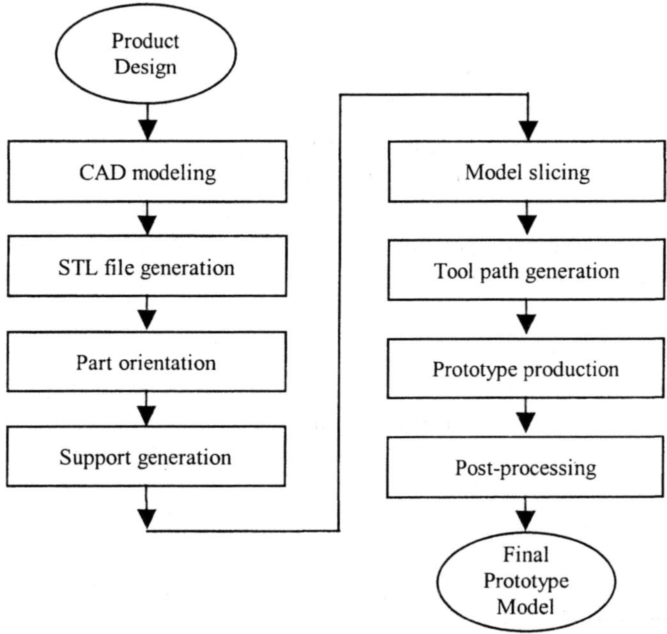

Figure 5-1. Overview of the entire process of rapid prototyping.

CAD data interfacing for rapid prototyping is a common issue during data processing for rapid prototyping. The de-facto standard for data interfacing is the STL interface specification initially proposed by 3D Systems, Inc. (1989). The STL interface specification is a file format for storing data of 3D objects represented by faceted triangular surface models. The format was extended from polyhedron-surface data structures widely used by the CAD and graphics community. The pros and cons of the STL interface have been widely discussed in the RP community over the past decade (Dolenc and Mäkelä 1992, Mäkelä and Dolenc 1993, Gilman and Rock 1995, Roscoe *et al.* 1995, Fadel and Kirschman 1996, Bohn 1997, Dutta *et al.* 1998, and Jurrens 1999, Pratt *et al.* 2002). Some alternative specifications were also proposed. However, the position of STL in the RP community has not been changed and it remains the de-facto standard for the RP industry today. Various issues concerning STL data interfacing will be discussed in Section 5.3.

Part building orientation is another issue affecting both the surface quality and the build time. Allen and Dutta (1994) described a method for determining the best orientation based on certain criteria related to part stability and the minimum area of part surface contact with support structures. Frank and Fadel (1994, 1995) developed an expert system that considered three important parameters for part fabrication: surface finish, build time, and the necessity for support structures. The system interacts with the user to recommend the best build direction. These parameters were also considered by Lan *et al.* (1997) and Xu *et al.* (1997) as criteria to find an optimal build orientation for stereolithography, one of the most successful RP processes. Xu *et al.* (1999) also addressed the issue of optimal build direction for other RP systems. Part orientation will be addressed in Section 5.4. In particular, a generic cost model extended from the work of Alexander *et al.* (1998) will be discussed in Section 5.4.2.1. It will be noted that the overall building time is the most important parameter affecting the overall prototyping cost. Criteria affecting part surface quality will be discussed in Section 5.4.2.2 whereas those minimizing the necessary support structures will be addressed in Section 5.4.2.3. Part build orientation determination is actually a multi-objective optimization problem. This aspect will be briefly discussed in Section 5.4.2.4. Some issues of specific interest to stereolithography will be presented in Chapter 6.

An inevitable consequence of building up a part layer by layer using a rapid prototyping technology is the staircase (or, stair-stepping) phenomenon. This phenomenon usually has a profound effect on surface smoothness. The influence is mainly determined by the layer thickness and the local part geometry. If the layer thickness is computed based on the local geometry, the staircase effect can be controlled to a user-specified tolerance level. In general, there are two slicing approaches applicable to the determination o f the layer t hickness: u niform s licing a nd a daptive s licing. Uniform s licing i s t he s impler approach s ince it involves s licing a p art a t equal intervals, so it is used in almost all commercially available rapid prototyping systems. However, there is a compromise with regard to layer thickness selection. If the layer thickness is made smaller, one obtains a smoother part. However, this may result in a larger number of layers and, hence, in a longer build time. On the other hand, if the layer thickness is made larger, the build time would be shorter, but one may end up with a part having a large staircase effect.

Adaptive slicing is a method of resolving the above problem. This method generates a variable layer thickness based on the local surface geometry and a pre-specified smoothness requirement. In literature, one can find a number of attempts directed towards adaptively slicing a part. Dolenc and Mäkelä (1994) discussed slicing issues for producing an accurate part surface with consideration of both tolerance and important peak features. The angle of the surface normal is used to predict a possible layer thickness. Sabourin *et al.* (1996) proposed another method by first subdividing the model space into uniform slabs. The thickness of each slab equals the maximum acceptable layer thickness of a particular RP process. Each of these slabs is then subdivided into uniform thinner layers such that the produced cusp height is within a given tolerance. A carefully designed adaptive slicing algorithm for STL models can be found in Tata *et al.* (1998). Cormier *et al.* (2000) and Mani *et al.* (1999) also used a region-based variable tolerance for adaptive slicing. All these approaches are based on the STL interface.

There are also approaches that, instead of going through the STL interface, directly operate upon the CAD surface or solid model without using the STL interface. Several researchers developed algorithms for direct slicing of a CAD model to maintain its original accuracy, i.e., algorithms of the direct slicing type. Suh and Wozny (1994) investigated direct and adaptive slicing procedures that take into account both peak features and tolerance requirement. Kulkarni and Dutta (1996) further developed an algorithm for direct and adaptive slicing of a parametric surface model. Normal curvature in the vertical direction was used to determine the maximum allowable layer thickness at each layer based on a pre-specified cusp height. To improve the computational performance, Lee and Choi (2000) proposed the use of both the skin contours and vertical character lines of the part for adaptive layer thickness computation. Jamieson and Hacker (1995), on the other hand, utilized a different algorithm to perform direct and adaptive slicing. They sliced the model using a standard layer thickness and compared the contours of the current layer with those of the previous layer to see if the layer thickness could be increased. If the difference between the contours on the two slices was small, the maximum layer thickness was first attempted. Otherwise, a middle slice between the standard slice and the previous slice was created. Once again, the contours on the newly created slice were compared to the previous one. The procedure was repeated until the maximum allowable layer thickness was found. With this approach, one may need to compute many slices in order to find an acceptable layer thickness. Hope *et al.* (1997) sliced the object using layers with sloping boundary surfaces that closely matched the part surface. Their slicing procedure can greatly reduce the staircase effect, but may not be suitable to commonly used commercial RP systems. Vuyyuru *et al.* (1994), Ma and He (1999), and Ma *et al.* (2003b) developed direct slicing algorithms based upon a NURBS surface model. The algorithm reported in (Ma and He 1999, and Ma *et al.* 2003b) is for direct and adaptive slicing and has been successfully implemented for FDM prototyping.

In addition to direct and adaptive slicing, Sabourin *et al.* (1997) presented an interesting approach for tackling the compromising issue of between smooth surface finish and fast build time. First, the object is uniformly sliced to create slabs with maximum allowable layer thickness of the R P s ystem s imilar t o the a pproach r eported in S abourin *et a l.* (1996).

Next, contour offsets are generated to separate the parts within each slab into exterior and interior regions. The exterior regions are further sliced into uniform thinner layers based on the local geometry and tolerance requirements. The exterior regions of the given part are, therefore, built with thin layers to achieve an acceptable surface finish, while its interior regions are built with fast, thick layers to reduce the overall build time. The slicing procedure is based on STL files. This approach has been further extended by Tyberg *et al.* (1998) by taking into account local features. Another approach, namely, selective hatching for layered manufacturing, was also reported in (Ma and He 1999 and Ma *et al.* 2003b) for direct and adaptive slicing of a NURBS-based surface model. The results have shown that a significant build time reduction can be achieved on general engineering parts with thick walls by combining adaptive slicing with selective hatching (Ma and He 1999 and Ma *et al.* 2003b).

Layer thickness determination and model slicing are important issues to be considered in this chapter. They will be covered in two sections. In Section 5.5, generic slicing algorithms for faceted surface models will be introduced. Direct and adaptive slicing algorithms will be presented next in Section 5.6. A selective hatching algorithm will also be briefly introduced in Section 5.7. Toll path generation will be covered in Section 5.8.

5.2 CAD MODEL PREPARATION

Rapid prototyping includes a class of fully automatic manufacturing technologies that are capable of producing prototype models of any shape provided a computer description of the object is available. With most of today's rapid prototyping systems, a valid STL model is required as input. To reliably produce a valid STL model, a non-ambiguous CAD solid model is often needed. The model surface must uniquely distinguish between 'inside' and the 'outside' of the object concerned. A solid model provides a complete, valid and unambiguous representation mostly suited for automatic processing, s uch a s i nterference a nalysis b etween i ndividual objects, m ass property calculation, automatic mesh generation for finite element analysis, and rapid prototyping.

A CAD surface model provides only geometric information. Surface representations are the main building blocks for solid modeling and a solid model is built upon surface representations. CAD surfaces created through reverse engineering techniques discussed in the previous chapter often need to be further processed and converted into a solid m odel for downstream applications, such as for rapid prototyping. To convert a CAD surface model into a solid model, one often needs to extend the CAD surfaces, find intersections between surfaces, apply chamfering and fillets, and finally form a closed watertight solid model.

Similar to surface modeling, there also exist a variety of schemes for solid modeling. Constructive solid geometry (CSG) is one of the important schemes for solid modeling (Hoffmann 1989 and Zeid 1991). CSG is built upon three key building blocks: solid primitives, transformation operations and Boolean operations. Solid primitives are standard solid features, such as a block, a cylinder, a sphere or a solid wedge. These primitives can be easily defined through just a few parameters. Boolean operations include union ($\cup$), intersection ($\cap$) and subtraction (-). Starting from simple solid primitives and combining with transformation operations such as translation, rotation, shearing, scaling and their combinations, one can gradually progress to very complex engineering parts. Figure 5-2 highlights the basic idea of CSG solid modeling represented as a CSG tree.

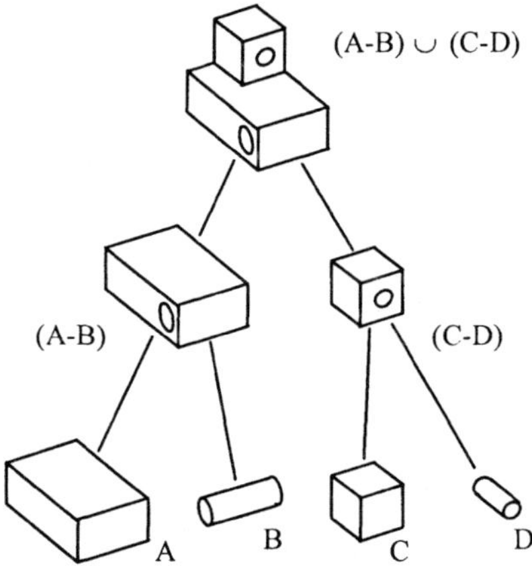

Figure 5-2. CSG solid modeling.

CSG is a very intuitive and user-friendly solid modeling scheme. However, depending on the particular CAD modeling system being used, the modeling capability might be limited owing to the limited availability of solid primitives. Boundary representation (B-rep) is an alternative solid modeling scheme that is entirely complimentary to CSG in solid modeling activities (Hoffmann 1989 and Zeid 1991). B-rep is a very powerful and flexible scheme that can be used to model any object found in the physical world. As the name of the scheme implies, B-rep defines an object by a set of boundary faces that can be either planar or freeform surfaces. Topologically, each boundary face is enclosed by a loop of boundary curves, each of which can again be a simple line segment or a freeform curve. Apart from geometric parameters, each boundary curve should have two end/boundary points. This leads to the well known topological data structures of the B-rep solid modeling schemes shown in Figures 5-3 and 5-4. While Figure 5-4 illustrates an object with only planar faces, as mentioned above, any of the faces and any of the edges can be freeform in nature. Figure 5-5 further highlights the data structure in the case of a freeform object. In this case, we only need to add a parameter node if a boundary face is a freeform surface or if a boundary curve is a freeform curve. If face F_1 is a B-spline surface defined by equation (4.3), e.g., the parameter node P_{f1} should be pointing to the memory where the surface definition parameters like the orders k_u and k_v, the number of control points $n = n_u \times n_v$, two sets of knots $\xi := \{\xi_i\}_{i=1}^{n_u+k_u}$ and $\varsigma := \{\varsigma_j\}_{j=1}^{n_v+k_v}$ and the control points $\{\{v_{ij} = [x_{ij}, y_{ij}, z_{ij}]^T\}_{j=1}^{n_v}\}_{i=1}^{n_u}$ can be found (Figure 5-6).

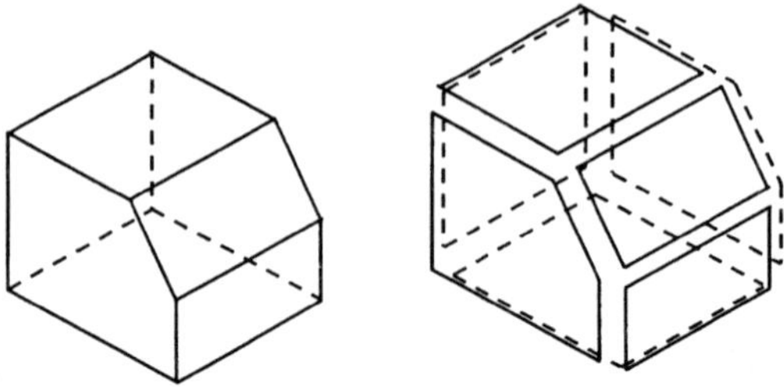

Figure 5-3. Boundary representation (B-rep) for solid modeling.

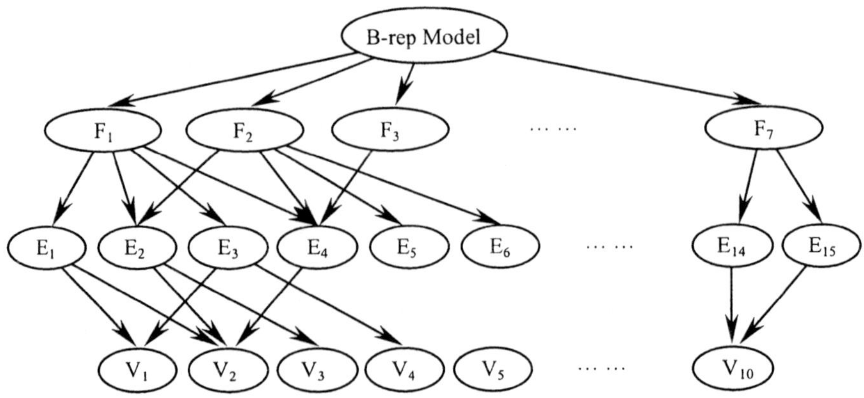

Figure 5-4. Topological data structure for B-rep solid modeling.

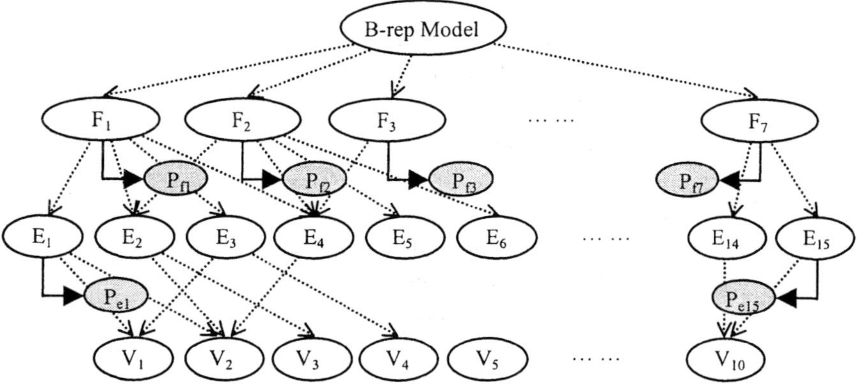

Figure 5-5. Topological data structure for B-rep solid modeling with freeform features.

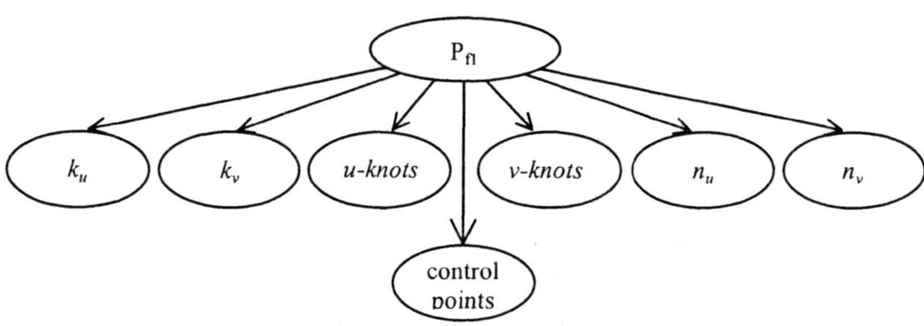

Figure 5-6. Parameter node for defining a B-spline surface.

While B-rep is undoubtedly a powerful s cheme for s olid modeling, it may not be very convenient for the user if one always starts from surface definition even for regular primitive features. To provide users both a friendly and intuitive interface and a powerful database, most of the computer-aided design and manufacturing (CAD/CAM) systems use a hybrid solid modeling scheme with B-rep for flexible model representation and CSG for user interfacing. To the user, the CAD/CAM system looks like a CSG solid modeling system with solid primitives and Boolean operations. However, for internal data representation, the B-rep data structure is used. Thus, all possible modeling operations applicable to a CSG solid modeler are available to the user. The user can also define any freeform features built upon surface modeling techniques. When equipped with other modeling operations such as sweep solid modeling, automatic filleting and chamfering, today's CAD/CAM systems offer powerful and sophisticated modeling tools for advanced solid modeling.

In addition to CSG and B-rep solid modeling, there are also some voxel based modeling schemes, such as cell decomposition and octree encoding, often used for scientific and medical visualization. These modeling schemes could be very useful for rapid prototyping of colored models.

5.3 DATA INTERFACING FOR RAPID PROTOTYPING

5.3.1 STL Interface Specification

STL interface is a file format specification for interfacing a CAD model with rapid prototyping equipment (3D Systems 1989). The specification was initially developed by 3D Systems Inc. for their stereolithography system and has become the de-facto standard input for all RP systems. The STL interface was developed based on polyhedral representations widely used in graphics kernels and solid modelers. Precise CAD surfaces are approximated with planar and linear primitive geometric elements into a tessellated (triangular) facet boundary surface model. The accuracy of conversion from a curved surface to a planar faceted surface is controlled by the number of facets whereas their organization is used to represent that surface. For a planar surface, the conversion is exact since each of the converted facets lies on the original planar surface.

A description of the STL specification is available in (3D Systems 1989). There are two representations based on the ASCII format and the binary format, respectively. Both representations provide a list of triangular facets that form the STL model and, for each of the triangles, the coordinates of the three vertices of the triangle in 3D space and the associated face normal pointing outside the model surface. Figure 5-7 illustrates a simple STL model with 4 facets. An ASCII STL description of this object is shown in Figure 5-8.

One of the important features of an ASCII STL file is the use of some key words to indicate the type of information. In addition to the key words used in Figure 5-8, one can use another keyword "`attribute`" followed by an attribute number for any of the vertices as shown in Figure 5-9 (Hull and Lewis 1991). The attribute record is however not popularly used.

In an ASCII STL file, the coordinates of a facet normal or an individual vertex can be floating point numbers, so "`23.325866`", "`2.3325866e+01`" or "`0.23325866E+02`" is acceptable. The order of vertices should follow the right hand rule with the thumb pointing outside the model surface. The normal vector should be a unit vector with magnitude equal to 1. If a STL file is produced for RP applications, the file must define a valid triangle mesh with each edge connecting two faces in the right topology with no flipped triangles and the entire mesh must define a closed watertight volume.

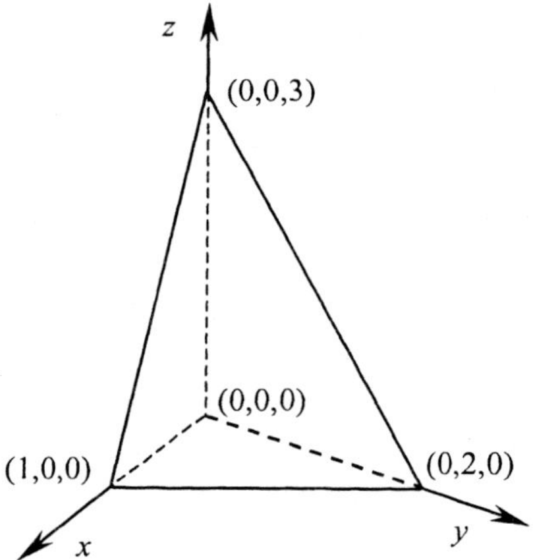

Figure 5-7. A faceted object with three faces and three vertices.

```
Solid irregular tetrahedron
  facet normal -1.0 0.0 0.0
    outer loop
       vertex  0.0   0.0   0.0
       vertex  0.0   0.0   3.0
       vertex  0.0   2.0   0.0
    endloop
  endfacet
  facet normal 0.0 -1.0 0.0
    outer loop
       vertex  0.0   0.0   0.0
       vertex  1.0   0.0   0.0
       vertex  0.0   0.0   3.0
    endloop
  endfacet
  facet normal 0.0 0.0 -1.0
    outer loop
       vertex  0.0   0.0   0.0
       vertex  0.0   2.0   0.0
       vertex  1.0   0.0   0.0
    endloop
  endfacet
  facet normal 0.85714286 0.42857143 0.28571429
    outer loop
       vertex  1.0   0.0   0.0
       vertex  0.0   2.0   0.0
       vertex  0.0   0.0   3.0
    endloop
  endfacet
endsolid irregular tetrahedron
```

Figure 5-8. STL representation of the object shown in Figure 5-7.

```
Solid irregular tetrahedron
  ... ...
  facet normal 0.0  -1.0  0.0
    outer loop
      vertex  0.0   0.0   0.0
      vertex  1.0   0.0   0.0
      vertex  0.0   0.0   3.0
      attribute 16
    endloop
  endfacet

  ... ...
endsolid irregular tetrahedron
```

Figure 5-9. Illustration of the field for attributes in a STL file.

While the ASCII format can be read and visually checked conveniently, the binary format results in a much smaller file size with a typical file size reduction of 85%. Due to its small file size and high speed for file transfer, the binary format is used almost exclusively in industry. The binary STL file structure is similar to that of the ASCII format. The file contains a list of facet records and each facet record contains information about the facet normal, coordinates of the three vertices and, optionally, some attributes. Similar to a STL file in ASCII format, the facet attributes are not used. The STL file specification in binary format has also been described in (3D System 1989). The file structure is shown in Figure 5-10.

In Figure 5-10, 1 byte equals to 8 bits. In ANSI C language, each character takes 1 byte and 80 bytes are equivalent to 80 characters (char). The index number of a facet record can be defined as an unsigned long integer that is equivalent to 4 bytes. The coordinates of the facet normal and vertices can be defined as floating numbers, each taking up 4 bytes. The facet attributes can be defined as an unsigned short integer that is equivalent to 2 bytes.

The STL interface specification provides a convenient input file format for all rapid prototyping systems. As will be seen in Section 5.5, the basic slicing algorithm for faceted surface models is very simple and easy to implement. This probably explains why STL is still the de-facto standard of the RP industry in spite of the various problems to be addressed in subsection 5.3.3.

```
(Top of file)
84 bytes - header record
            80 bytes - unformatted general information
                       such as file name, part name and comments.
             4 bytes - number of facet records
                       each facet record defines one triangle.
50 bytes - first facet record
            12 bytes - facet normal vector
                        4 bytes - i coordinate
                        4 bytes - j
                        4 bytes - k
            12 bytes - first vertex
                        4 bytes - x coordinate
                        4 bytes - y
                        4 bytes - z
            12 bytes - second vertex
                        4 bytes - x coordinate
                        4 bytes - y
                        4 bytes - z
            12 bytes - third vertex
                        4 bytes - x coordinate
                        4 bytes - y
                        4 bytes - z
             2 bytes - optional facet attributes
50 bytes - second facet record
50 bytes - third facet record
    :
    :
    :
50 bytes - the last facet record
(End of File)
```

Figure 5-10. Binary STL file specification.

5.3.2 STL Data Generation

When a CAD solid model is available, all CAD systems provide an option to export it in STL file format. The conversion from a precise solid model to an approximate facet model is controlled through a tolerance specification. Some CAD systems allow direct control of an absolute chordal tolerance. Figures 5-11 and 5-12 illustrate how the chordal tolerance is used for producing a tessellated facet surface.

Some systems allow the user to provide a quality parameter for controlling the final model resolution. For instance, Pro/Engineer allows users to specify the degree of resolution on curved surfaces by entering a quality value through its user interface (Jacobs 1992a). The value ranges from 1 to 10, with 10 creating a tessellated facet model with the highest resolution. As one might imagine, a high resolution value could produce an unduly large STL file. Some other systems allow users to specify a relative chordal tolerance, i.e., a percentage parameter that equals the allowed maximum deviation divided by the maximum facet dimension.

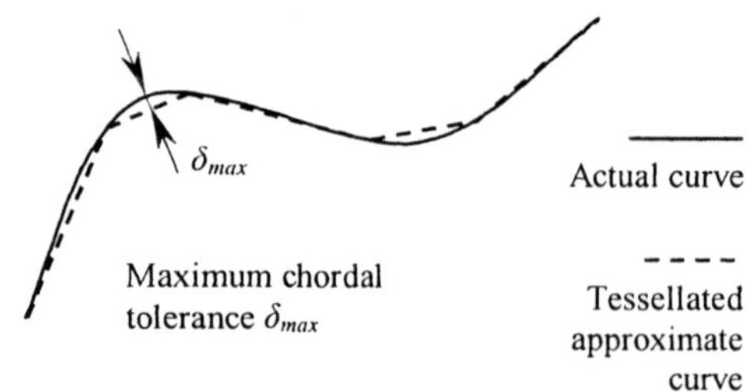

Figure 5-11. Illustration of maximum chordal tolerance control by a 2D curve.

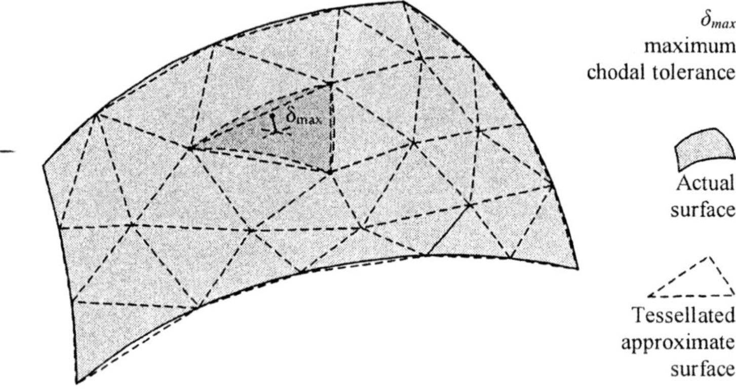

Figure 5-12. Illustration of maximum chordal tolerance control for surface tessellation.

Model tessellation is a well-known problem associated with surface visualization and finite element analyses. With a B-rep solid model, the tessellation process is quite straightforward since a mosaic of boundary surfaces is already defined explicitly within the CAD data structure. One may first produce surface meshes of individual boundary faces and finally assemble the individual surface meshes as a single surface mesh. To avoid invalid and degenerate facets, coordination among individual surfaces is needed when producing individual surface meshes. For each prodividual surface, whether trimmed or non-trimmed, one may first produce a triangulation in the 2D parametric domain. The 2D mesh can be further mapped to 3D through the parametric definition of the corresponding boundary face. Mesh correction in 3D might be needed in order to produce a smooth surface mesh.

For a CSG solid model, one may need to first convert the solid model in CSG representation to a B-rep representation. Regular surface meshing techniques can then be applied for further model tessellation.

5.3.3 STL Data Manipulation

It has been known for a long time that the STL interface is associated with a number of potential problems during practical application (Dolenc and Mäkelä 1992, Mäkelä and Dolenc 1993, Gilman and Rock 1995, Roscoe *et al.* 1995, Fadel and Kirschman 1996, Jacobs 1996, Bohn 1997, Dutta *et al.* 1998, and Jurrens 1999, Pratt *et al.* 2002). The quality of a STL file also depends on the system used for exporting the STL file. This subsection addresses some common problems that need to be corrected before an STL file can be used for prototype fabrication.

5.3.3.1 Data redundancy

The most obvious problem is the large file size. As one may see from the file specification, the notion of surface normal is entirely unnecessary provided that the facet vertices are ordered properly. In addition, the file repeats the coordinates of an individual vertex as many times as it appears in the number of triangular facets. For a regular triangle mesh, each of its vertices will have six neighboring triangles. This implies that a further data reduction of six to one can be achieved.

5.3.3.2 Topological problems

In a STL file, all vertices, edges and triangles must satisfy valid topological rules as shown in Figure 5-13. All triangles around a vertex must satisfy a valid two-manifold condition with the vertex being located at the topological center. All face normals must point towards the same side. Each of the edges must be shared by exactly two triangles (except for an open mesh, and if the edge is located on the boundary). Figure 5-14 highlights some of the problematic cases that may appear with a bad STL file. All problems must be corrected through face flipping, local re-triangulation, and edge, triangle and vertex reconnection, insertion and deletion, as appropriate.

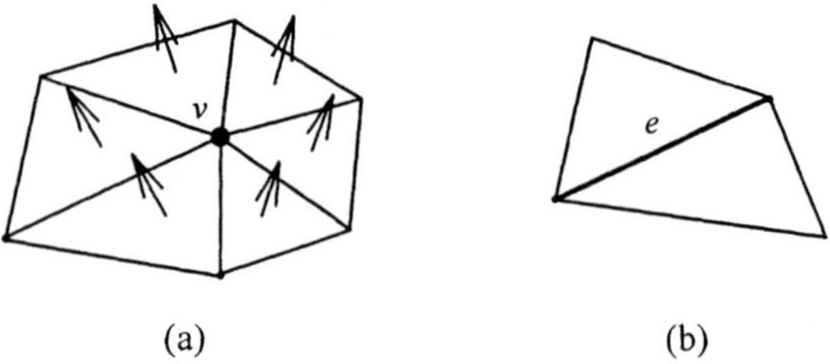

Figure 5-13. Topological rules for a valid STL model: (a) vertex-face condition; (b) edge-face condition.

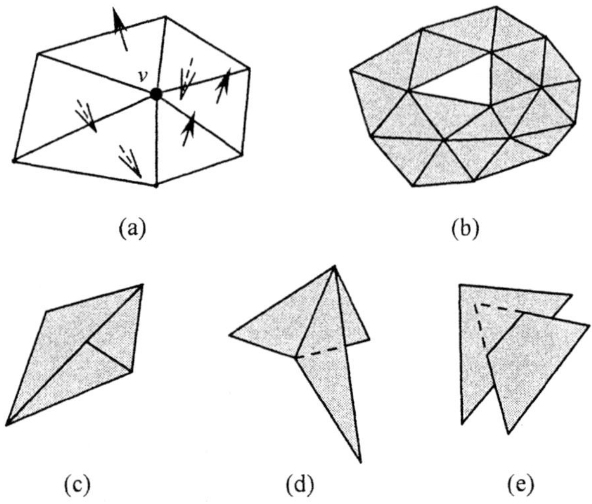

Figure 5-14. Common topological problems: (a) flipped triangles; (b) missing triangles; (c)-(d) invalid edge sharing; and (e) misplaced triangles.

5.3.3.3 Geometric problems

Even if a CAD system produces a topologically valid surface mesh, inappropriate h andling o f t olerances a nd f loating p oint a ccuracy m ay a lso lead to a variety of geometric problems. Figure 5-15 highlights several such cases of geometric inaccuracy. Such geometric problems can often be corrected by deleting degenerate triangles through valid edge contraction algorithms or by vertex repositioning.

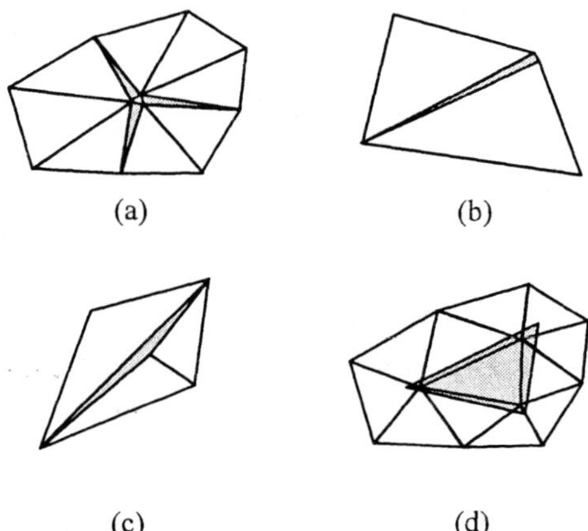

Figure 5-15. Geometric problems: (a)-(c) degenerate triangles; and (d) face overlapping.

5.3.4 Alternative RP interfaces

Some researchers have studied other options, such as the use of IGES, STEP, VRML and some other new file format proposals (Dolenc and Mäkelä 1992, Gilman and Rock 1995, Bohn 1997, Dutta *et al.* 1998, and Jurrens 1999, Pratt *et al.* 2002). Many researchers also suggested direct slicing of a CAD model for RP applications (Rajagopalan *et al.* 1992 and Jamieson and Hacker 1995 and references therein). In spite of all the previously mentioned problems (Mäkelä and Dolenc 1993, Roscoe *et al.* 1995, Fadel and Kirschman 1996 and Jurrens 1999) and various efforts made so far for alternative RP interfaces (see above for references), the approximate faceted STL interface has been adopted by almost all RP system developers and it remains the de-facto standard of the RP industry today. As mentioned earlier, one of the important reasons is its simplicity for RP system manufacturers in processing a faceted model for slicing and support generation. It would be unrealistic for every RP system manufacturer to develop their own processing software based upon high-level CAD models, not mentioning the vast number of schemes available for surface a nd s olid m odeling. A nother i mportant r eason m ust b e d ue t o t he leading market position of 3D Systems and ready support from CAD software developers to the STL interface. Possibly, other RP systems manufacturers are just following in order to minimize their resources for key system developments. We should however not ignore these problems in connection with the STL interface specification and should strive for an informative, complete, logical, and simple interface specification for the RP industry of today and the future.

5.4 PART ORIENTATION AND SUPPORT GENERATION

Part orientation and support generation are two closely related issues in layered m anufacturing. B y s electing a n o ptimal p art orientation for m odel prototyping, it is possible to shorten build time and minimize the overall prototyping cost.

5.4.1 Factors Affecting Part Orientation

Many a uthors have a ddressed t he p roblem o f p art o rientation. A mong others, Frank and Fadel (1994, 1995), Cheng *et al.* (1995), Lan *et al.* (1997), Xu et al. (1997), Xu et al. (1999), and Pham *et al.* (1999) studied methods for determining part orientation in stereolithography prototyping. Thompson and Crawford (1993) and Hur *et al.* (2001) proposed methods for SLS prototyping. Masood *et al.* (2000) and Masood and Rattanawong (2002) proposed an approach by estimating the volumetric e rrors for FDM rapid prototyping. Hu *et al.* (2002) and Yang *et al.* (2002b) investigated the issue for the case of hybrid prototyping incorporating machining processes. Allen and Dutta (1995) and Alexander *et al.* (1998) explored methods applicable to to several RP processes. Some related geometric optimization problems were addressed by Majhi *et al.* (1999a and 1999b).

The basic methodology for building parts layer by layer is common to most RP processes. By identifying the factors common to all such processes, an approach that i s n ot b ound to a ny one process can be developed. Part orientation has a significant effect on the final part quality and prototyping cost. The general part orientation characteristics are as follows (see also the discussion in (Alexander *et al.* 1998)):

- The main envelope of the part in 3D is one of the important factors. For example, with regard to the stereolithography process, the switching b etween individual layers takes a significant p art o f the overall building time and hence must be properly optimized. For the FDM process on the other hand, there is no difference in terms of the total building time for adding the modeling materials, but it is still better, where possible, to orient the part such that the part would firmly stand on the platform during the building process.

- For processes t hat need support s tructures, part orientation s hould also be optimized such that it would require minimal support. There are two key parameters to be optimized in this respect. While ensuring that the part is firmly supported during the entire prototyping process, the overall support contact area should be minimized. This helps in minimizing the influence of the support on the surface q uality o f the prototype. I t a lso reduces further e fforts during post-processing. We should also make good use of the allowable overhang angle that needs no further support. The total support volume should also be minimized to save time and material(s) for building the support structures.

- The external surfaces produced should be as smooth as possible. As RP models are built in a layered fashion, a staircase effect is unavoidable, but it should be minimized. This can be achieved by reducing the number and areas of inclined faces, i.e., by trying to position the part such that most of the faces or, at least, most of the important faces are positioned either vertically or horizontally without support, to the extent possible.
- Other factors, such as trapped volume should be avoided in the case of stereolithography rapid prototyping. Wall and ceiling thickness values and possible curling should also be considered.

5.4.2 Various Models for Part Orientation Determination

5.4.2.1 A generic cost model

Based on the above considerations and a generic cost model proposed by Alexander *et al.* (1998), we may use the following consolidated generic cost model for decision-making. It is assumed that a good STL model is available and the entire evaluation starts from model preparation. The consolidated generic cost model is defined as

$$C_{tot} = C_{pre} + C_{build} + C_{mat} + C_{post} \tag{5.1}$$

where the cost components are:

- C_{pre}, direct cost related to pre-processing;
- C_{build}, machine utilization cost for building the prototype model;
- C_{mat}, material cost including part modeling material and support material, if different; and
- C_{post}, post-processing cost.

The pre-processing cost is determined by a number of factors including

- model positioning time, T_{pos};
- support generation time, T_{sup};
- model slicing time, T_{sli};
- tool path generation time, T_{path}; and
- machine set up time, T_{setup}.

The overall pre-processing cost is then defined by

$$C_{pre} = \left(T_{pos} + T_{sup} + T_{sli} + T_{path}\right) \cdot \left(R_{staff} + R_{comp}\right) + T_{setup} \cdot \left(R_{staff} + R_{mach}\right) \quad (5.2)$$

In the above equation, the related costs are as follows:
- R_{staff} , staffing cost, that can be based on an hourly rate;
- R_{comp} , cost related to computing facilities and rapid prototyping software; and
- R_{mach} , machine utilization rate.

The machine utilization cost can be further estimated based on the initial machine cost and annual maintenance cost after considering machine depreciation. If the machine life is T, the initial machine cost is C, annual maintenance is $M = 15\% \cdot C$, i.e., 15% of the initial machine cost, and it is estimated that on average the machine is utilized 15 hours per day, the hourly machine cost C_{mach} can then be estimated as

$$R_{mach} = \left(M + C\!\!\left/\!\!_T\right.\right)\!/(365*15) \quad (5.3)$$

The second cost component is directly proportional to build time, T_{build} , and is given by

$$C_{build} = T_{build} \cdot R_{mach} \quad (5.4)$$

The material cost is further divided into two parts

$$C_{mat} = C_{mat-modelling} + C_{mat-support} \quad (5.5)$$

Post-processing cost includes two components

$$C_{post} = T_{post} \cdot R_{staff} + C_{misc} \quad (5.6)$$

where C_{misc} stands for miscellaneous costs that could include any additional equipment utilization, such as a UV-oven for post-curing of a stereolithography part or an ultrasonic tank for the removal of the water soluble support for FDM prototyping. In addition to the above generic cost model, (Pham *et al.* 1999) also reported results for building time and build-cost estimation for the stereolithography process. A brief summary can be found in next chapter on stereolithography (Section 6.6.4).

Build time estimation varies from process to process and it dictates the overall c ost. In t he case o f s tereolithography, i t d epends on the l aser s pot dimension, the laser power that determines the scanning speed, and the hatching pattern. Further, the switching time between two consecutive layers also contributes to a significant part of the overall building time (Pham *et al.* 1999). One can also find discussions regarding build time prediction for stereolithography in (Chen and Sullivan 1996). As for the SLS process, the building time mainly depends on the laser drawing speed. In the case of FDM, the overall building time for a given modeling material is usually not sensitive to part orientation, but part orientation will largely affect the consumption of support material. Part orientation will thus directly influence the overall building time and the post-processing time.

In many situations, one can produce multiple parts in a single setup. This is particularly useful for processes such as stereolithography and SLS that involve substantial times while switching between layers. In addition to part orientation, another technical issue concerns how one should pack the part in 3D in order to achieve maximum utilization of the machine capacity (Hur *et al.* 2001).

5.4.2.2 Part orientation for surface quality improvement

While minimizing the overall cost, the surface quality of the final prototype model must also be observed. For a fixed building orientation, the minimization of build time, and hence the overall cost, is an optimization issue r elated t o s urface q uality i mprovement. If a sm all l ayer t hickness i s used, one would obtain a fine part surface, but the overall building time will be longer and hence the surface quality is improved at a cost. It is however possible by adjusting the part orientation to obtain a n optimal solution at which we gain the maximum possible surface quality at a bearable cost.

Several authors have reported on attempts to approximately quantify the error for minimization (Majhi *et al.* 1999a, Majhi *et al.* 1999b, Masood *et al.* 2000, Rattanawong *et al.* 2001, and Masood and Rattanawong 2002). Majhi *et al.* (1999a) proposed several alternative criteria for determining an optimal part orientation based on staircase error evaluation (Figure 5-16):

- To minimize the maximum staircase error of all of its facets of a particular feature for a given direction.
- To minimize a weighted staircase error, as some features might be more important than others.
- To minimize the sum of staircase errors.

To evaluate the staircase error, Masood *et al.* (2000), Rattanawong *et al.* 2001, and Masood and Rattanawong (2002) proposed an approximate volumetric error for primitive features. Figure 5-17 illustrates the case of a cylinder for which the volumetric error is estimated as

$$V_e = \sum_{i=1}^{n} A_i \cdot l_i = \sum_{i=1}^{n} \left(\frac{1}{2} \cdot d^2 \tan\theta \cdot l_i \right) \tag{5.7}$$

where, d is the layer thickness (also called layer pitch L_p), l_i is the perimeter at layer i, n is the total number of layers of the primitive feature.

The perimeter length at layer i can be estimated according to the orientation of the individual feature and the region a particular layer falls in. For just one feature, the answer would be obvious at $\theta = 0$, but in the case of multiple features, a compromise is needed between these features and hence an optimal orientation.

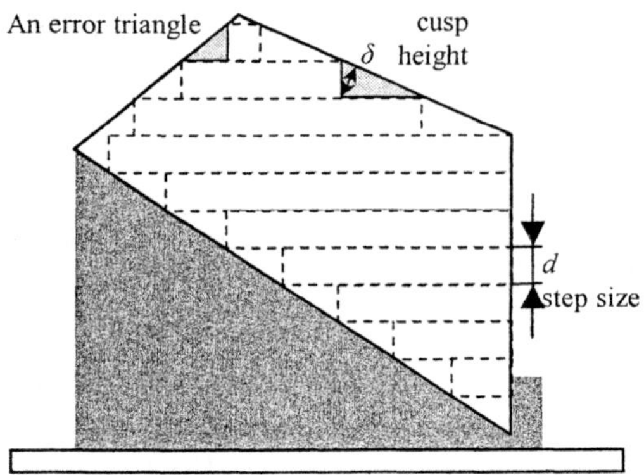

Figure 5-16. Illustration of staircase errors for minimization.

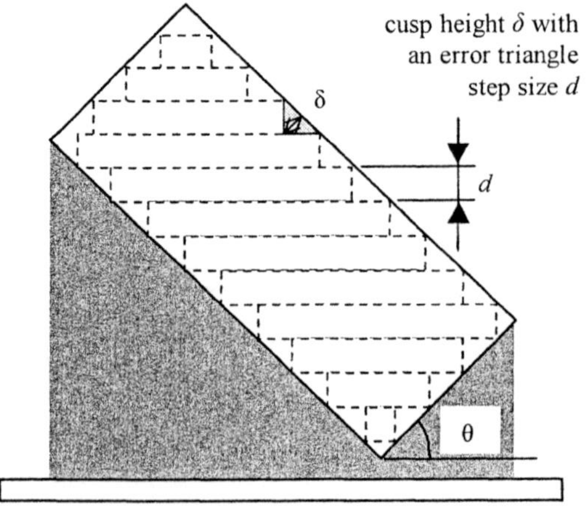

Figure 5-17. Illustration of staircase errors for minimization.

5.4.2.3 Part orientation with minimum support

Part orientation will affect the volume of material required for model support and hence extend building time for some RP processes. In addition, overhanging part areas having direct contact with support will have a poor surface finish and require more post-processing. Several researchers have proposed approaches to compute the volume of supports for determining part orientation (Majhi *et al.* 1999a, Majhi *et al.* 1999b, and Pham *et al.* 1999).

The volume of support includes all areas under down-facing and overhanging facets, whether located on the outer part surface or the inner surface surrounding an internal void. For a given facet of the STL model, if the facet normal is pointing downward and the inclination angle between the facet and the horizontal slicing plane is smaller than 25° (otherwise the facet is considered a self-supporting facet), the facet is considered as a down-facing and overhanging facet while calculating the support volume. Figure 5-18 illustrates the volume of support in shading. It can be computed as

$$V_{\text{sup}} = \sum_{i=1}^{n} V_i = \sum_{i=1}^{n} A_i \cdot h_i \tag{5.8}$$

where n is the total number of down-facing and overhanging facets of the STL model requiring external support, A_i the project area of facet i onto the horizontal slicing plane, and h_i the average support height evaluated from the corresponding facet to either the platform or the next intersection facet when projected downward toward the platform, whichever is less. The volume V_i can also be directly computed on the basis of the geometry of the corresponding facet and the projected geometry onto either the platform or another part area.

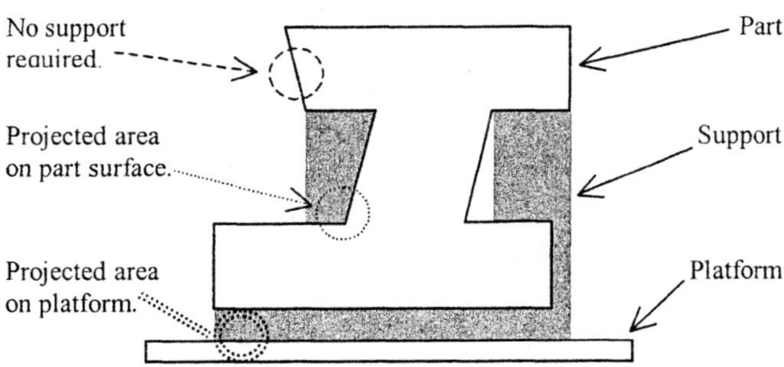

Figure 5-18. Illustration of support structures.

When e valuating t he a mount o f s upport m aterial, the g eometry o f t he support structures to be discussed in the following section must be considered.

5.4.2.4 Part orientation with multiple objective optimization

In addition to the criteria discussed in the previous subsections, there are many factors that need to be considered when searching for an optimal part orientation. The entire problem for part orientation determination can be formulated as a constrained optimization problem, i.e., as a problem involving the optimization of the most important criterion while satisfying other constraints (Xu *et al.* 1999). It can also be formulated as a multi-objective optimization problem with an appropriate weighting for each of the individual criteria (Cheng *et al.* 1995 and Majhi *et al.* 1998). As summarized by Jacobs (1992) and Cheng *et al.* (1995), a part should be oriented to meet the following criteria:
- maximize the number of perpendicular surfaces;
- maximize the number of up-facing horizontal surfaces;
- maximize the number of holes with their axes in the slicing direction;
- maximize the number of curved cross sections drawn in the horizontal plane;
- maximize the area of the base surface;
- minimize the number of sloped surfaces;
- minimize the total area of overhanging surfaces; and
- minimize the number of trapped volumes.

In addition, in order to minimize the building time, a part should also be oriented to minimize the total number of slices and to minimize the height of the required support structures (Cheng *et al.* 1995). One may also find approaches based on expert systems for build direction determination (Frank and Fadel 1994 and 1995 and Pham *et al.* 1999a).

5.4.3 The Functions of Part Supports

Depending on the nature of a particular RP process, there will be a need for part supports while implementing the prototyping process. One example is the stereolithography process where, without proper support, the overhanging structure would fall down while in the resin vat. For some other processes, such as selective laser sintering, no special support is required since the powder itself actually supports the part during the building process. The following summarizes some important functions of part supports (Hull and Lewis 1991 and Jacobs 1992a). The studies are based on the stereolithography process, but the functions of part supports for other processes, such as fused deposition modeling, are similar:

- *To separate parts from the platform*: The use of supports will make it easier to safely remove the part from the platform after model production. It will also be easier to control the layer thickness and surface quality of the bottom layers. Marks on the platform would not be printed on the final part.
- *To provide support to hanging structures*: The function is twofold. First of all, it provides support to hanging structures and prevents such structures from collapsing. In addition, it can also strengthen overhanging regions for the prevention of deformation and curling for stereolithography as well as other processes. In the case of stereolithography, curl occurs when shrinkage of layer $n+1$ pulls on layer n, similar to bimetallic strip effect
- *To provide a collision avoidance buffer*: The supports also provide a collision avoidance buffer between the platform and other components of the machine since a different building strategy is applied for building the support structures compared with building the main prototype model.
- *Process improvement*: To improve liquid flow in and around the part for the stereolithography process, especially during the initial building stage and while switching between individual layers.

Figure 5-19 shows several cases where supports are required. Figure 5-20 illustrates how part supports could prevent curling and deformation during the prototyping process.

5.4.4 Support Structure Design

Depending on the design and the application, a support structure may be decomposed into three functional areas with different building strategies for practical applications. For an area connecting to part surfaces, the support structure should be easily removable while providing sufficient support. The support structure for this area is often defined as sierras or needles with minimum contact with the part surface. The main support should be strong enough to withstand both the vertical weight and other horizontal disturbances. A different strategy may be applied for areas between the main support and the platform for easy removal while providing a stable base support. The structure should be designed such that its total weight is minimized. Thus, the three functional areas are:

- sierras or needles: connections between the main support and the part;
- supports: the main support structure, and

- separators: connections between the main support and the platform. Figure 5-21 illustrates the three functional areas of a support structure.

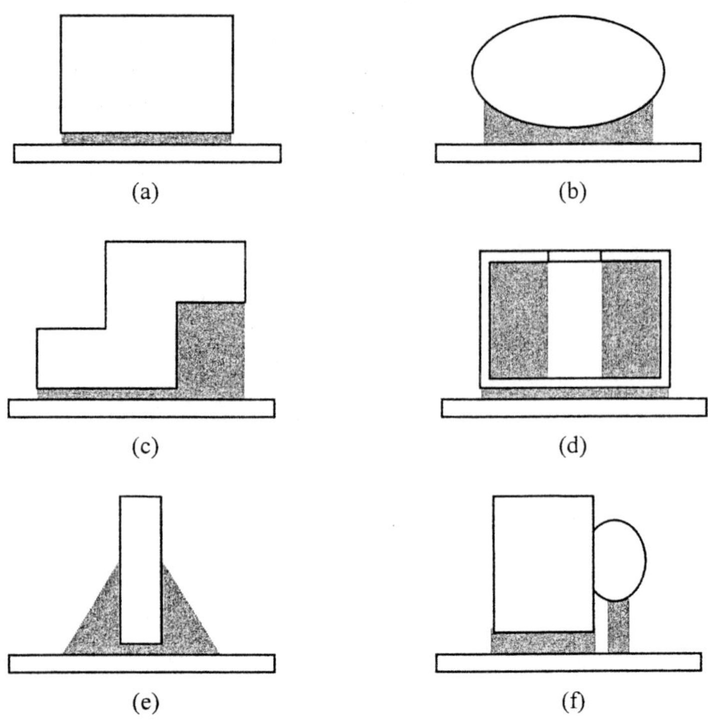

Figure 5-19. Incidences where supports are needed: (a) separation between part and platform; (b) down facing regions below the equator of the surface normal curvature; (c) supports for hanging structures; (d) internal supports; (e) support for part stability; and (f) supports for islands.

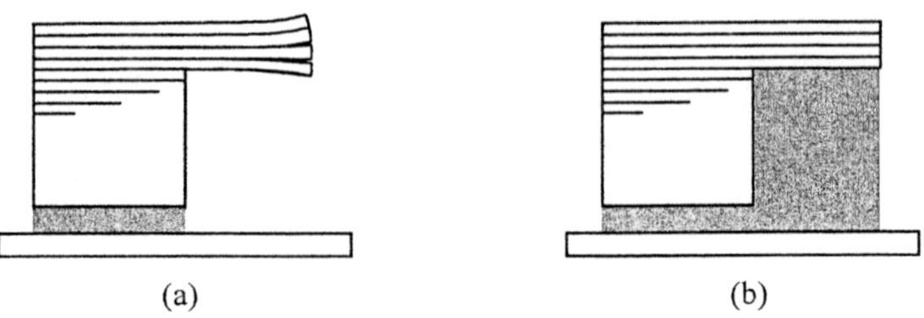

Figure 5-20. The prevention of deformation and curling through part supports: (a) unsupported cantilevered beams or similar structures may deform without support; (b) properly attached layers under support.

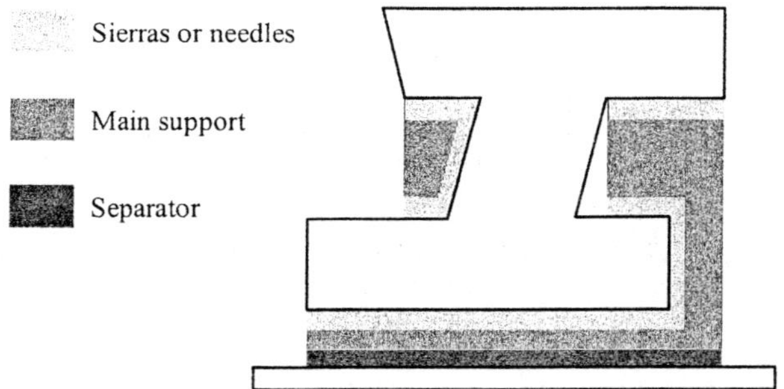

Figure 5-21. Illustration of support structures..

There exist a variety of designs for the main support structure. Some of the approaches can be found in (Hull and Lewis 1991, Jacobs 1992a, Swaelens *et al.* 1995 and 1997, Chalasani *et al.* 1995, and Hur and Lee 1996). The following are some commonly used support structures:

- *Gussets*: As illustrated in Figure 5-22a, gussets (a single one or a set) are used to support lightweight overhang areas during the part building process and attach to a vertical wall near the overhang areas. Gussets provide the optimal support for overhang areas while requiring minimal resources during the building process. The supports are also easily identified during cleanup.
- *Projected feature edges* (Figure 5-22b): The edges of unsupported lightweight areas where gussets cannot reach are projected downwards to provide support. Projected feature edges support the edges of the feature and provide excellent control against curling and warpage.
- *Single webs* (Figure 5-22c): Thin walls can be supported by single webs produced by projecting the center-line along the narrow side of the thin walled feature. Cross members are added to provide stability of the support structure.
- *Webs* (Figures 5-22d, 5-22e): Large unsupported areas maybe supported by various web structures, such as those shown in Figures 5-22d and 5-22e. Contact of such support structures with vertical part walls should be avoided to protect the final part surface. To minimize support material consumption, perforated walls may be used in the web structure as shown in Figure 5-22f. Swaelens *et al.* (1995 and 1997) proposed a number of perforated wall structures for stereolithography prototyping.
- *Scaffolding*: Various scaffolding structures used in the construction and tissue engineering can also be designed as support structures

for rapid prototyping. The perforated support structure reported in Swaelens *et al.* (1995) and Vancraen *et al.* (1994) is actually a scaffolding structure if simple sticks are used to compose a wall for the web structure instead of perforating the walls. Chua et al. (2003) developed a scaffold structure library for rapid prototyping in tissue engineering.

- *Honeycomb*: Other sophisticated support structures similar t o t he honeycomb s tyle f or h ollowing master prototype m odels initially developed for QuickCast 2.0 (Hague *et al.* 2001) may also be used as support structures.
- *Columns*: For isolated small islands, column type support structures can be used as shown in Figure 5-22g. For large islands, columns defined by other web structures may also be used.
- *Zigzag and perimeter support*: Delicate support structures are most suitable for processes such as steoreolithography as the laser beam can be easily blocked anytime anywhere to prevent the resin from solidification. For processes such as FDM that use nozzles for material injection, a continuous path is preferred whether for support generation or for building the part. The zigzag and perimeter support structure shown in Figure 5-22h is most suited for FDM prototyping with a continuous path for each layer.

For all the above mentioned support structures, the thickness of the thin webs can be just a single cured line (usually 0.18mm to 0.3mm thick) in the case of laser lithography or a single road width (often two times the layer thickness) in the case of fused deposition modeling. Sierras or needles for connecting the part surface, if used, should penetrate into part surface by a few layer thicknesses. The intersection will ensure that the supports physically connect to the part features.

5.4.5 Automatic Support Structure Generation

Given a STL model and a building direction, all support structures can be automatically generated based on the inclination angle of a particular facet (Hur and Lee 1996). If the inclination angle between the facet under consideration and the slicing plane is larger than the face angle specified by the user, the corresponding face is considered to be a self-supporting face, so there is no need to provide any extra support. Otherwise, the face needs to be counted as a face requiring support. The areas that require support for the entire part will be the union of all individual areas identified with the face selection angle. Figure 5-23 shows an arbitrary part and indicates areas requiring supports.

When the areas requiring support have been identified for the entire part, support patterns can first be designed on a projection plane parallel to any of the slicing planes. Ray projection algorithms often used in the graphics community (Mantyla 1988, Foley *et al.* 1990 and Zeid 1991) can then be applied to extrude the support structures in 3D. Further details of the support structure, such as separators, sierras or needles for connecting to part surfaces and features for perforated structures, can be added afterwards. Figure 5-24 illustrates a part supported by perforated structures (Swaelens *et al.* 1995 and Vancraen *et al.* 1994) and with needles connecting the part surface.

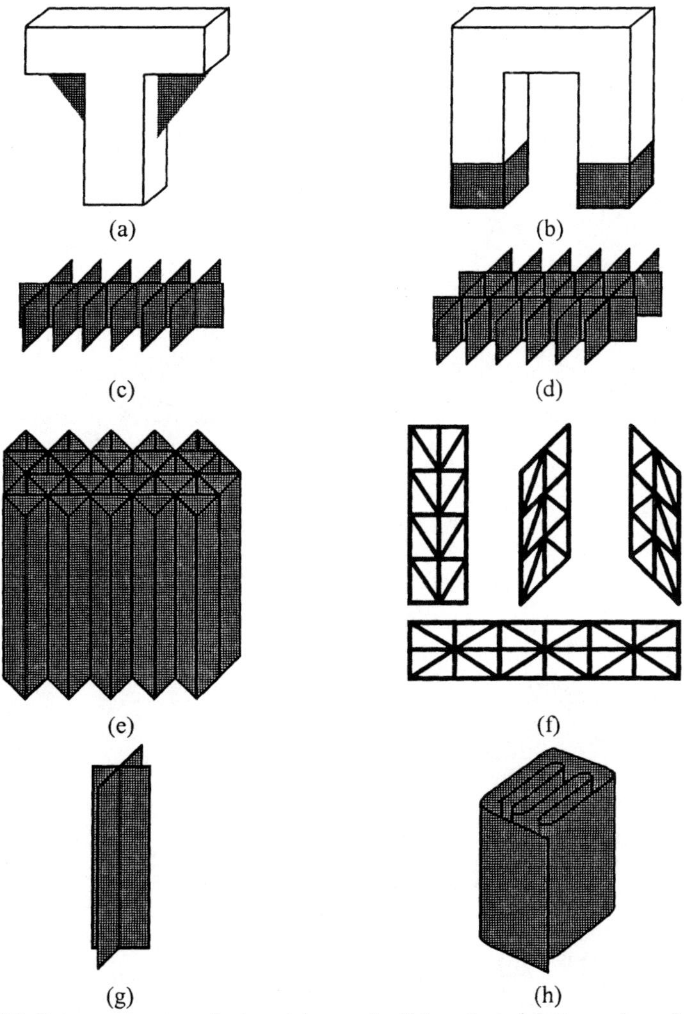

Figure 5-22. Support structure design: (a) gussets; (b) projected feature edges; (c) single webs; (d) webs; (e) triangle webs; (f) perforated wall structures for various web-based support design; (g) columns; and (h) zigzag and perimeter support.

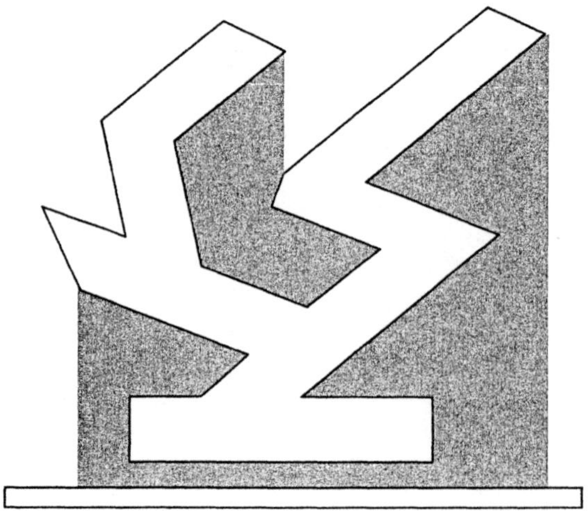

Figure 5-23. Illustration of part sections requiring support.

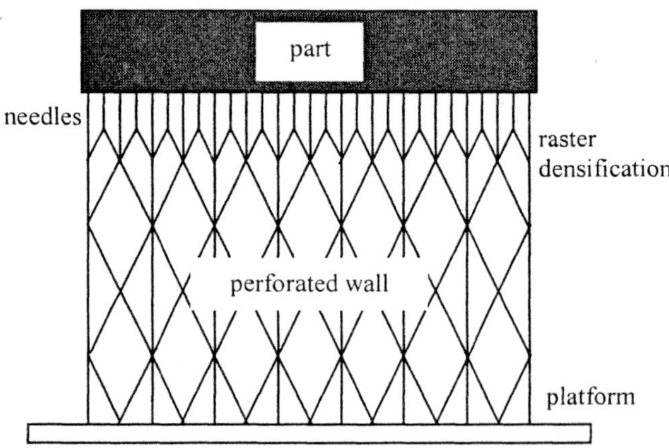

Figure 5-24. Illustration of perforated wall with needles connecting to part surface.

In some cases where there is no STL information, such as with direct interfacing of CT or MRI data for rapid prototyping, support structures can also be produced entirely on the basis of contour data (Swaelens *et al.* 1995). The support structures can also be automatically produced by comparing the surface contour information associated with two successive layers and estimating the steepness of the part surface locally. Wherever the surface is not steep enough, a support structure should be used.

In the past, support structures were also designed during the CAD session. With today's RP software, however, this is no longer the case. Using RP software, support structures can now be easily and, often, automatically produced based on either STL information or contour data. The algorithms based on STL or contour data are much simpler compared with those based on CAD surface or solid models.

5.5 MODEL SLICING AND CONTOUR DATA ORGANIZATION

A number of algorithms have been reported for slicing faceted models with uniform layer thickness (Luo and Ma 1995 and Liao and Chiu 2001a and references therein). Some researchers have also explored adaptive slicing using a variable layer thickness (Tata *et al.* 1998 and Tyberg and Bohn 1998). This section focuses on the basic slicing algorithms applicable to a faceted surface model with uniform slicing. Adaptive slicing will be discussed in the next section together with direct slicing. To efficiently perform model slicing for rapid prototyping applications, the input STL geometry must be organized in a concise data structure.

5.5.1 Model Slicing and Skin Contour Determination

A STL facet model used for rapid prototyping applications contains a collection of planar faces. These faces define an approximate boundary representation for the object. During subsequent tool path generation, we need to slice the model based on either uniform layer thickness or adaptively variable layer thickness. In this section, we use uniform slicing to illustrate the slicing algorithm. As for adaptive slicing, one only needs to determine the corresponding adaptive layer thickness and the slicing algorithms are the same. A generalized discussion of adaptive slicing will be presented in the next section.

Based on a user-entered layer thickness, a sequence of parallel slicing planes can be defined for model slicing. As a convention, we assume that the model has been properly oriented such that the z-axis will be the building direction. Let d be the layer thickness and n be the total number of slicing planes excluding the bottom plane with $z = Z_{min}$ that are not used during the slicing procedure. Further, let Z_{max} and Z_{min} be the extreme z-coordinates of the STL model. The total number of layers (valid slicing planes) required is then defined by the following equation

$$n = \frac{Z_{max} - Z_{min}}{d} \tag{5.9}$$

The term $(Z_{max} - Z_{min})$ defines the dimension of the object along the z-axis. In this case, the slicing planes are defined as planes parallel to the xy-coordinate plane as follows

$$z_i = z_{i-1} + d \qquad \text{for} \qquad i = 1, 2, \cdots, n \tag{5.10}$$

with $z_n = Z_{min}$. For each of the slicing planes, a slicing procedure is performed. Figure 5-25 highlights a generalized procedure for slicing faceted models. To efficiently use this algorithm, contour points should be sorted first in the slicing direction, i.e., sorted following the z-coordinates. For contour point computation, i.e., for determining the intersection between one of the face edges and the slicing plane, a simple intersection algorithm may be used. Assuming that the two end points of the edge are defined by $\mathbf{p}_1 = x_1\mathbf{i} + y_1\mathbf{j} + z_1\mathbf{k}$ and $\mathbf{p}_2 = x_2\mathbf{i} + y_2\mathbf{j} + z_2\mathbf{k}$ (in vector form), the intersection point $\mathbf{P}$ at plane z_i is given by

$$\mathbf{p} = \left[x_1 + \frac{(z_i - z_1)(x_2 - x_1)}{(z_2 - z_1)}, y_1 + \frac{(z_i - z_1)(y_2 - y_1)}{(z_2 - z_1)}, z_i \right]^T \tag{5.11}$$

In the case of stereolithography, the layer thickness for the first layer must be determined based on the curing depth of the machine. In other words, if the curing depth of a particular machine is h, the z-height of the first layer should be $z_1 = Z_{min} + h$ instead of $z_1 = Z_{min} + d$.

The contour points computed above as simple intersections between face edges and a cutting plane are surface contour points. For RP applications, one further needs to convert the surface contour points into skin contour points for later tool path generation. The determination of skin contours is mainly based on the tolerance requirement. Figure 5-26 illustrates several cases with different tolerance requirements.

- Figure 5-26a shows a model produced using a 'top-down' approach. All the computed surface contours are directly used as skin contours for the current layer for model prototyping.
- Figure 5-26b shows a model produced using a 'bottom-up' approach. The computed surface contours are directly used as skin contours of the next layer.
- Figure 5-26c shows a model produced with negative tolerance. The produced model is always smaller than the actual computer model.

- Figure 5-26d shows a model produced with positive tolerance. The produced model is always larger than the computer model.

With regard to both negative and positive tolerances, one needs to determine a surface contour point should be used as a skin contour point of the current or the next layer based on face inclination, i.e., facing up or facing down, of the model surface (Liao and Chiu 2001a). For simplicity, we still call the computed skin contours as surface contours in the rest of this section.

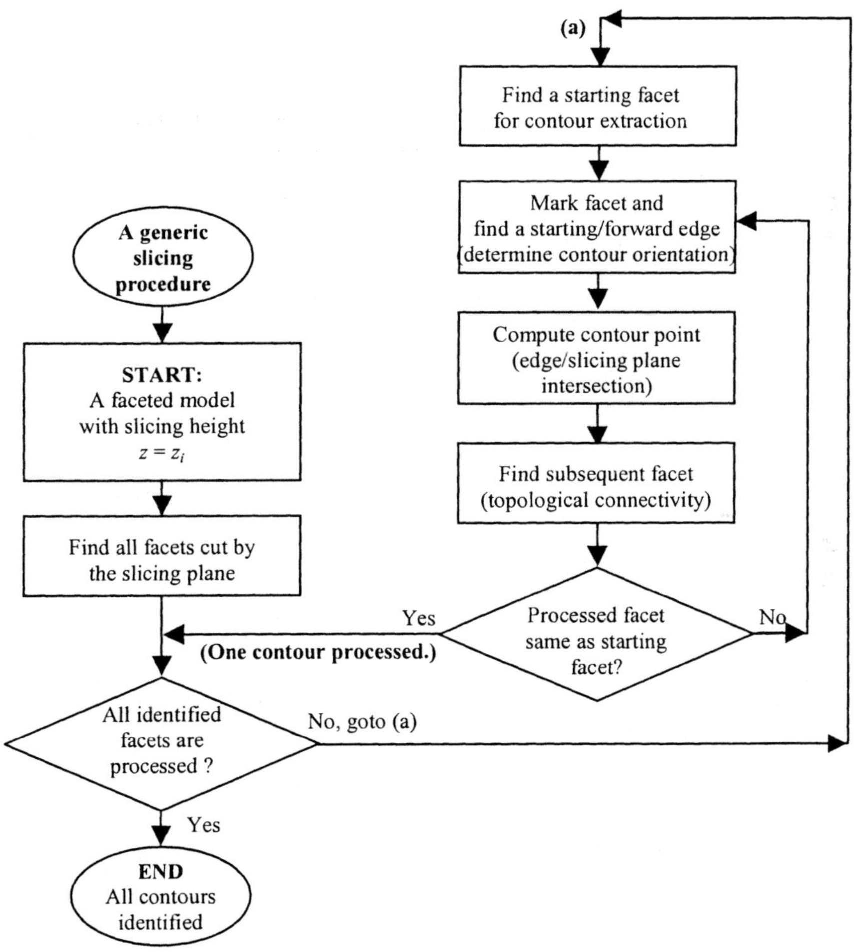

Figure 5-25. A general slicing procedure for STL models.

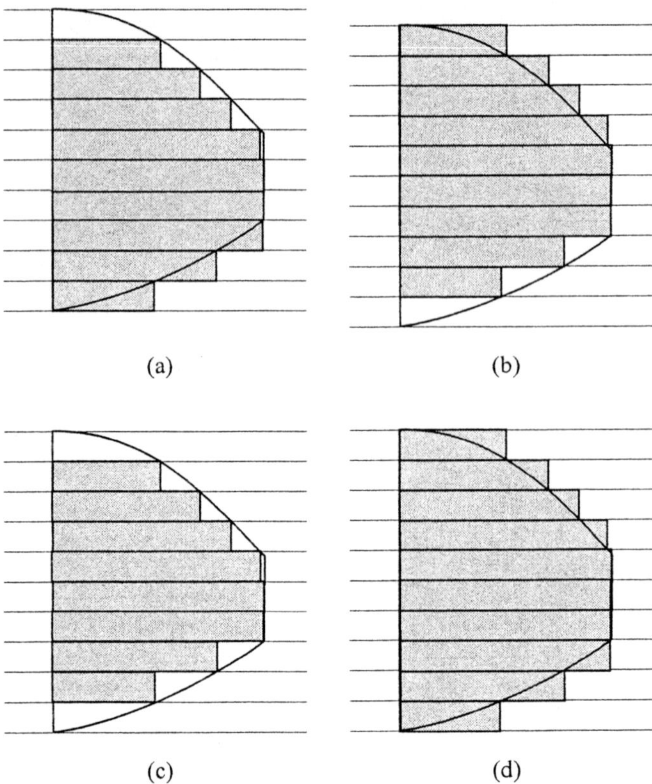

Figure 5-26. Skin contour computation based on tolerance requirement: (a) A 'top-down' approach; (b) a 'bottom-up' approach; (c) negative tolerance; and (d) positive tolerance.

After completing model slicing and having determined the skin contours, we should have arrived at a list of skin contours. Each of the contours is defined by a list of chain contour points. One can use an arbitrary starting point and the direction of ordering, i.e., clockwise or counter clockwise, is also arbitrary at this point and will be decided later. Figure 5-27a shows an object being cut by a slicing plane and Figure 5-27b shows the individual triangles with cutting lines. Figures 5-27c and 5-27d show the surface contours (scaled) produced after model slicing with two alternative orientations.

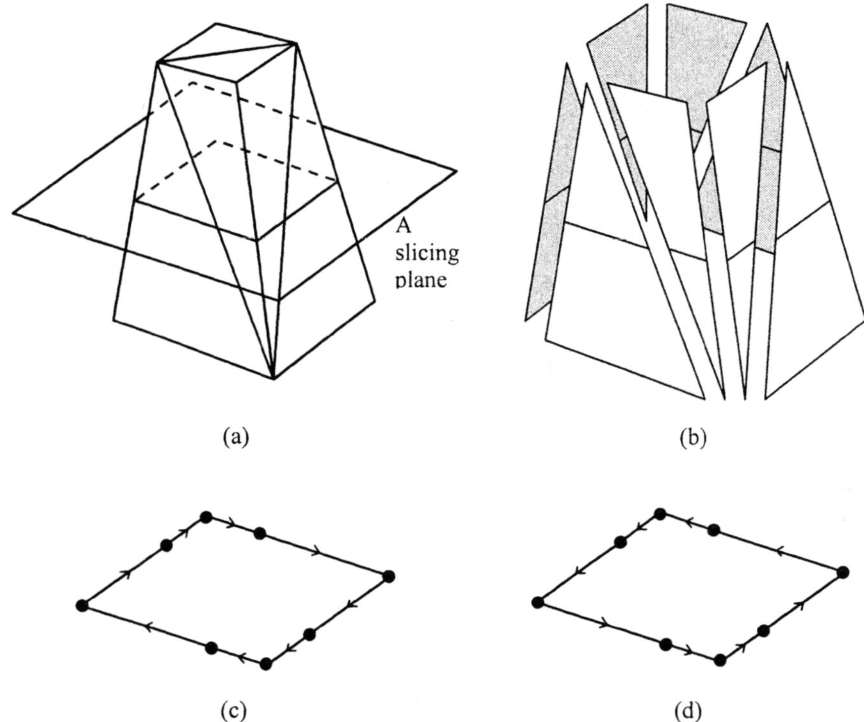

(a)   (b)

(c)   (d)

Figure 5-27. Facet model slicing and contour data initial sorting: (a) A STL model with a slicing plane; (b) related individual triangles cut by the slicing plane; (c)-(d) the identified contour (scaled) after initial contour points sorting with two possible orientations.

5.5.2 Identification of Internal and External Contours

The contour data identified during the previous step include both external contours and internal contours as shown in Figure 5-28. Any continuous solid area is defined by one external contour and one or more immediate internal contours.

The algorithm for the identification of the internal and external contours can be interpreted as follows. It is also a kind of ray casting algorithm. For each slice, a line located on the slicing plane is drawn across the contours as shown in Figure 5-28a. Since all slicing planes are parallel to the *xy*-coordinate plane, the line can either be parallel to the *x*-axis or the *y*-axis. The intersection points between the line and the contours are then computed. If no single line satisfies the purpose, more lines should be constructed until intersection occurs for all contours as shown in Figure 5-28b. If there is any contour with no intersection, one can simply add another line going through one of its contour points. In theory, one may also use inclined lines as shown in Figure 5-28b.

For each of the casting lines, the intersections are sorted and registered from one side, such as from left to right for horizontal lines or downwards for vertical lines. For each of the contours, a registration code is reserved for indicating the status of a particular intersection, with an odd number indicating the 'in' condition and an even number indicating the 'out' condition. An initial registration code of "0" is assigned to all the intersecting contours before processing. The registration code of a contour will be incremented by one when the intersection line meets an intersection point of that contour during the identification process. Each of the intersection points will also have a unique registration code equaling the registration code of the corresponding contour at that intersection point.

Based on the above conventions, we may adopt the following algorithm for contour type classification:

- As shown in Figure 5-28a, starting from the left end, the contour to which the first intersection point (point #1) belongs is classified as an external contour (contour #A) and the registration code of the identified contour is incremented by one.
- If the next intersection point is on a previously classified contour (points #6-9, #11-14, #16 and #19-24), there is no need to classify the contour again and the registration code of the corresponding contour is simply incremented by one.
- Otherwise, if the next intersection point belongs to an unclassified contour, the following algorithm can be used for classification:
    - If the registration code of its previous intersection point is an odd number (points #2-5 and #18), the contour type will be the inverse of the contour type of the previous intersection point;
    - If the registration code of its previous intersection point is an even number (points #10, #15 and #17), the contour type will be the same as that of the previous intersection point.

In the case of a tangent point, such as the intersection point T between the vertical line and the circle, the registration code should be incremented by two instead of one.

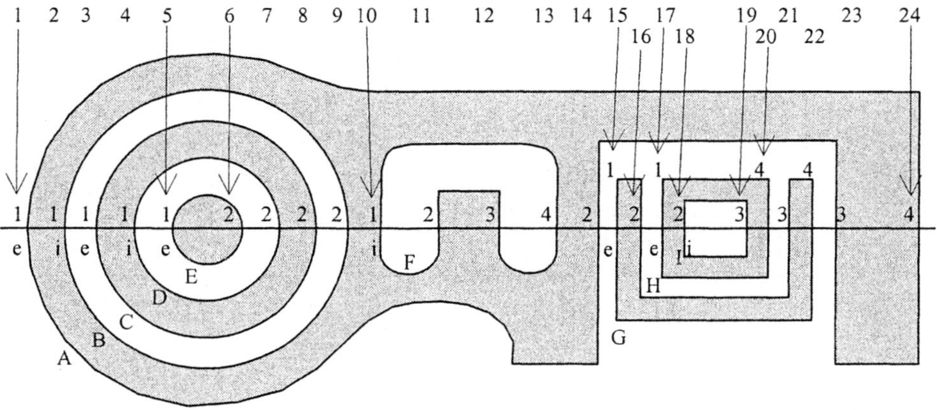

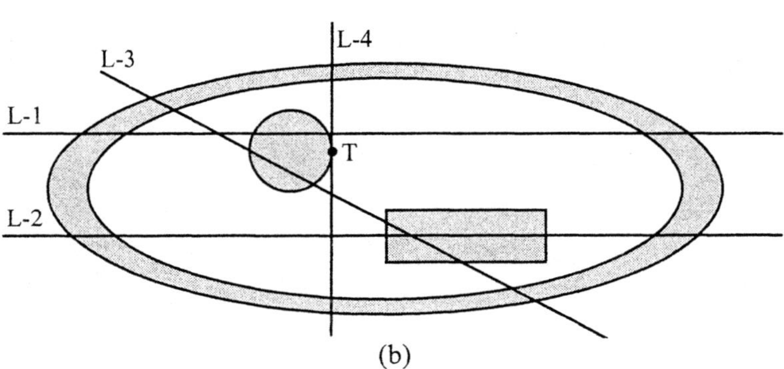

Figure 5-28. Contour type identification: (a) identification with a single horizontal line; (b) identification with two horizontal lines (L-1 and L-2) or with a single inclined line (L-3).

5.5.3 Contour Data Organization

A variety of contour data interfaces are being used by the RP community. One of the commonly used contour data interfaces is the Common Layer Interface (CLI) developed through a Brite-Euram project (Materialise 1994 and Vancraen *et al.* 1994). In addition, individual RP machine manufacturers also use their proprietary contour data interfaces. Typical examples include SLC from 3D Systems, Inc. (3D Systems 1994) and SSL (Stratasys Slicing Language) from Stratasys, Inc. (Stratasys 1997). Figure 5-29 illustrates a data structure adapted from the CLI interface for contour data organization. Figure 5-30 shows an example indicating how the directions of internal and external contours are defined in a CLI file.

With the CLI contour data interface, all contours are organized layer by layer. Strictly speaking, contour formats only include closed contours. Sometimes, open contours also need to be supported in order to represent support structures or hatching. Closed contours could be either external or internal contours as discussed in the previous subsection. The internal and external contours are arranged clockwise and counter clockwise, respectively, in the direction of the slicing axis, i.e., in the direction of the z-axis if the slicing planes are parallel to the xy-plane.

```
Contour data organization:
n_layer                  : total number of layers;

// Contour data of the first layer
layer_index              : layer number
z_cord                   : z coordinate of the current layer
n_contours               : number of contours of this layer

    // Information of the first contour
    contour_index    : contour number
    contour_closed   : closed or open contour
    contour_external : external or internal contour
    father_contour   : contour pointer, 0 if outmost contour

    n                : number of contour points
    x(1), y(1), z(1) : coordinates of the 1st contour point
    x(2), y(2), z(2) : coordinates of the 2nd contour point
        :
        :
        :
    x(n), y(n), z(n) : coordinates of the last contour point

    // Information of the second contour
        :
        :
        :
    // Information of the last contour

// Contour data of the second layer
    :
    :
    :
// Contour data of the last layer
```

Figure 5-29. Parameters for contour data organization.

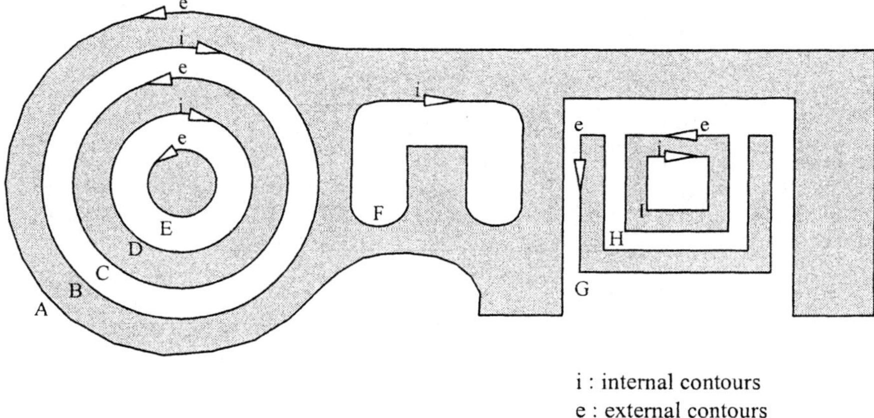

i : internal contours
e : external contours
A-I : contour index number

Figure 5-30. Contour data organization: external contours - counter clockwise; internal contours - clockwise.

5.6 DIRECT AND ADAPTIVE SLICING

This section addresses adaptive slicing for obtaining a smooth surface finish while ensuring high building speed. Instead of working with a STL model, d irect s licing a lgorithms a re p resented, i.e., t he a lgorithms d irectly work with a CAD model. The discussion is mainly adapted from Ma *et al.* (2003b) on direct slicing of a NURBS-based surface model, but is applicable to all parametric surface or B-rep solid models. The general approach is also based on the work reported in (Kulkarni and Dutta 1996) on generic adaptive slicing and (Dolenc and Mäkelä 1994) on adaptive slicing of a STL model. The procedure is subdivided into the following major steps.

- *Peak feature point identification*: When producing prototype models with uniform layer thickness, there i s n o guarantee t hat important features of an object, such as horizontal features and other important feature points, are properly reproduced. With adaptive slicing, one can place a layer anywhere and hence all the peak features can be reproduced on the prototype model. In order to do so, all the peak features in a CAD model are first identified from the model surfaces as a set of feature points. These feature points subdivide the CAD model into slabs along the slicing direction, i.e., the z-direction.

Feature points are sorted according to the slicing direction and will be used as inputs for the adaptive slicing algorithms.

- *Adaptive slicing with arbitrary tolerance control*: An adaptive slicing algorithm based on surface curvature along the vertical direction at the reference level/points is applied to each of the slabs with a pre-specified cusp height tolerance, and the minimum and maximum layer thickness. The skin contours on each layer are obtained from the allowable layer thickness, the local geometry information, and the given tolerance.

The issue of maintaining the desired tolerances is often called a containment problem in layered manufacturing (Kulkarni and Dutta 1996) and either a positive or a negative tolerance can be used depending on the applications of the prototype model. For maintaining the desired tolerance among layers, Kulkarni and Dutta (1996) used a 'bottom-up' and a 'top-down' approach for determining the skin contours. A direct method (Ma *et al.* 2003b) is presented here to compute the skin contours with the computed adaptive layer thickness and required cusp height tolerance. With this algorithm, the cusp height tolerance can be either positive, negative or an arbitrary c ombination o f these, so t he a lgorithm i s c apable o f p roviding a satisfactory solution to the containment problem.

The use of adaptive slicing with a variable layer thickness can yield the minimum number of layers along a given direction that satisfy the cusp height requirement or other tolerance criterion. The build time is thus reduced. At the same time, direct slicing of a CAD model can avoid potential problems related to the STL interface and thus improve the slicing accuracy. A major drawback with the use of direct slicing is the complexity of slicing algorithms, which could be a major reason why the STL interface is still the de-facto standard of the RP industry. However, the algorithms presented here can also be extended to process models represented in STL format.

In addition, all the algorithms presented here, including the selective hatching strategy discussed in the next section, have been implemented and tested on a Fused Deposition Modeling (FDM) rapid prototyping machine (Ma *et al.* 2003b). Given the same tolerance requirement, the results show that significant build time reduction can be achieved with these adaptive slicing and selective hatching algorithms.

5.6.1 Identification of Peak Features

Peak features of an object are easily missed when performing slicing in layered m anufacturing (Dolenc and Mäkelä 1 994). Figure 5 -31 i llustrates two slicing strategies. All the peak features are preserved with the second strategy shown in Figure 5-31b. This is however not the case for the first strategy shown in Figure 5-31a.

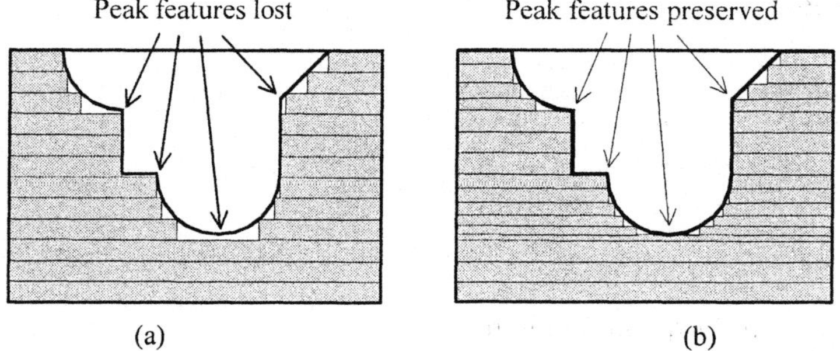

Figure 5-31. Two slicing procedures: (a) peak features lost with uniform slicing; (b) peak features preserved with adaptive slicing.

A peak feature may occur at places including a corner, a boundary curve, and an interior area of a surface. A peak feature may also be a horizontal edge or a horizontal face. For the purpose of layered manufacturing, all peak features can be handled with characteristic/feature points on the individual surfaces. These characteristic points can be subdivided into three classes:
- peak/corner points P_p;
- boundary/curve points P_c; and
- interior/surface points P_s.

Horizontal feature edges or feature faces can be represented by any point on the edge or face. In the following algorithms, they are marked by one of their corner points. In the following discussion, the characteristic/feature points are also called extreme points.

To proceed further for the extreme points, let

$$\mathbf{S}(u,v) = \left[x_s(u,v), y_s(u,v), z_s(u,v)\right]^T \qquad 0 \le u,v \le 1 \qquad (5.12)$$

represent a parametric surface in 3-dimensional space and let

$$\mathbf{C}(t) = \left[x_c(y), y_c(t), z_c(t)\right] \qquad 0 \le t \le 1 \qquad (5.13)$$

stand for a parametric curve defining the boundaries of the surface. For non-trimmed surfaces, equation (5.13) can be directly extracted from equation (5.12). For trimmed surfaces, the boundary curves are defined as curves on the original surface. In this case,

$$\mathbf{C}(t) = \mathbf{S}[u(t), v(t)] \qquad 0 \le t \le 1 \tag{5.14}$$

where $\chi(t) = [u(t), v(t)]^T$ is defined as a planar B-spline curve in the parametric *uv*-space of $\mathbf{S}(u, v)$.

Using the notations summarized in the previous paragraphs, all the corner points of a surface are first taken as candidate peak feature points. They are further examined with tighter criteria. A corner point is considered as a peak feature point only if it is a sharp corner, or it is a smooth corner, but the surface normal direction at the corner position is parallel to the slicing direction, i.e., the z-axis. The latter case automatically covers feature faces since, for any point on such a face, the surface normal is always parallel to the slicing direction.

The extreme points on a boundary curve are those where

$$\frac{dz_c(t)}{dt} = 0 \tag{5.15}$$

and can be determined through the minimization of $F(t) = [z_c'(t)]^2$. This is again an overestimation. A further verification is applied to keep only those points where the boundary curve is a crease feature, or where the surface is smooth, but the surface normal direction is parallel to the slicing direction. In a similar way, all the extreme points on a surface are those where

$$\begin{cases} \dfrac{\partial z_s(u, v)}{\partial u} = 0 \\[2mm] \dfrac{\partial z_s(u, v)}{\partial v} = 0 \end{cases} \tag{5.16}$$

Equation (5.16) can again be solved by minimizing $F(u, v) = \sqrt{Z_u^2 + Z_v^2}$.

For trimmed surfaces, it is possible that an extreme point found by equation (5.16) might not be on the trimmed area. For this reason, all extreme points $P_s(u,v)$ are further verified to check whether they are lying on the trimmed surface or not. Those extreme points not within the trimmed surface are rejected. The verification is done in the 2D parametric space. Since the boundaries of a trimmed surface are defined as 2D curves in the parametric *uv*-plane defined by equation (5.14), it is possible to determine whether an extreme point should be accepted or rejected by comparing the location of the extreme point relative to the regions formed by the 2D parametric boundary curves in the *uv* plane. For example, Figure 5-32 shows a spherical surface after trimming and the parameter region enclosed by the boundaries on the *uv* plane. If the parameter (u,v) of an extreme point is within the gray region shown in Figure 5-32b, then the extreme point is accepted. One can draw a line in the *uv*-plane passing through (u,v) along the direction of the *u*-axis, and count the number of intersection points of the line with the boundary curves in the 2D space. If the number of intersection points on one side of (u, v) is an odd number, the extreme point is then on the trimmed surface.

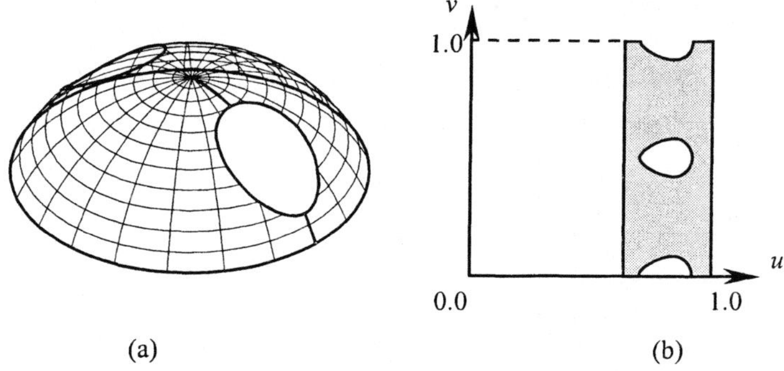

(a)   (b)

Figure 5-32. A trimmed spherical surface and its parameter space: (a) the trimmed surface; (b) the parameter region (u,v).

It should be noted that there might be multiple extreme points on a single curve or surface. The minimization of $F(t)$ or $F(u,v)$ should therefore be applied to all curve segments and surface patches, which can be realized by using multiple starting points when solving the minimization problem. It is also possible that an extreme point on a boundary curve is the same as an extreme point found on a surface. To eliminate duplicate points, all extreme points P on the NURBS surface model are examined and the final extreme points can be represented as

$$\frac{dz_c(t)}{dt} = 0 \qquad (5.17)$$

$$\mathbf{P} = \left(\cup \, \mathbf{P}_p\right) \cup \left(\cup \, \mathbf{P}_c\right) \cup \left(\cup \, \mathbf{P}_s\right) \qquad (5.18)$$

When the extreme points are obtained from equation (5.18), the object can be subdivided into slabs according to the *z*-coordinates of the extreme points for further adaptive slicing.

5.6.2 Adaptive Layer Thickness Determination

The tolerance requirement of a RP part is an important input for layer thickness determination. The tolerance distribution can be either negative, positive or a combination of both as illustrated in Figure 5-33. The negative tolerance shown in Figure 5-33a can be used for situations where the fabricated RP part acts as a core pattern. The positive tolerance shown in Figure 5 -33b i s o ften u sed i n s ituations w here a p ost-treatment o peration, such as polishing, can be entertained. Both the cases of negative and positive tolerances are addressed in (Kulkarni and Dutta 1996) as a containment problem. In other general situations, a mixed tolerance as shown in Figure 5-33c might be desirable for faithfully representing the model shape or for satisfying other fitting conditions. This section provides a generalization of the containment problem where the user may use an arbitrary combination of both negative and positive tolerances. It is therefore possible for users to control the fitting conditions irrespective of whether the RP model is to be used as a core pattern or is followed by some post treatment operations. In this case, the same distribution of the mixed tolerance occurs at all layers. This f eature is d ifferent from t he s ituation p ointed o ut i n (Sabourin *et a l.* 1996, Kulkarni and Dutta 1996) where the RP part is skewed.

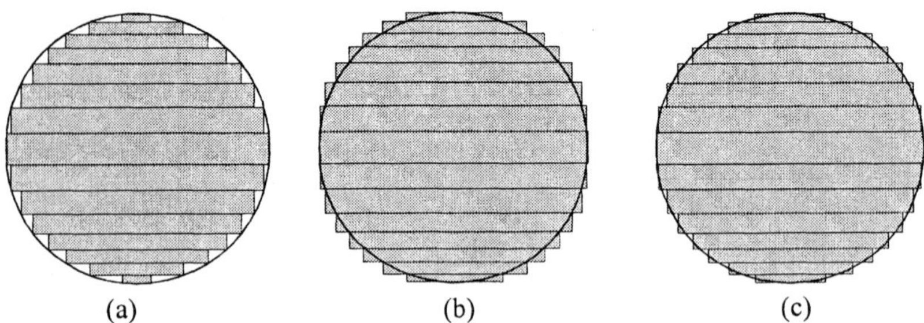

Figure 5-33. Tolerance distribution: (a) negative tolerance only; (b) positive tolerance only; and (c) mixed tolerance.

The acceptable layer thickness at a point on a surface is often determined by an allowable cusp height and the local surface shape in the vertical direction. This shape at a point is represented by an intersection curve between the surface itself and a plane passing through the point. The plane normal is determined by a vector product of the surface normal and the z-axis. For determining the layer thickness, the curve is locally approximated by a circle whose radius is determined by the curvature of the curve at that point, the so called normal curvature in the vertical direction in (Kulkarni and Dutta 1996).

Given a point P=$\{x, y, z\}$, we let δ_n, δ_p and $\delta = \delta_n + \delta_p$ be the positive, negative and total allowed cusp heights respectively, ρ the radius of the normal curvature κ_n at a point P of the present layer along the vertical direction, d the next layer thickness, N the normal vector of the surface at P and $n=\{n_x, n_y, n_z\}$ the unit vector of N. The determination of layer thickness with both negative tolerance and positive tolerance can be found in (Kulkarni and Dutta 1996). For mixed tolerances, Figure 5-34 can be used to illustrate various cases and derive the corresponding formulas for determining the layer thickness. The results for the eight cases are the following:

(a) $\qquad d = +\dfrac{\delta}{n_z}$ $\hfill$ (5.19)

(b) $\qquad d = -\dfrac{\delta}{n_z}$ $\hfill$ (5.20)

(c) $\qquad d = -\rho n_z + \sqrt{\rho^2 n_z^2 + \delta(2\rho + \delta_p - \delta_n)}$ $\hfill$ (5.21)

(d) $\qquad d = +\rho n_z + \sqrt{\rho^2 n_z^2 + \delta(2\rho - \delta_p + \delta_n)}$ $\hfill$ (5.22)

(e) $\qquad d = -\rho n_z - \sqrt{\rho^2 n_z^2 - \delta(2\rho + \delta_p - \delta_n)}$;

$\qquad$ if $\rho^2 n_z^2 > \delta(2\rho + \delta_p - \delta_n)$ $\hfill$ (5.23)

(f) $\qquad d = +\rho n_z - \sqrt{\rho^2 n_z^2 - \delta(2\rho - \delta_p + \delta_n)}$;

$\qquad$ if $\rho^2 n_z^2 > \delta(2\rho - \delta_p + \delta_n)$ $\hfill$ (5.24)

(g) $\qquad d = -2\rho n_z$; if $\rho^2 n_z^2 \le \delta(2\rho + \delta_p - \delta_n)$ $\hfill$ (5.25)

(h) $\qquad d = +2\rho n_z$; if $\rho^2 n_z^2 \le \delta(2\rho - \delta_p + \delta_n)$ $\hfill$ (5.26)

Considering equation (5.23) derived from Figure 5-34e, if $\rho^2 n_z^2 - \delta(2\rho + \delta_p - \delta_n) \leq 0$, one obtains $n_z^2 \leq \dfrac{\delta(2\rho + \delta_p - \delta_n)}{\rho^2}$ or, approximately, $n_z^2 \leq \dfrac{2\delta}{\rho}$ since $\delta = \delta_p + \delta_n << \rho$. Similarly, for equation (5.24) derived from Figure 5-34f, if $\rho^2 n_z^2 - \delta(2\rho - \delta_p + \delta_n) \leq 0$, one obtains $n_z^2 \leq \dfrac{\delta(2\rho - \delta_p + \delta_n)}{\rho^2}$ or, approximately, $n_z^2 \leq \dfrac{2\delta}{\rho}$. These two cases imply that n_z is very small or that point P lies very close to the equator. In such cases, the layer thickness is defined by equations (5.25) and (5.26). Figure 5-34g and Figure 5-34h show the two special cases when the layer thickness is symmetric about the equator. General information on surface normal and curvature evaluation can be found in (Hoschek and Lasser 1993 and P iegl a nd T iller 1 995). F urther i nformation f or e valuating t he n ormal curvature κ_n in the vertical direction and the corresponding radius ρ can be found in (Kulkarni and Dutta 1996 and Ma *et al.* 2003b).

Equations (5.19)-(5.26) are formulas for layer thickness determination at a point. To find the layer thickness for the entire slice, the minimum layer thickness at all contour points on that layer needs to be used. This leads to the following notational minimization problem:

$$
\begin{cases}
\underset{u,v}{\text{minimize}} & d(u,v) \\
\text{Subject to} & (u,\ v) \in \text{the corresponding trimmed surface} \\
& \&\ z(u,\ v) = \text{constant}
\end{cases}
\qquad (5.27)
$$

If there is more than one surface on the current slice, all the resulting layer thickness values must be compared and the smallest should be used. To find the minimum layer thickness, we check all the discrete sample contour points o n t hat l ayer t o d etermine t he m inimum l ayer t hickness. W hen the adaptive slicing distance is available, contours of the next layer are computed as the intersection contours between the slicing plane and the parametric surface model.

5.6.3 Skin Contours Computation

Depending on the tolerance requirement, the skin contours might not be the same as the sliced contours as shown in Figure 5-34. A method that integrates the 'top-down' and the 'bottom-up' (Figures 5-26a and 26b, respectively) approaches are used in (Kulkarni and Dutta 1996) for determining the skin contours in the cases of pure positive or pure negative tolerances. With this approach, the skin contours can be readily determined when the vertical coordinate of the surface normal maintains its sign on any given contour. Otherwise, the upper slice and the lower slice of the layer are projected onto a common horizontal plane to determine an inner and an outer closed contour. The inner and outer contours are then regarded as the skin contours with negative and positive tolerance, respectively.

We now present a direct method for computing the skin contours based on the parameters of the current layer. The approach is similar to that described earlier for the determination of the layer thickness. It computes skin points with layer thickness d and is applicable to all cases with either negative, positive or mixed tolerance requirements. Since the surface curvature information (the curvature radius ρ) at the present layer is readily available after computing the layer thickness, the skin contours can be effectively computed based on the various cases shown in Figure 5-34.

Let us assume that at a point $P=\{x,\ y,\ z\}$, the coordinates of the corresponding skin point is $\{x_s,\ y_s,\ z+d\}$, and

$$\begin{cases} n'_x = \dfrac{n_x}{\sqrt{n_x^2 + n_y^2}} \\[4mm] n'_y = \dfrac{n_y}{\sqrt{n_x^2 + n_y^2}} \end{cases} \tag{5.28}$$

where $n=\{n_x,\ n_y,\ n_z\}$ is the unit vector of the surface normal at P. When the layer thickness d is obtained for each layer, we can further determine the skin contour points according to the actual cusp height. Let $\delta_n{'}$ be the actual 'negative' cusp height, $\delta_p{'}$ the actual 'positive' cusp height, and $\delta' = \delta_n' + \delta_p'$ the total actual cusp height, where

$$\delta_n' = \frac{\delta_n}{\delta_n + \delta_p}\delta' \tag{5.29}$$

and

$$\delta_p' = \frac{\delta_p}{\delta_n + \delta_p}\delta' \tag{5.30}$$

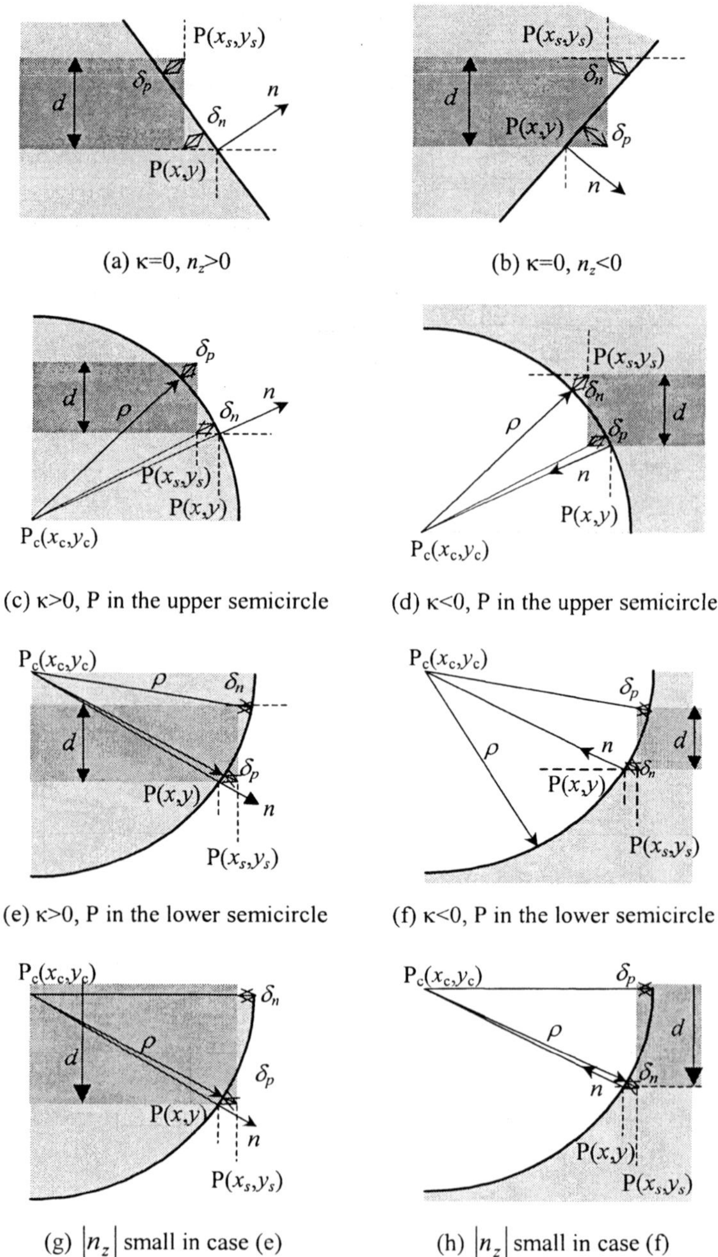

Figure 5-34. Determination of layer thickness and skin contours.

The formulas for determining the skin point at point P can also be derived from Figure 5-34 and can be represented in terms of either δ'_n or δ'_p as follows. In terms of δ'_n, the skin contours can be defined as

(a)
$$\begin{cases} x_s = x - \dfrac{n'_x}{\sqrt{1 - n_z^2}}\,\delta'_n \\[4mm] y_s = y - \dfrac{n'_y}{\sqrt{1 - n_z^2}}\,\delta'_n \end{cases}$$
(5.31)

(b)
$$\begin{cases} x_s = x - \dfrac{n'_x}{\sqrt{1 - n_z^2}}\,(dn_z + \delta'_n) \\[4mm] y_s = y - \dfrac{n'_y}{\sqrt{1 - n_z^2}}\,(dn_z + \delta'_n) \end{cases}$$
(5.32)

(c)
$$\begin{cases} x_s = x - \rho n_x + n'_x \sqrt{(\rho - \delta'_n)^2 - \rho^2 n_z^2} \\[3mm] y_s = y - \rho n_y + n'_y \sqrt{(\rho - \delta'_n)^2 - \rho^2 n_z^2} \end{cases}$$
(5.33)

(d)
$$\begin{cases} x_s = x + \rho n_x - n'_x \sqrt{(\rho + \delta'_n)^2 - (\rho n_z - d)^2} \\[3mm] y_s = y + \rho n_y - n'_y \sqrt{(\rho + \delta'_n)^2 - (\rho n_z - d)^2} \end{cases}$$
(5.34)

(e)
$$\begin{cases} x_s = x - \rho n_x + n'_x \sqrt{(\rho - \delta'_n)^2 - (\rho n_z + d)^2} \\[3mm] y_s = y - \rho n_y + n'_y \sqrt{(\rho - \delta'_n)^2 - (\rho n_z + d)^2} \end{cases}$$
(5.35)

(f)
$$\begin{cases} x_s = x + \rho n_x - n'_x \sqrt{(\rho + \delta'_n)^2 - \rho^2 n_z^2} \\[3mm] y_s = y + \rho n_y - n'_y \sqrt{(\rho + \delta'_n)^2 - \rho^2 n_z^2} \end{cases}$$
(5.36)

(g)
$$\begin{cases} x_s = x - \rho n_x + n'_x (\rho - \delta'_n) \\[2mm] y_s = y - \rho n_y + n'_y (\rho - \delta'_n) \end{cases}$$
(5.37)

(h)
$$\begin{cases} x_s = x + \rho n_x - n'_x \sqrt{(\rho + \delta'_n)^2 - \rho^2 n_z^2} \\ y_s = y + \rho n_y - n'_y \sqrt{(\rho + \delta'_n)^2 - \rho^2 n_z^2} \end{cases} \tag{5.38}$$

In terms of δ'_p, the skin contours can also be defined as

(a)
$$\begin{cases} x_s = x - \dfrac{n'_x}{\sqrt{1 - n_z^2}}(dn_z - \delta'_p) \\ y_s = y - \dfrac{n'_y}{\sqrt{1 - n_z^2}}(dn_z - \delta'_p) \end{cases} \tag{5.39}$$

(b)
$$\begin{cases} x_s = x + \dfrac{n'_x}{\sqrt{1 - n_z^2}}\delta'_p \\ y_s = y + \dfrac{n'_y}{\sqrt{1 - n_z^2}}\delta'_p \end{cases} \tag{5.40}$$

(c)
$$\begin{cases} x_s = x - \rho n_x + n'_x \sqrt{(\rho + \delta'_p)^2 - (\rho n_z + d)^2} \\ y_s = y - \rho n_y + n'_y \sqrt{(\rho + \delta'_p)^2 - (\rho n_z + d)^2} \end{cases} \tag{5.41}$$

(d)
$$\begin{cases} x_s = x + \rho n_x - n'_x \sqrt{(\rho - \delta'_p)^2 - \rho^2 n_z^2} \\ y_s = y + \rho n_y - n'_y \sqrt{(\rho - \delta'_p)^2 - \rho^2 n_z^2} \end{cases} \tag{5.42}$$

(e)
$$\begin{cases} x_s = x - \rho n_x + n'_x \sqrt{(\rho + \delta'_p)^2 - \rho^2 n_z^2} \\ y_s = y - \rho n_y + n'_y \sqrt{(\rho + \delta'_p)^2 - \rho^2 n_z^2} \end{cases} \tag{5.43}$$

(f)
$$\begin{cases} x_s = x + \rho n_x - n'_x \sqrt{(\rho - \delta'_p)^2 - (\rho n_z - d)^2} \\ y_s = y + \rho n_y - n'_y \sqrt{(\rho - \delta'_p)^2 - (\rho n_z - d)^2} \end{cases} \tag{5.44}$$

$$\text{(g)} \quad \begin{cases} x_s = x - \rho n_x + n'_x \sqrt{(\rho + \delta'_p)^2 - \rho^2 n_z^2} \\ y_s = y - \rho n_y + n'_y \sqrt{(\rho + \delta'_p)^2 - \rho^2 n_z^2} \end{cases} \quad (5.45)$$

$$\text{(h)} \quad \begin{cases} x_s = x + \rho n_x - n'_x (\rho - \delta'_p) \\ y_s = y + \rho n_y - n'_y (\rho - \delta'_p) \end{cases} \quad (5.46)$$

5.7 A SELECTIVE HATCHING STRATEGY FOR RP

The use of the variable layer thickness approach improves the surface quality and shortens the build time of layered manufacturing. The process may however produce very thin layers when slicing objects having curved surfaces and the build time may still be very long. To further improve the building speed without sacrificing surface accuracy, a selective hatching strategy can be incorporated with adaptive slicing (Ma and He 1999 and Ma *et al.* 2003b). Only the skin regions will be scanned/solidified at all adaptive layers to ensure a smooth part surface. The internal areas will be scanned only when the accumulated layer thickness approaches the maximum allowable layer thickness o f a particular process s o a s to maintain a h igh building speed. Based on this idea, we may subdivide a layer into skin and internal regions. In other words, the skin regions of a layer are built with the adaptive layer thickness while the internal or the kernel regions are built with the maximum possible layer thickness of a RP process. The idea is illustrated in Figure 5-35.

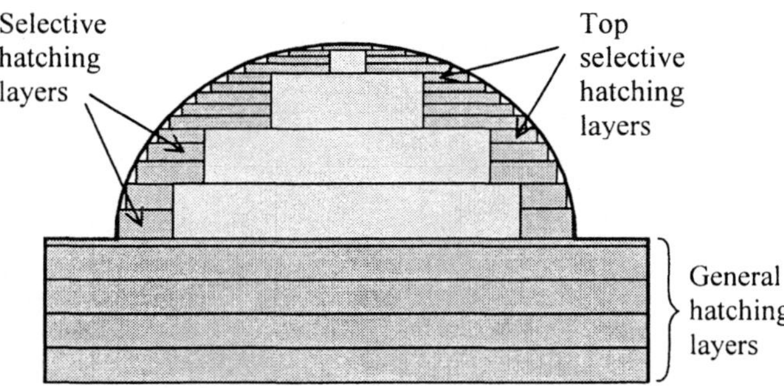

Figure 5-35. Thin layers for the skin regions and thick layers for the kernel regions.

Let us assume that the adaptive layer thickness at a level z_i is d_i. One can find an integer n such that:

$$\sum_{i=0}^{n-1} d_i \leq D < \sum_{i=0}^{n} d_i \qquad\qquad (5.47)$$

where D is the maximum layer thickness allowed by the given RP system/process. These n layers are then grouped as a hatching unit for determining the skin and the kernel regions. To find the common separation contours between the skin and the kernel regions for all these layers, the skin contours of these n layers are first projected onto a common horizontal plane. The innermost contours of the external skin contours and the outermost contours of the internal skin contours are further identified. The separation contours between the skin region and the kernel region are then the offset contour segments of these extreme skin contours, i.e., the innermost and the outermost skin contours of a hatching unit, with an offset direction pointing towards the material side of the part surface. The offset distance is called the (minimum) wall thickness for the skin region. Figure 5-36 illustrates various contours and the skin and kernel regions. For both Figures 5-36and 5-37 and later discussions, the innermost contours of the external contours and the outermost contours of the internal skin contours are simply marked as 'internal contours' and 'external contours', respectively, for simplicity. At the same time, the partition contours between skin and kernel regions are marked as 'offset contours'. As shown in Figure 5-35, the computed partition contours between the skin and the kernel regions are applied to all layers of the hatching unit.

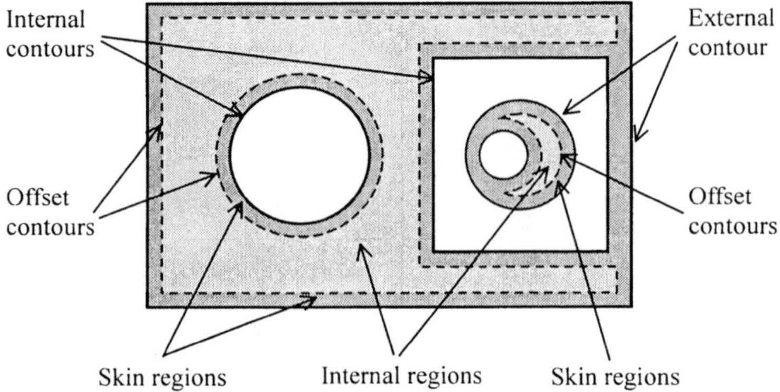

Figure 5-36. The skin and kernel regions of a single slice.

To determine the skin and kernel regions, the simple three-phase approach shown in Figure 5-37 is adopted. During the first phase, all the offset contours (Figure 5-37a), both internal and external, are calculated according to the given wall thickness. During the second phase, internal and external contours are processed independently for the removal of overlapping areas. For internal offset contours, all contour segments falling within any other internal offset contours with intersections are removed and the remaining internal offset contour segments are reorganized (Figure 5-37b). For external offset contours, all contour segments falling outside other external offset contours with intersections are removed and the remaining external offset contour segments are reorganized (Figure 5-37b). During the third phase, all external contour segments falling inside any other internal contours with intersections and all internal contour segments falling outside any external contours with intersections are removed. The remaining contours and contour segments after reorganization are the partition contours for the skin and kernel regions (Figure 5-37c). One may also first determine the skin and kernel regions for each layer in a hatching unit. The kernel regions of the entire hatching unit are then the common areas of the n layers. Since, after adaptive slicing, all contours are defined by discrete contour points, both contour offsetting and offset contour processing are built upon contour points/segments.

In literature, one can also find a similar approach independently developed for accurate exterior and fast interior layered manufacturing (Sabourin *et al.* 1996 and 1997). The main objective of the above selective hatching strategy (Ma and He 1999 and Ma *et al.* 2003b) and that of the approach reported in (Sabourin *et al.* 1996 and 1997) is the same. Both of the approaches use a fine layer thickness for external regions for achieving an accurate part surface and a large layer thickness for internal regions for reducing the build time. However, their approaches exhibit the following differences:

- The algorithms reported by Ma and He (1999) and Ma *et al.* (2003b) directly operate upon a NURBS-based CAD surface model, while the approach of Sabourin *et al.* (1996, 1997) works with STL models.
- Ma *et al.* subdivide the object into slabs according to global peak features. The approach reported by Sabourin *et al.*, on the other hand, subdivides the object according to the maximum layer thickness, although the subdivided slices are also called slabs.
- The approach of Ma *et al.* uses an adaptive layer thickness within each of the individual slabs. The approach of Sabourin *et al.*, however, uses a uniform layer thickness within individual slabs, although the layer thickness for different slabs may be different.
- In the approach of Ma *et al.*, a direct method is used to calculate the skin contours of the next layer based on the layer thickness, tolerance requirement and the exact geometric parameters of the

current layer. The skin contours of Sabourin *et al.* are computed from the contours of that layer and tolerance requirement.

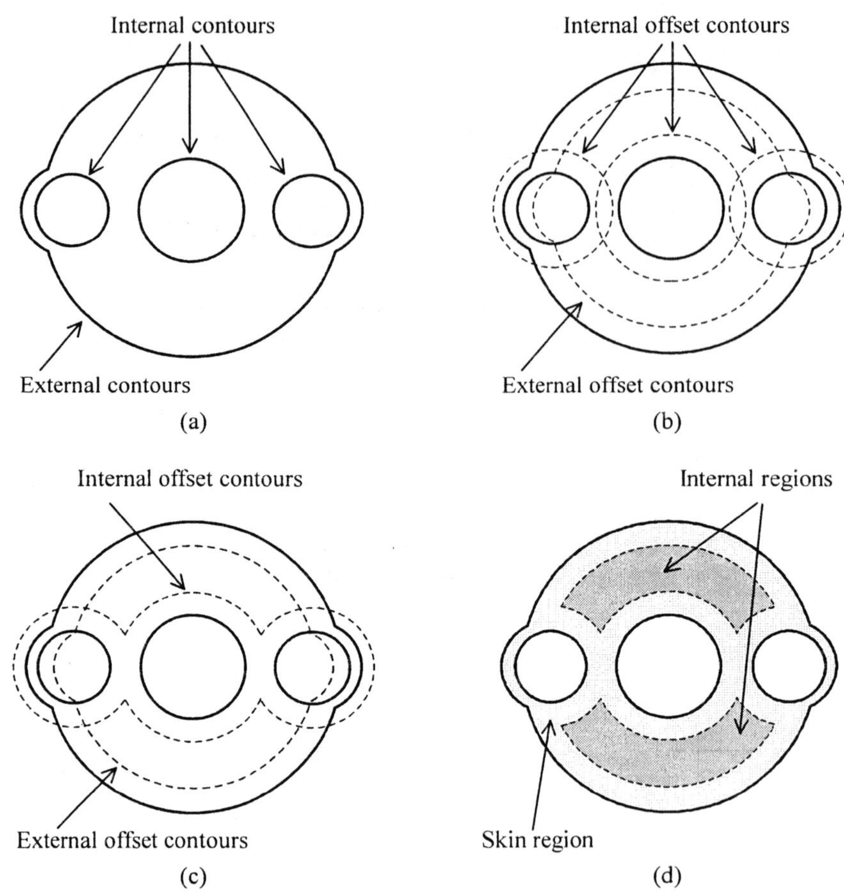

Figure 5-37. The determination of the skin and kernel regions of a selective hatching layer: (a) initial surface contours; (b) produced offset contours; (c) offset contours after Boolean operation; and (d) identified skin and kernel regions.

5.8 TOOL PATH GENERATION

Once the slicing contour data are ready, one can start addressing process-dependent issues for tool path generation. In particular, as most of the RP processes are layer-based processes, one can produce the tool path layer-by-layer starting from the bottom layer. Figure 5-38 shows a sample part with one layer of sliced surface contours for illustrating various tool paths (Figure 5-39) used for rapid prototyping applications. In general, one can classify tool paths for all RP processes into the following basic categories:

- *Raster scanning*: Raster scanning refers to scanning along one coordinate axis for model solidification. This is the simplest tool path for layered manufacturing. The scanning strategy can be applied to processes such as stereolithography (SL), selective laser sintering (SLS) a nd s ome other 3 D p rinting p rocesses f or i nternal hatching. Figure 5-39a illustrates a typical scanning pattern produced by two orthogonal raster scanning operations along the x- and y-axis for internal area solidification.

- *Perimeter scanning*: The perimeter scanning approach shown in Figure 39b (illustrating only a single perimeter) is often used for producing external surfaces (Stratasys 1997, Wasser *et al.* 1999). This approach can also be turned into a contouring approach with multiple perimeters and their offsetting contours. The method is applicable to almost all RP processes involving skin region solidification. It is a lso the m ain s canning s trategy u sed f or p aper cutting in the LOM process. Figure 5-39c shows a scanning pattern produced by single perimeter scanning plus o rthogonal i nternal x - and y-hatchings. This example illustrates the basic pattern of several scanning styles such as the WEAVE and the STAR-WEAVE patterns used on SLA machines of 3D Systems (Jacobs 1992a) (see also Section 6.6.3 for further discussions on these building styles).

- *Directional scanning*: Sometimes, raster scanning may produce a large number of scanning vectors and it might be advantageous to perform scanning along arbitrary paths, such as inclined linear and other contouring paths (Crawford 1993, Beaman *et al.* 1997). In certain other cases, we may also need to use directional scanning for improving the mechanical properties of the produced model prototype. Fig. 5-39d illustrates a scanning pattern with a single perimeter and three groups of scanning lines at 0°, 60° and 120° angles to the x-axis. This is the basic pattern used for the 'Tri-Hatch' scanning style that was once used on stereolithography machines of 3D Systems. It is also the basic pattern adopted by QuickCast 1.0 (Pang and Jacobs 1993 and Jacobs 1996).

- *Zigzag tool path*: Zigzag tool path is often used for FDM prototyping, 3D welding and other extruding type RP systems (Stratasys 1997, Wasser *et al.* 1999 and Qiu *et al.* 2001). It has also been used for some drop-on-demand and point-by-point 3D printing processes such as the 3D Printing (3DP) process developed by MIT and Ballistic Particle Manufacturing (BPM). Figure 5-39e illustrates a zigzag tool path for internal solidification with a single perimeter for surface scanning. Figure 5-39f shows an inclined zigzag path for internal filling with three perimeters for skin area solidification.

- *Contouring and spiral paths*: Contouring and spiral tool paths can also be used for model prototyping (Crawford 1993, Farouki *et al.* 1997, Beaman *et al.* 1997, Yang *et al.* 2002a). For parts with certain geometries, these approaches may produce parts with improved mechanical properties. Figure 5-39g illustrates a contouring tool path for the example part.

- *Line by line scanning*: Line by line scanning is often used for some inkjet type printing processes, such as the process used by ThermoJet 3D printer of 3D Systems. Each layer is produced through single sweeping of a line component along the principal scanning direction.

- *Area by area solidification*: Some RP processes directly solidify the object area by area. Typical examples include photopolymer-based processes such as the Cubital's stereolithography process. As shown in Figure 5-39i, a mask is first developed based on the sliced contours and a thin layer of photopolymeric resin is then selectively solidified by exposing to a flash of UV light.

- *Boundary cutting tool paths (a variant of perimeter scanning)*: In laminated object manufacturing (LOM), a boundary cutting strategy is used. For cutting waste and supporting materials during the LOM process, a basic orthogonal *xy*-hatching pattern is often used. Such an approach is illustrated in Figure 5-39j.

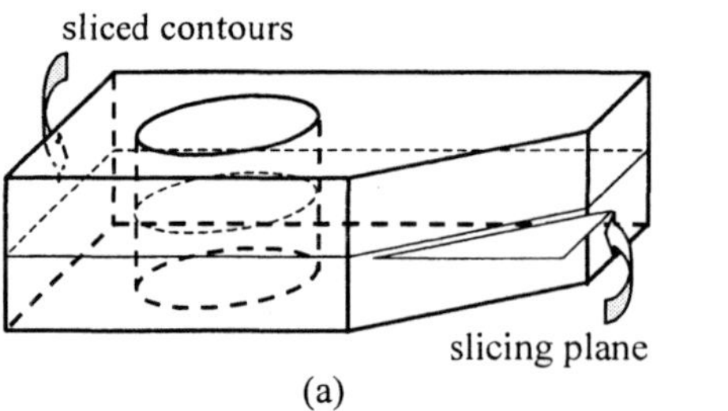

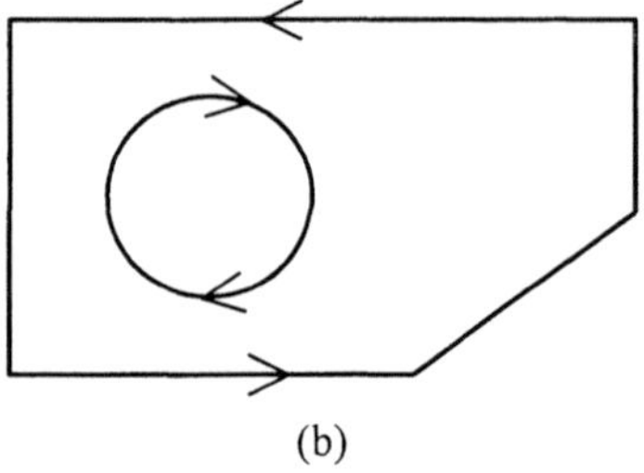

Figure 5-38. An example illustrating various types of tool paths for model prototyping: (a) the original part with a slicing plane and sliced contours; (b) the sliced contours on the *xy*-plane.

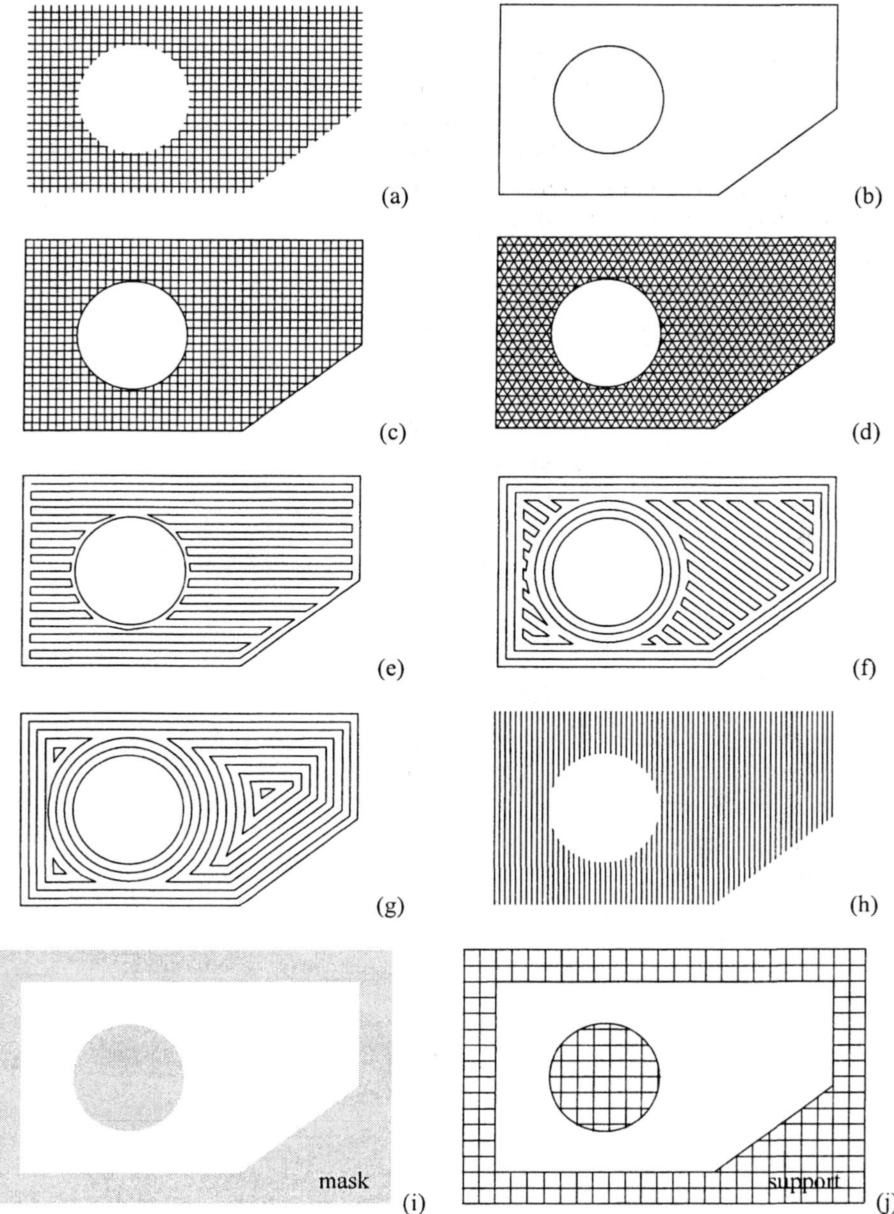

Figure 5-39. Typical tool paths for model prototyping: (a) *x*- and *y*-raster scanning for internal hatching (may be used separately); (b) perimeter scanning; (c) a typical tool path with a single perimeter surface scanning and orthogonal internal *xy*-hatching; (d) single perimeter surface scanning with internal directional hatching; (e) single perimeter with internal horizontal/vertical zigzag paths; (f) three perimeters with internal inclined zigzag paths; (g) contouring or equidistant paths; (h) line-by-line scanning; (i) area-based solidification using masks; and (j) boundary cutting with orthogonal cross-cutting for LOM.

In addition to the basic tool paths discussed above, there are also a variety of well-studied scanning strategies/styles capable of rapidly producing prototype models for various applications. In addition to Tri-Hatch, WEAVE and STAR-WEAVE mentioned earlier, Figures 5-40, 5-42 and 5-43 illustrates three scanning styles used for QuickCast, a popular scanning strategy for producing hollow stereolithography master models for quick tooling. The finally produced master model is a shell model with internal supports specially designed for easy drainage of residual resins after the building process and for easy burn out of the support structures during the tooling process. Figure 5-40 illustrates the pattern for QuickCast 1.0. During the first level of the building process, the basic pattern of Figure 5-40a is used. During the second level of the building process, the pattern shown in Figure 5-40b that is produced by offsetting the pattern of Figure 5-40a is used. The offsetting is produced such that the vertices of the pattern for the even levels will be located at the center of the triangles of the odd levels. The entire process is then repeated till the completion of the QuickCast master model (Pang and Jacobs 1993, Jacobs 1996 and Hague *et al.* 2001). The entire internal volume is composed of interconnected triangular cells and hence the resin contained within can be easily released after the building process. A similar approach is used for QuickCast 1.1 (Jacobs 1996) with orthogonal grid cells as shown in Figure 5-41. Figure 5-42 illustrates the pattern used for QuickCast 2.0 (Hague *et al.* 2001). The build style produces hexagons in three levels. The three levels of this structure use the same scanning pattern with a rotation of 120° as shown in Figures 5-40a-c. Together, the three levels produce a complete hexagon when seen from the top, often called a honeycomb structure. Further discussions regarding the QuickCast style can be found in Section 9.2.

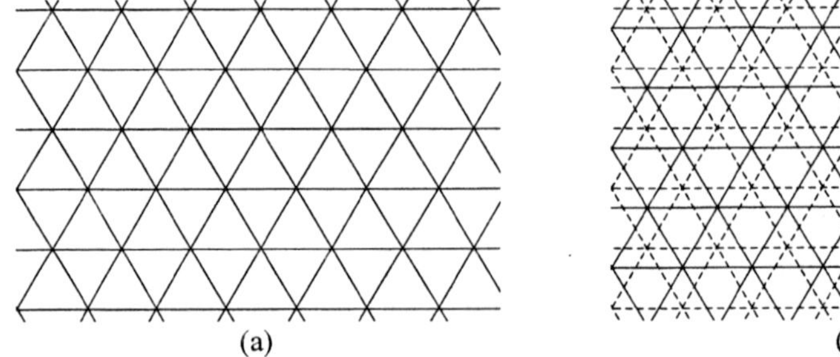

(a)   (b)

Figure 5-40. Special scanning patterns used for QuickCast 1.0: (a) first level scanning pattern; (b) top view of the scanning pattern after several levels with solid lines indicating odd levels and broken lines indicating even levels.

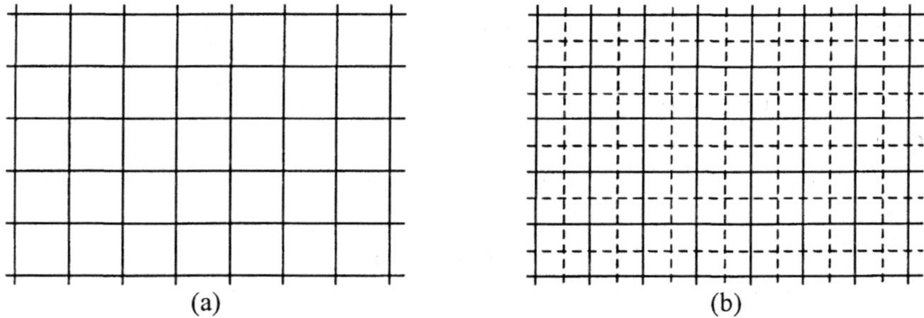

Figure 5-41. Special scanning patterns used for QuickCast 1.1: (a) first level scanning pattern; (b) top view of the scanning pattern after several levels with solid lines indicating odd levels and broken lines indicating even levels.

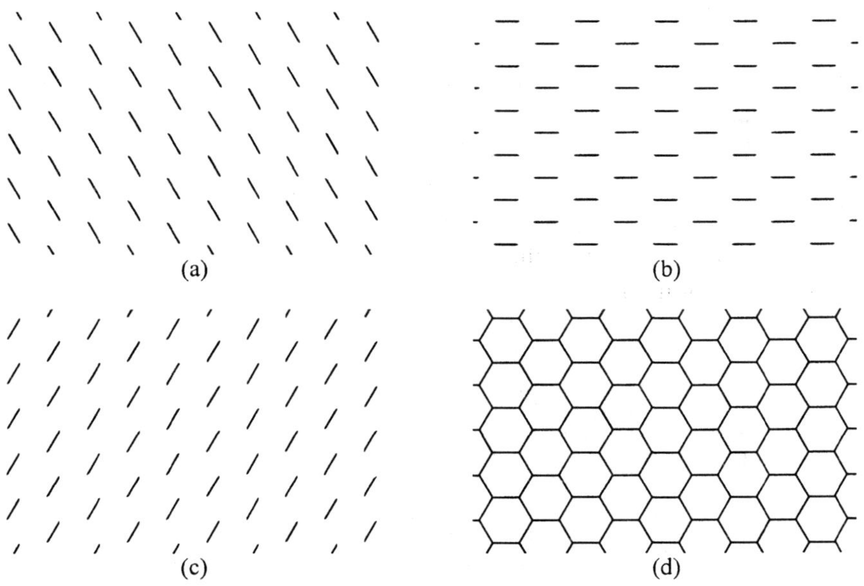

Figure 5-42. Special scanning patterns used for QuickCast 2.0: (a) first level scanning pattern; (b) second level scanning pattern; (c) third level scanning pattern; (b) top view of the scanning pattern after more than three levels.

While most of the present RP systems use vectors for various scanning paths discussed earlier, direct contouring based on exact parametrization, such as circular movements, may also be used (Crawford 1993, Beaman *et al.* 1997). Similar to material removal processes, tool compensation should also be considered when producing tool paths for surface scanning. This is sometimes a difficult problem especially for laser based processes and when the laser beam is not vertically directed towards individual layers as the projected laser spot is elliptical, i.e., it is not circular in this case anymore

(Fadel and Kirschman 1996). Post processing is another issue that has to be addressed with respect to almost all processes. One has to convert the computed tool path into control codes of the individual axis of a particular prototyping process.

For computing the individual tool paths, the intersection points between the scanning vectors and the surface contours need to be determined. Again, this can be accomplished by the aid of ray casting algorithms (Mantyla 1988, Foley *et al.* 1990 and Zeid 1991). In addition, the offset curves (planar) of the initial sliced contours need to be produced for approaches using perimeter scanning, contouring approaches and spiral approaches. Offset curves have been well-studied in CAD/CAM literature on surface contouring or pocketing operations for material removal processes (Takashi 1999 and Hoschek and Lasser 1993).

Chapter 6

STEREOLITHOGRAPHY (SL)

Stereolithography (SL), also called Laser Photolithography, is a 3-dimensional (3D) printing process that uses a laser beam directed by a computer onto the surface of a photo-curable liquid (resin) to produce physical copies of solid or surface models. It is the most widely used among the available RP processes because of its superior part building accuracy.

Stereolithography (SL) builds three-dimensional prototypes layer by layer through selective photo-polymerization of a liquid polymer held in a vat. 'Stereo' signifies 3D object realization, 'litho' indicates solidification, and 'graphy' suggests printing. Photonic energy is supplied by scanning a CNC controlled laser beam across the liquid polymer surface.

This chapter describes the basic principles of stereolithography and similar processes. Issues concerning process planning for SL will also be discussed.

6.1 THE STEREOLITHOGRAPHY (SL) PROCESS

6.1.1 Part Building Using SL

This section aims to provide a simple overview of the part-building process using SL with a view to serving as an introduction to later sections where we will discuss specific issues of practical significance.

Figure 6-1 illustrates the operational principles common to several SL machines including those produced by 3D Systems, Inc. The entire machine is a sealed system (to prevent the fumes from escaping) consisting of several subsystems. A liquid photopolymer resin is held in a vat. A laser device senses the level of the resin. A servo-controlled vertically movable table (elevator) holds a base-plate (platform) on which the prototype would be built. Prior to building each layer of the part, the table is positioned below the resin surface so that a layer of fresh (uncured) resin of controlled thickness (layer pitch, L_p; denoted by d in Chapter 5) flows over the previously cured layer. However, since the resin is usually highly viscous, it takes quite sometime for a uniformly thick fresh layer to be established. To reduce the time thus wasted, the liquid above the previously cured part section is smoothed with the help of a horizontally movable recoater blade (also called the squeegee). This process is called recoating. A separate laser system is used to selectively polymerize the liquid resin while each layer is built. The wavelength, λ, of laser radiation is usually in the visible to ultra-violet range and, therefore, much larger than the molecular dimension, d_m, of the resin, i.e., $d_m<<\lambda$. The laser subsystem includes mechanisms for controlling beam power and beam size (lens system). A pair of mirrors (x and y mirrors) enables the control of the position where the beam strikes the resin surface. The mirrors are driven by fast and highly controllable servomotors. Each layer is built by moving the laser beam over the surface of the liquid photopolymer to trace the geometry of the cross-section (as obtained from CAD data) of the object in that layer. The tracing and recoating steps are repeated until the object is completely fabricated.

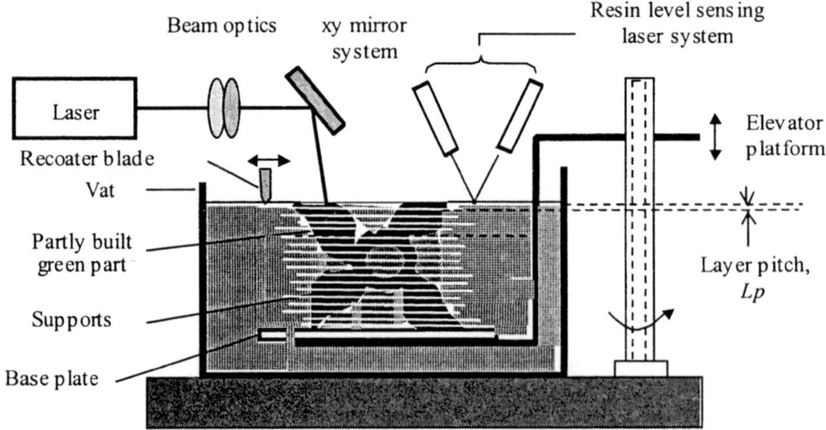

Figure 6-1. The principle of stereolithography (SL).

Current SL machines are so sophisticated that over 90% of the build time is spent in actually building the part and its supports. The rest of the time usually arises because of file errors, power failures, and manual checking activities.

After a part has been built, it (along with supports, if any) is gently removed from the base-plate and drained. Excess resin is swabbed manually.

However, owing to reasons to be described later, the part is usually not in a fully and uniformly cured state. Hence, it may be necessary to further cure the 'green' prototype in a post-cure apparatus (PCA) flooded with ultra-violet radiation. The support structures are then cut off from the fully cured part so that the prototype is ready for use.

6.1.2 Post-build Processes

Once a last layer on the part has been built, the operator requests the computer to elevate the platform and the part to reach z-position above the resin level. At this moment, quite an amount of resin is attached to the part. The part is therefore left to stand for sufficient time to allow excess resin to drain back into the vat. Further, many parts contain cavities capable of trapping liquid resin. Some of the trapped resin may be drained by manually tilting the platform over its support arms. SL resins are fairly expensive. So, it is of importance to save the excess resin by letting it flow back into the vat.

After draining the part, the platform and the part are manually removed from the machine using rubber gloves and placed in a shallow-rimmed stainless steel tray or on to cellulose padding. The removed components are then wiped with a paper towel. Narrow gaps may be wiped using 'Q-tips'. The components are then subjected to solvent cleaning in a special apparatus. One method is to remove excess resin using TPM (tripropylene glycol monomethyl ether). Next they are rinsed with water followed by rinsing with isopropyl alcohol.

The cleaned components are then dried using low-pressure compressed air so that any residual resin and solvent layers sticking to the part and platform are removed. The removal needs to be thorough. Residual resin stuck in the narrow passages, corners, or blind holes of the part can adversely affect part accuracy by getting polymerized in the subsequent post-curing operation. In contrast, the residual solvent layer cannot post-cure. This can result in a tacky surface. Further, residual solvent can dilute the resin and render the photo-initiator ineffective.

Next, the part is removed very carefully from the platform using a flat bladed knife, fine scissors, and/or a 'X-acto' knife. At this stage, the part is in a green state, i.e., it is not fully cured. To reach a fully polymerized state, the green part is placed in a post-curing apparatus (PCA). A PCA subjects the green part to broadband or continuum ultra-violet radiation. Some PCAs are capable of optimizing the wavelength of UV radiation for the given part material (resin). Most parts can be post-cured within a couple of hours. Large parts may require as much as 10 hours.

6.1.3 Pre-build Processes

Although we have started with a description of the part-building process first, several preparatory steps need to be executed before a part can be built. Actually, the SL process sequence starts with a solid or surface model of the part created in a suitable CAD system. Successful prototype building using SL critically depends on how good the CAD system used and the CAD engineer using it are. Although systems utilizing the solid model directly are under development, most c urrently available SL systems require the solid model to be tessellated and presented as an .STL file. Tessellation is the process of representing the surfaces in terms of a large number of tiny triangles.

SL builds parts layer by layer in the vertical direction. The number of layers, the geometry of the cross-section of each layer as well as the overall part accuracy and surface finish depend on the orientation of the part relative to the vertical. As a result, total part build time can be quite sensitive to part orientation. It is therefore advisable to determine the optimal part orientation. This is usually done manually or with the aid of a suitable software module.

The objects to be built may have overhangs or undercuts that need to be supported during the part fabrication process. Supports are usually incorporated for three reasons. Firstly, it is necessary to avoid collision between the recoater blade and the part being built. Secondly, one needs to make sure t hat a ny small d istortions of t he platform on which t he part i s being built do not adversely affect the part building process. Finally, the presence of supports makes it e asier f or t he part to b e removed from the build platform. The way supports appear will depend on the part orientation. Hence part orientation decisions should consider how to minimize the time needed to build the supports. Sometimes, multiple parts are concurrently built on the same platform. In such situations, there may be a need for additional supports. Supports are usually designed manually or with the aid of suitable software and stored in a separate CAD file.

After generating support structures, both the part and supports are mathematically 'sliced' by the computer into a series of parallel horizontal planes. D uring t his s tep, t he S L p rocess p lanner w ill a lso s elect t he l ayer thickness, the building style, the hatch spacing, the cure depths, the line width compensation value, and the shrinkage compensation values. The significance of these process-planning parameters will be explained later. The .STL file is then modified to incorporate the supports.

The next step is to merge the part and the supports. This step is followed by t he s election of operational parameters s uch a s t he n umber o f r ecoater blades per layer, the sweep period, and z-wait period (these terms will be explained later). The final pre-build process involves ensuring that the level of the liquid resin in the vat is at the proper z-level for optimum laser focus.

6.2 PHOTO-POLYMERIZATION OF SL RESINS

6.2.1 SL Polymers

The polymeric resins being used in SL originally grew out of the coating industry. M any o f t he e arly c oatings h ad b een c ured t hrough t he input o f thermal energy. Such curing was highly inefficient. Subsequently efforts were made to develop photo-curable resins. These turned out to be 50 to 100 times more energy efficient. In coating as well as SL, resins are applied in thin layers in both cases. Hence, spreading and viscosity properties are important in both cases. Likewise, issues related to toughness, speed and shrinkage are of particular significance.

Typically, UV lamps in the wavelength range 200-400nm are used in the conventional photo-curable coating industry. Many SL lasers output radiation in the UV range although lasers emitting other wavelength are not ruled out (Hug 1992). SL machines using He-Cd lasers work at a wavelength of 325nm. Machines adopting Argon-ion lasers use laser wavelengths equal to 351 and/or 364nm.

Almost all SLA use a photo-curable liquid resin as the part material. There are many types of liquid photopolymers that can be solidified by exposure to electromagnetic radiation, including wavelength in the gamma rays, X-rays, UV and visible range, or electron-beam. The major photopolymers used in 3D Systems' SLA machines are curable in the UV range. UV-curable photopolymers are resins formulated from photo-initiators and reactive liquid monomers. There is a large variety of them. Some contain fillers and other chemicals to mechanical strength requirements. The process through which photopolymers are cured is called photo-polymerization. A great deal of research into photo-curable prototyping resins has been conducted on a commercial basis by Ciba (e.g., SL 5410, SL 5510), Allied-Signal (e.g., a low viscosity resin called Exactomer), Du Pont (e.g., SOMOS 2100, 6110, 7100, and 8120), Zeneca (e.g., several colorable a nd clear resins), Coates Brothers, DSM Desotech (e.g., Somos® ProtoComposites™), Loctite (e.g., epoxies), Asahi Denka (e.g., HS 663, and 773), Japan Synthetic Rubber (e.g., SCR 200, and 310), Teijin Seiki (e.g., TSR 752), and other companies. In the beginning, the resins were mainly acrylic-based, although urethane acrylate, vinyl ether, and epoxy acrylate materials were also developed (Figure 6-2). However, most of the newer resins are epoxies. The greatest beneficiary of these developments has been 3D Systems, Inc. 3D Systems recommends different resins to suit different SLA machines. Each type of resin has its own characteristics and mechanical properties such as achievable accuracy, speed of photo-polymerization, temperature resistance, humidity resistance, optical clarity, and color. These may be obtained from the catalogs of the respective vendors.

6.2.2 Radical Photo-polymerization

The first polymer system used for SL had relied on photo-polymerization via free radicals. Photo-curable SL resins relying on free radicals usually consist of 2 to 5% of a photo-initiator with the rest consisting of approximately equal parts of a reactive liquid pre-polymer and a diluent to help reduce resin viscosity (Hoyle 1990). Diluents may also act as lubricants, preservatives, and dyes.

Pre-polymers may consist of monomer, oligomer, or other types of molecules. Usually, monomers have a number molecular weight of several hundreds (after polymerization). Some monomers also increase the sensitivity of the process and prevent swelling. In a vinyl monomer, there exist C-C double bonds. The vinyl group can get attached to other molecules. Typically, this occurs with a carbon atom from another monomer molecule.

In SL practice, both vinyl-based and acrylate-based monomers are often used (Figure 6-2). A vinyl group has the chemical formula $CH_2=CHX$ where X is a pendant substituent. The latter may be polymeric and contain additional vinyl groups. In many resins, the monomer molecules are end-capped with an acrylate group. In fact, the first photopolymer system used for SL was acrylate-based (Hunziker and Leyden 1992). Acrylates form a subset of the vinyl family. They are highly reactive and versatile because of the carboxylic acid group (–COOH) stuck to the C-C double bond (Figure 6-2a). The group allows a variety of other chemical groups to be attached thus resulting in resins with a wide variety of properties.

Oligomers are low-molecular-weight polymers typically with two to five monomer units. However, they have higher molecular weights than single monomers. Urethane and epoxy acrylates are among the more popular oligomers. For instance, an SL resin called SCR-310 uses urethane acrylate that results in a molecular weight of several thousands after polymerization.

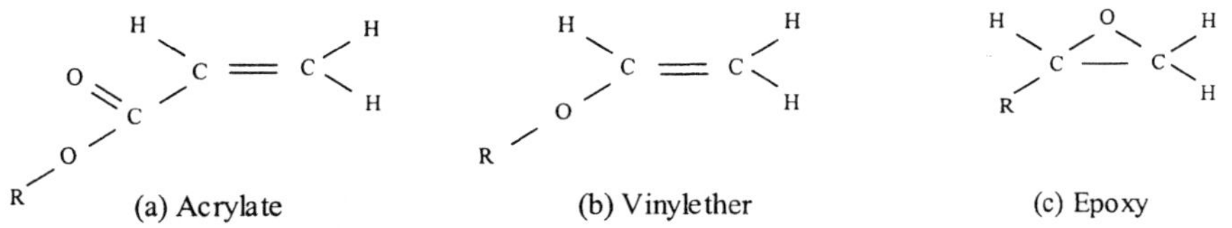

(a) Acrylate   (b) Vinylether   (c) Epoxy

Figure 6-2. Acrylates, vinylethers, and epoxies.

The principle of SL is based on the possibility of initiating polymerization through laser radiation. Each output photon has energy equal to h/λ where h is Planck's constant and λ is the wavelength of laser radiation. A photo-initiator generates free radicals upon exposure to UV radiation. Aromatic carbonyl compounds are among the most frequently used photo-initiators. Since different photo-initiators exhibit different absorption wavelength spectra, the photo-initiator selected must be compatible with the laser used by the machine.

Any monomer that can be polymerized by free radical polymerization can also be polymerized by photo-polymerization or photosensitized polymerization. Except for the initiation phase, the steps involved in photo-polymerization are similar to those described in Chapter 4. In SL, the initiation is done by actinic photons with wavelength in the UV range (250 to 500nm). Since most vinyl monomers are incapable of absorbing light in this wavelength range, the process needs to be assisted by photo-sensitizers capable of absorbing light in the desired wavelength range and break up into free radicals. An example of such a photo-sensitizer is azoisopropane (Figure 6-3). Although this material does not dissociate thermally over a wide range of temperature, it can readily dissociate when irradiated with UV light in the wavelength range 300-400nm (Allcock and Lampe 1990).

When radical polymerization is activated by laser exposure, the potential energy becomes high. The photons striking the molecules start opening up the C=C bonds and the corresponding molecules move to an excited state. Once polymerization is complete, the molecules return to the stable basic state. The energy difference between the excited and stable states is released as 'reaction heat' and fluorescent light. For every two actinic photons ($\approx 2 \times 10^{-18}$ J), one radical is produced that could polymerize up to 1000 acrylate monomers that, in turn, releases 50 to 100 hundred times more energy. Thus, this process is highly energy efficient and exothermic. In free radical polymerization, the exothermic action stops very soon after exposure to laser radiation has stopped.

The photo-speed of acrylate-based SL polymers is generally quite high. However, owing to oxygen inhibition, their polymerization could result in incomplete curing, so tacks may appear in areas with insufficient irradiance. Further, acrylates can exhibit high volumetric shrinkage of up to 5% owing to the conversion of each double bond into a pair of single bonds.

In contrast to conventional polymerization by thermal curing, modeling of photo-polymerization is complicated by the fact that exposure $E(y, z)$ decreases exponentially as depth z increases. This is called the Beer-Lambert Law. In such a situation, the rate of photo-initiation, R_i, may be expressed as a function of photo-power per unit volume, dP/dV:

$$R_i = 2n_c \, dP / dV \qquad\qquad (6.1)$$

where n_c (= 0 to 1) is the number of chains initiated per quantum of light absorbed, and the factor 2 results from the assumption that two radicals are generated.

$$Me2CHN{=}NCHMe2 + h/\lambda \quad\longrightarrow\quad 2Me2CH + N2\,\bullet$$

$$I + light \quad\xrightarrow{kd}\quad 2R\bullet + nitrogen\ gas$$

Figure 6-3. Photo-dissociation of azoisopropane (Beaman 1997).

Now, by considering the energy balance across a thin slice of fluid of thickness dz at depth z, it can be shown from equation 6.1 that

$$\frac{dP}{dV} = \frac{E(y,0)}{D_p} \exp\left(\frac{-z}{D_p}\right) \qquad\qquad (6.2)$$

Combining equations 6.1 and 6.4,

$$R_i = \frac{2n_c \, E(y,0)}{D_p} \exp\left(\frac{-z}{D_p}\right) \qquad\qquad (6.3)$$

Irrespective of whether termination occurs by combination, disproportionation, or occlusion, the rate of termination, R_t, may be expressed as

$$R_t = 2k_t[\text{M}\bullet]^2 \tag{6.4}$$

Usually, R_t is at least 10,000 times larger than R_i. As a result, shortly after the reaction has started, radicals form and get destroyed at the same rates. Hence, we may equate the right hand sides of equations 6.3 and 6.4 to determine the steady state concentration of radicals, $\text{M}\bullet$, as

$$[\text{M}\bullet] = \sqrt{\frac{n_c}{k_t D_p} E(y,0)\exp\left(\frac{-z}{D_p}\right)} \tag{6.5}$$

Clearly, the rate of polymerization, R_p, must be equal to the rate of monomer consumption, so

$$R_p = -\frac{d[\text{M}]}{dt} = k_p[\text{M}][\text{M}\bullet] = k_p[\text{M}]\sqrt{\frac{n_c}{k_t D_p} E(y,0)\exp\left(\frac{-z}{D_p}\right)} \tag{6.6}$$

The average number of monomer units polymerized per chain initiated is called the 'kinetic average chain length,' v_0. This parameter is determined as the rate of propagation per rate of initiation that, in steady state, is equal to the rate of termination:

$$v_0 = \frac{R_p}{R_t} = \frac{k_p[\text{M}][\text{M}\bullet]}{2k_t[\text{M}\bullet]^2} = \frac{k_p[\text{M}]}{2k_t[\text{M}\bullet]} = \frac{k_p[\text{M}]}{2\sqrt{\frac{k_t n_c}{D_p} E(y,0)\exp\left(\frac{-z}{D_p}\right)}} \tag{6.7}$$

The degree of polymerization, DP_0, is the non-dimensional number equal to the average molecular weight of the polymer divided by the molecular weight of the repeat unit. It is possible to express DP_0 as

$$DP_0 = v_0(1 + y'') \tag{6.8}$$

where y'' is the ratio of the rate of termination by coupling to that by all other termination mechanisms. Thus, if coupling dominates, $DP_0 = 2v_0$.

It follows from equations 6.7 and 6.8 that a compromise needs to be sought between conflicting desires for high polymerization rates (hence, higher part build rates) and high molecular weight of the cured part. While the former increases with increasing exposure, the latter decreases.

6.2.3 Cationic Polymerization

We have already noted the general characteristics of cationic polymerization in Chapter 4. Three classes of cationic photo-initiators are in commercial use for SL: diaryliodonium salts, triarylsulfonium salts, cyclopentadiene-Fe-Arene complexes. 3M (Smith 1975) and General Electric (Crivello 1976a, Crivello 1976b) were among the first companies to introduce diaryliodonium salts. These salts exhibit high yields and efficiencies. Hence they are used with most cationically polymerizable monomers. They exhibit weak absorption at wavelengths greater than 350nm. However, they can be sensitized to absorb higher wavelengths by using dyes such as acridine orange, benzoflavine (Crivello *et al.* 1977), anthracene, or some anthracene derivatives. Triarylsulfonium salts are commercially available from several companies. The performance of these salts is comparable to that of diaryliodinium salts. Cyclopentadiene-Fe-Arene complexes are of more recent origin (Meier 1986). A popular commercial brand is Irgacure 261 of Ciba-Geigy.

6.2.4 Vinylethers and Epoxies

Vinylethers (Figure 6-2b) and epoxies (Figure 6-2c) are two main classes of cationically polymerizable resins in use. Vinylethers were introduced earlier than epoxies. Early versions were exclusively developed for SL-190 and SL-250 machines of 3D Systems that utilized He-Cd lasers operating at a wavelength of 325nm. Around 1993, Allied Signal introduced Exactomer^{TM} 2201 (Sitzmann *et al.* 1992) and 2202 resins (Anon. 1993) on a commercial basis. A potential disadvantage of these resins is the high value of critical exposure, E_c (25 to 50mJ/cm^2).

Vinylethers have carbon double bond structures that can be opened up cationically owing to the strong tendency of the adjacent oxygen atom to donate an electron (Figure 4.2b). They generally have low viscosity and, hence, are easy to clean. They also exhibit low curl. Early vinylethers such as Exactomer 2201 had suffered from low photo-speeds. This disadvantage was overcome subsequently, e.g., through the development of Exactomer 2202 SF.

Epoxies are widely used in industry as two-part adhesives, etc. The first epoxy-based SL resin was Cibatool SL5170 introduced by 3D Systems Inc. in 1993. This material was designed for use with He-Cd lasers. Subsequently, SL-5180 epoxy resin was introduced for use with argon-ion based laser systems adopted in SLA-500 machines.

Most conventional epoxy systems utilize thermal curing instead of cationic polymerization. However, it is possible to polymerize epoxies through reactions that open the OCC rings (see Figure 6-2b). This can be made to occur in the presence of cationic photo-initiators that are capable of generating Lewis acids, Broenstedt acids, or protons efficiently (Odian 1981). Ring-opening reactions involve minimal volume-change since the types and numbers of bonds remain unchanged during reaction. Hence shrinkage can be as low as 2 to 3% (Pang 1993). They also exhibit negligible curl, improved flatness, almost zero swelling, and minimal creep distortion in the green state. It was the development of epoxy resins that made direct shell investment casting of functional metal prototypes from stereolithography patterns (e.g., by applying QuickCast 2.0 technique to be described in Chapter 9).

6.2.5 Developments in SL Resins

New resins are being constantly developed to expand the usefulness of stereolithography. For instance, most conventional SL materials soften at temperatures above 100°C, whereas, higher softening temperatures are desirable in SL applications such as injection molding, wind tunnel testing, and under-the-hood automotive components. One way of obtaining glass-transition temperatures of the order of 140°C is to use certain liquid crystal (LC) resins that have appeared recently (Chartoff *et al.* 1998) (Ullett *et al.* 1999) (Ullett *et al.* 2 000). L C materials c ontain s tiff r od-like s egments i n their molecules that can be pre-aligned with the help of an externally applied magnetic field to produce the desired anisotropy of properties. Photo-polymerization locks in the properties. Consequently, it is possible to minimize the thermal expansion of liquid crystal parts over a wide temperature range by rotating the molecular alignment 90° from one layer to the next (Ullett *et al.* 1999). Further, the fracture toughness of liquid crystal coupons can be about twice that of isotropic coupons manufactured from same resin. Figure 6-4 shows the structure of one type of LC diacrylate monomer whose photo-polymerization can be initiated by I rgacure 569 of Ciba-Geigy.

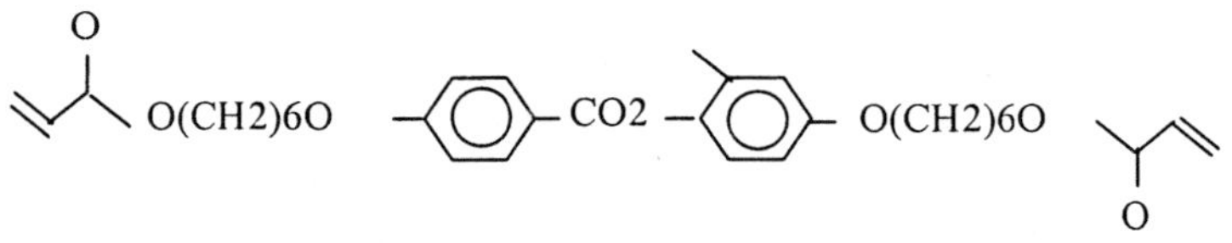

Figure 6-4. A liquid crystal diacrylate monomer.

A major problem with SL is that overhanging part features need to be supported through appropriate structures. The support structures are added to the part before the part-building process starts and removed after the part has been built. The determination of the optimum support structures requires much domain knowledge although software packages are available to support decision-making. Further, the supports need to be removed manually, thus leading to possible accuracy loss. One method to reduce the need for support structures is to increase the viscosity of the photopolymer by mixing it with 3–6% of a fine powder such as Aerosil B812 (Nee *et al.* 2001). Aerosil is a commercially available, highly dispersed, pyrogenic, amorphous silica powder with primary particles of spherical shape with average diameter from 7 to 40nm. The actual properties are, however, determined by the agglomerates formed by the primary particles. Further, owing to the very high viscosity of the Aerosil-mixed photopolymer, the conventional SL technique of relying on a vat filled with liquid resin will not suffice, so it is necessary to extrude the resin mixture by means of a specially designed coating head. A side advantage of using Aerosil mixture is that the overall part shrinkage is significantly diminished while the cure depth is increased.

Improved SL materials are continually appearing in the market in response to consumer demand for ever-superior materials from several competing viewpoints: high optical clarity with refractive index of the order of 1.5 (water-clear resins), high modulus with superior elongation at break, high characteristic photo-speed, a combination of high temperature resistance with high resolution, superior water and humidity resistance, ability to take on distinctive colors, etc.

Research is in progress at several locations to enable new applications through new SL resins and hardware strategies (Wohlers and Grimm 2002). Micron Technology, Inc. has recently published several new SL applictions in electronic component manufacture: semiconductor marking, protective layers for exposing contact pads, protective structures for bond wires, die-to-die interconnection methods, and the generation of conductive elements and assemblies. Tsighua University in Beijing, China, is developing an SL system for the fabrication of clear plastic aligners to straighten human teeth. SRI International, CA, is working on a process called Direct Photo Shaping that is mechanically similar to stereolithography, but uses a deformable mirror device (DMD), instead of a laser, to expose an entire layer at one time. Unirapid of Japan is offering an SL machine that uses a lamp and fiber optics to deliver the UV light. Materialise of Belgium is building what is presently the largest SL machine with a capacity equivalent to five SLA 7000 machines.

Research is in progress at several locations to enable new applications through the development of new SL resins. Development of temperature and flame resistant resins is expected to lead to direct applications in producing under-the-hood parts for the automobile industry. Chemical industry can benefit from prototypes made from resins with higher solvent resistance. Electrical circuits could be directly fabricated once electrically conductive resins are developed. Photo-elastic resins would be of use in nondestructive testing. Medical applications will be accelerated once biocompatible resins are developed. It becomes possible to produce functional prototypes through SL once superior composite resins are developed. For tooling applications, it is useful to develop resins with improved temperature and abrasion resistance.

6.3 ABSORPTION OF LASER RADIATION BY THE RESIN

6.3.1 Beam Size and Positioning over the Resin Surface

In many RP processes, it is necessary to traverse the laser beam across the work area. This is usually achieved through a pair of galvanometric mirrors placed centrally and over the area. Each mirror controls the beam position in one orthogonal direction (x or y) across the work area. When the beam is directed normal to the resin surface, the Gaussian laser spot on the resin surface would be circular with a radius (Gaussian half width), W_0, given by

$$W_0 = \frac{\lambda M^2 L_s}{\pi a_{ga}} \tag{6.9}$$

where M is the mode purity number, L_s is the distance of the closest galvanometer mirror from the center of the work area, and a_{ga} is the radius of the galvanometer aperture. However, when the beam strikes the resin surface at an angle other than $90°$, the spot would be elliptical.

The minimum possible feature size on the part depends upon the focused laser spot on the photopolymer surface. On the other hand, the resolution of a part depends upon laser power, the angular accuracy of the galvanometers, and radiation distribution over the beam cross-section at the photopolymer surface.

Beam positioning accuracy is a function of the dynamics of the galvanometer mirror system. Systems with higher inertia exhibit smaller acceleration and deceleration. Machine operators should take all available measures to tune the system. For instance, some machines have the facility for adjusting the delay times associated with laser ON/OFF (LO and LF) commands. These control the delay between times when the ramp signal is sent and when the laser is turned on. The optimized tuning parameters can be determined by applying different build parameter sets. Several Design of Experiments (DOE) approaches including the Taguchi method have been used to determine such parameter sets.

6.3.2 Laser Scanning Patterns

Equation 5.4 (see Chapter 5) describes the irradiance distribution at the surface of a resin exposed to a stationary laser beam. However, in SL, the laser beam is invariably scanned across the surface of the liquid resin in a certain pattern so that the resin gets hardened through polymerization in the vicinity of the areas where the beam strikes the surface (selective curing). The nature of the cured layer strongly influences the strength, accuracy, and surface finish of the green part.

There are two basic laser scanning patterns used in contemporary SL practice: hatching pattern, and filling pattern. Usually, the interior of the part is cured using a hatching pattern whereas sections belonging to the skin of the part are cured using a filling pattern. Both patterns consist of a set of linear scanning paths resulting in strands of cured layers. In a hatching pattern, adjacent strands are spaced apart whereas, in a filling pattern, they touch or slightly overlap. Hence it is useful to start modeling the SL process by examining what is likely to happen when a laser beam of certain size traverses along a straight path over a photo-polymerizable resin (see Figure 6-5a). In the following analysis, we will assume that the distribution of irradiation across the beam is Gaussian (equations 5.4 to 5.6 in Chapter 5).

6.3.3 Total Exposure from a Single Laser Scan

In Figure 6-5a, the scanning line of the laser beam is along the x-axis from $x = -\infty$ to $+\infty$. Hatch spacing is along direction y whereas axis z is directed normal to the resin surface into the resin. The scanning velocity is V_s. A point inside the resin is specified by coordinates $x, y,$ and z. Beam size at the resin surface is indicated by the $1/e^2$ Gaussian half width, W_0.

Clearly, the total exposure (units of energy received per unit area) received by a particle located on the resin surface (z =0) at distance y from x-axis (the trajectory of the center of the laser beam) may be calculated as

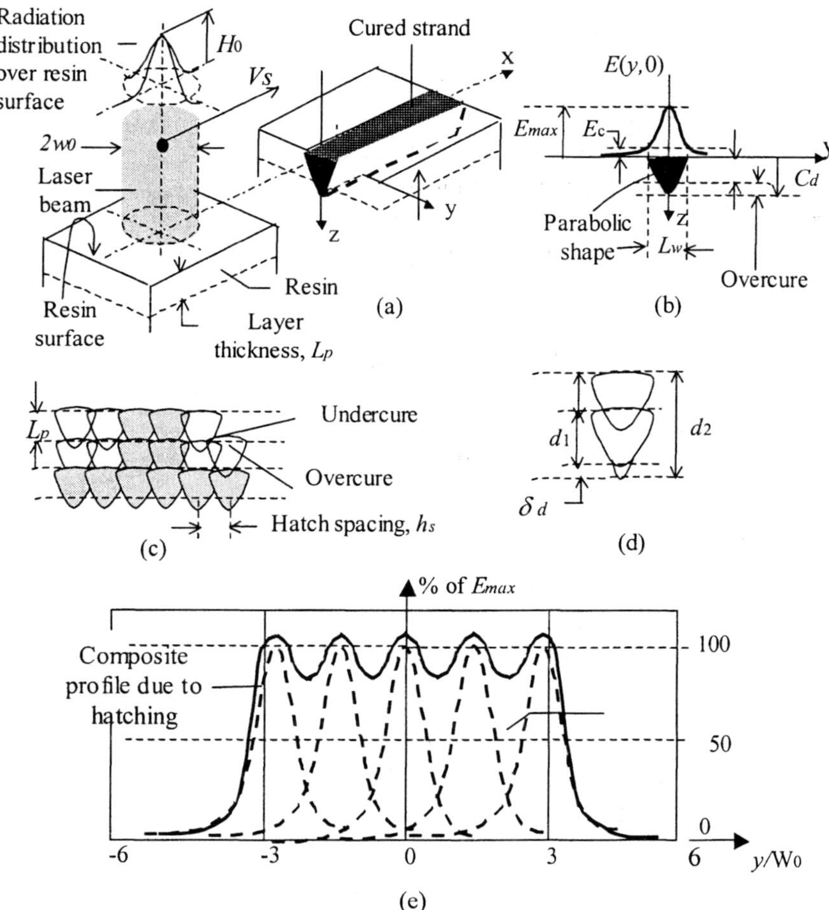

Figure 6-5. Single line and multi-line scanning and t6e corresponding cured lines.

$$E(y,0) = \int_{t=-\infty}^{t=\infty} H[r(t),0]dt \tag{6.10}$$

where $r(t)=[x(t)^2+y^2]^{1/2}$ is the instantaneous distance between the point of interest $(x,y,0)$ on the resin surface and the center of the laser spot on the resin surface at time, t.

Now noting that $V_s = dx(t)/dt$, substituting the expression for $H[r(t),0]$ implied by equation 5.6, and performing the integration over $t = -\infty$ to ∞, it can be shown that (Jacobs 1992a)

$$E(y,0) = \sqrt{\frac{2}{\pi}} \frac{P_L}{W_0 V_s} \exp\left(\frac{-2y^2}{W_0^2}\right)$$

(6.11)

where P_L is the total power of the laser beam and W_0 is the $1/e^2$ Gaussian half width of the laser spot as it strikes the resin surface.

6.3.4 Total Exposure of Interior Resin Layers

When a laser beam strikes the resin surface, its energy is progressively absorbed by resin layers just below the area of incidence. Consequently, as depth, z, increases, the energy available decreases. According to Beer-Lambert law, the exposure, $E(y,z)$, at depth $z>0$ is related to the exposure, $E(y,0)$, at the surface ($z = 0$) as (Jacobs 1992a):

$$E(y,z) = E(y,0)\exp\left(\frac{-z}{D_p}\right)$$

(6.12)

where D_p is the 'depth of penetration' defined as the depth of the resin at which the irradiance has reduced to a value equal to $1/e$ ($= 37\%$) of the surface irradiance. Note that, for a given laser wavelength, D_p is a characteristic of the resin. Further, for most resins, D_p is smaller than the $1/e^2$ diameter, $2W_0$, which, in turn, is smaller than the zone of influence, R_i (see equation 5.7), i.e., $D_p<2\ W_0< R_i$. Many SL resins such as Ciba-Geigy's XB5131, XB5134-1, XB5139, and XB5143 exhibit D_p values between 0.14 and 0.18mm (Jacobs 1992a).

The time, t_l, taken for a photon to pass through a single resin layer of thickness, L_p, is equal to L_p/c_r where c_r is the velocity of light in the resin. Typically, t_l is of the order of a picosecond (1×10^{-12}s). Note that this time scale is of the same order as the times needed for the molecular absorption of the photons and the generation of free radicals. However, the process of cross-linking that follows free radical generation takes much longer—the $1/e$ time, t_K, for cross-linking in conditions typical of SL is of the order of microseconds.

Combining equations (6.11) and (6.12),

$$E(y,z) = \sqrt{\frac{2}{\pi}} \frac{P_L}{W_0 V_s} \exp\left(\frac{-2y^2}{W_0^2}\right)\exp\left(\frac{-z}{D_p}\right)$$

(6.13)

This result shows that "the line spread function of a Gaussian laser beam is itself Gaussian in the surface coordinate (system) orthogonal to the scan direction (Jacobs 1992a)."

The time needed for the laser exposure to reach 99.99% of its value when a Gaussian beam is scanned at a constant velocity past a given point is called the characteristic exposure time, t_e. The typical range of this time in SL is from 70 to 2000μs, so t_e is much larger than t_l and t_K.

6.3.5 Shape of a Cured Strand

In general, a photo-polymerizable resin remains liquid when the total exposure is below a threshold value, E_c (see Figure 6-5b). Many classical SL resins exhibit E_c values between 4.3 and 7.6mJ/cm^2 (Jacobs 1992a). In the case of free radical polymerization, the threshold arises mainly because of oxygen inhibition. In other words, the resin gets cured only at locations where $E(y,z) \geq E_c$. Typical values of E_c are between 5 to 50% of the applied exposure, E. The time required for the exposure to reach E_c can range from a few microseconds to a millisecond.

When the exposure is equal to E_c, the resin is said to be at its 'gel' point corresponding to the transition from the liquid to the solid phase. We may determine the boundary, (y^*,z^*), between the liquid and solid phases by setting $y=y^*$, $z=z^*$, and $E(y,z) = E_c$ in equation 6.13. Thus,

$$E_c = \sqrt{\frac{2}{\pi}} \frac{P_L}{W_0 V_s} \exp\left[-\left(\frac{2y^{*2}}{W_0^2} + \frac{z^*}{D_p}\right)\right] \tag{6.14}$$

Taking natural logarithms of both sides of the above equation and rearranging,

$$\frac{2y^{*2}}{w_0^2} + \frac{z^*}{D_p} = \ln\left[\sqrt{\frac{2}{\pi}} \frac{P_L}{W_0 V_s E_c}\right] \tag{6.15}$$

It follows from the above equation that, for a given resin (i.e., for constant values of D_p and E_c) and given laser scan (i.e., for constant values of P_L, W_0, and V_s), the boundary between the uncured and cured parts of the resin is of parabolic shape (see Figure 6-5b).

6.3.6 Cure Depth and Width

Owing to the symmetry of the parabolic boundary about plane $y = 0$, the maximum value of z^* occurs when $y^* = 0$. Hence substituting $y^* = 0$ in equation 6.13, we obtain the maximum cure depth, C_d, as

$$C_d = D_p \ln\left(\frac{E_{max}}{E_c}\right) \qquad (6.16a)$$

where

$$E_{max} = E(0,0) = \sqrt{\frac{2}{\pi}} \frac{P_L}{W_0 V_s} \qquad (6.16b)$$

Combining the above two equations,

$$V_s = \sqrt{\frac{2}{\pi}} \left(\frac{P_L}{W_0 E_c}\right) \exp\left(-\frac{\text{the desired } C_d}{D_p}\right) \qquad (6.17)$$

The magnitudes of D_p and E_c for a given SL resin may be determined using the WINDOWPANE^{TM} procedure developed by 3D Systems Inc. (Nguyen *et al.* 1992). The resin is scanned by a laser beam at a prescribed series of different velocities. The resulting C_d values are plotted against maximum exposure, E_{max}, on a logarithmic scale (usually, $\log_{10} E_{max}$ is used). Since the magnitudes of D_p and E_c are constants for a given resin and laser output wavelength, it follows from equation 6.16 that this semi-logarithmic plot should yield a straight line. Using this straight line, D_p and E_c are easily determined. Indeed, experiments conducted over a wide range of photo-curable resins have yielded such straight-line plots within experimental error. These plots are called the 'working curves' of the respective resins.

The working curve of a resin is of particular importance since cure depth has a profound effect on the quality of SL prototypes. The curve is usually determined through single strand laser curing experiments conducted in the range $D_p \leq C_d \leq 4D_p$. Trying to obtain a cure depth smaller than the penetration depth is usually undesirable owing to weakly developed properties of the cured layer. On the other hand, the working curve usually becomes non-linear when a cured depth much larger than $4D_p$ is attempted.

Figure 6-6 shows the experimentally obtained working curve of a UV curable resin called SCR-300 of Nippon Synthetic Rubber Ltd. (Anon. 1995). Note that the working curve is essentially logarithmic in nature. It can be surmised from the equation for the working curve that $D_p \approx 195\,\mu m$ and $E_c \approx 1.86\,mJ/cm^2$ for this resin.

Some resins exhibit s uperlogarithmic behavior in a certain range, i .e., the cured depth rises with increasing exposure faster than that suggested by a logarithmic relationship. One reason for such behavior is 'optical bleaching' where the partially cured photopolymer may have a reduced absorption coefficient. Another possible reason is 'optical self focusing' resulting from resin shrinkage due to polymerization. When the resin shrinks, the density increases and, hence, the index of refraction increases. This effect is greater near the center of the cured zone than at the outer edges.

If the working curve rises slower than a logarithmic curve, the behavior is said to be sublogarithmic. A possible reason for such behavior is 'optical scattering' arising from voids and other microscopic imperfections within the resin. Such scattering can result in the actual cured depth being smaller than that expected otherwise.

In addition to the WINDOWPANE^{TM} test, 3D Systems, Inc. has also developed a diagnostic test called the 'Reverse WINDOWPANE^{TM}' that enables the assessment of "how close a laser beam is to Gaussian mode", "how well does the system determine the laser power", "how consistent is the resin from batch to batch", and "how well is the system calibrated" (Nguyen *et al.* 1992).

The maximum width of a cured strand of photopolymer is called the 'cured line width', L_w. An expression for this parameter may be obtained by setting $z^* = 0$ and $2y^* = L_w$ in equation 6.13. Adopting such a procedure, it can be shown that (Jacobs 1992a)

$$L_w = W_0 \sqrt{2\ln\left(\frac{E_{max}}{E_c}\right)} = W_0 \sqrt{\frac{2C_d}{D_p}} \tag{6.18}$$

Note that the cured line width is directly proportional to the size of the laser beam. Note also that L_w is proportional to the square root of maximum exposure that, in turn, is inversely proportional to the scan speed. Usually, L_w ranges from $1.4W_0$ to $2.8W_0$.

The above expressions for cured depth and width have been derived on the assumption that the irradiance distribution across the laser beam is essentially Gaussian—as in the case of argon-ion lasers. However, many lasers (e.g., He-Cd lasers) used in SL equipment are multi-modal and, hence, could deviate from the Gaussian condition. Further, note that the derivation of equation 6.16 has involved setting $y^*=0$. This, in effect, ignores the radial distribution of irradiance across the laser beam. Hence, in practice, equations

6.16 and 6.18 are found to be reasonably effective in the case of a multi-modal (non-Gaussian) laser beam too.

In contrast to cure depth, cured line width can be quite sensitive to the multi-modal characteristics of the laser beam. As a result, the actual line width is usually larger than the corresponding Gaussian estimate. The difference can be of significance when certain laser scanning patterns (such as WEAVE or Star-WEAVE to be discussed later) are employed.

The laser energy required to solidify a unit volume of a resin is called the specific energy (J/cm^3). This parameter is indicative of the effectiveness of a laser beam in solidifying the resin. The relationship between specific energy and exposure (E) is usually nonlinear while exhibiting a minimum (optimum) at a certain E-value. One usually strives to use the optimum value so as to maximize part building efficiency. For instance, for Ceiba-Geigy XB5081-1 resin, the minimum specific energy is of the order of 1 J/cm^3 and this occurs when the exposure is about 15mJ/cm^2. Interestingly, at least for this resin, this exposure value corresponds to the condition $C_d \approx D_p$.

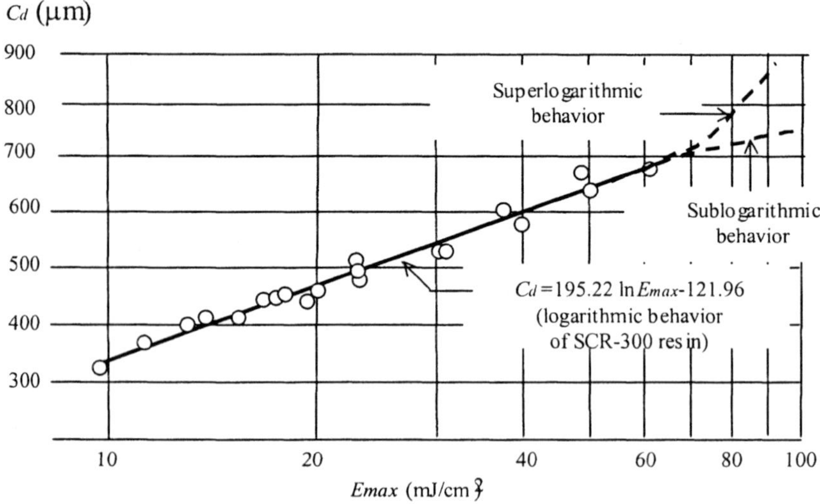

Figure 6-6. Log-linear working curve of a resin and two types of deviations.

6.3.7 Multi-layer Part Building, Overcure, and Undercure

Figure 6-5c shows the consequence of building a part in multiple layers with layer pitch equal to L_p (usually in the range 0.12 to 0.5mm) and a hatch spacing h_s smaller than the laser beam's 1/e^2 diameter ($2W_0$). Figure 6-5d shows the corresponding parameters in the z-direction. Figure 6-5e shows a radiation intensity distribution resulting from multi-line scanning (hatching) as a composite of single line Gaussian distributions spaced at hatch spacing, h_s. The composite profile can be obtained by superposing individual line profiles with the aid of a computer program. As h_s is decreased, the 'ripple' becomes flatter. At a small enough value of h_s, the exposure at all intermediate points exceeds E_c and the whole area is solidified. However, the surface area will be uneven owing to the ripple. This is called the composition error. As h_s is further decreased, the composite profile becomes almost flat and the exposure uniform and equal to the average exposure, E_{av}, for all practical purposes. Consequently, the difference between the cure depths at the peaks and the valleys can become smaller than 5μm. This condition is called 'planar curing'.

Computer analyses of planar curing have shown that, the average exposure is given by

$$E_{av} = \frac{P_L t_d}{A_s} = \frac{P_L}{V_s h_s} \tag{6.19}$$

where t_d is the total laser drawing time, and A_s part surface area (Jacobs 1992a).

Usually, E_{av} is set to be in the range $2E_c$ to $50E_c$. Lower values will lead to uncured resin. At higher values the working curve is unlikely to be linear, hence leading to uncertainties. For many classical SL resins, E_{av} is set to be between 200 and 300mJ/cm^2.

It must be noted that, owing to the parabolic shape of each cured strand, there could be uncured regions even when h_s is small. The size of the uncured region decreases as overlapping of scan lines increases. However, this is achieved at the expense of part building speed.

Ideally, one would like to have uniform curing throughout the current layer and no curing below the current layer. Unfortunately, this is not possible in practice owing to the exponentially decreasing irradiation through layer depth (Beer-Lambert law). Hence, laser power (P_L) and layer pitch (L_p) are so adjusted as to result in a cure depth (C_d) larger than L_p. The difference, (C_d-L_p), is called 'overcure' (Figure 6-6b). An idealized model of overcure resulting from multiple passes is available in (Lu *et al.* 2001).

6.4 RECOATING ISSUES

6.4.1 Recoating Cycle

Recoating is the process of establishing a new layer of fresh resin over the previously cured layer. Although the recoating mechanisms in machines manufactured by different manufacturers differ somewhat (Anon. 2001) (Schwartz 1993), most stereolithography machines use a blade (knife-edge) and an elevator as shown in Figure 6-1 (Shellabear 1994).

A successful recoating step is one that is capable of establishing a fresh layer of liquid resin of thickness exactly equal to the desired thickness, L_p, within a reasonably short time. An incorrect layer thickness will adversely affect the accuracy of the product. However, owing to effects related to the viscosity and surface tension properties of the liquid resin, it is not easy to meet these requirements. As a result, one needs to adopt quite a complex recoating cycle involving several stages. Figure 6-7 schematically illustrates a recoating cycle involving six stages: (a) to (e). The goal is to be able to go from stage (a) to stage (f) in the shortest possible time.

At the start of the recoating cycle (Figure 6-7a), the surface of the previously completed part is level with the resin surface level. In the next stage, the elevator fully immerses the previously completed part under computer control to allow the resin to flow over the part (Figure 6-7b). Hence this step is sometimes called 'deep dip'. However, the resin surface will not become level immediately. Owing to the high viscosity of the resin, the resin will take quite some time before it assumes a level surface over the entire part. Hence, there will be a depression in the resin surface above the part. The depression can be quite significant over regions with trapped resin volumes. Such volumes occur when there are pockets of liquid resin within the part interior that do not have connecting passages to the rest of the liquid in the vat.

In time, the resin fills the depression under the influence of gravity and the resin surface becomes level once again. The time, T_{cl}, taken for the depression to be thus 'closed' has been studied by means of viscous flow analyses and experimental studies. It is found that

$$T_{cl} \propto R_{cr}\mu / h_d^2 \qquad\qquad (6.20)$$

where R_{cr} is the so-called 'critical circle radius' for that cross-section, μ is the viscosity of the resin and h_d is the depth of the depression. Deeper the dip, larger is h_d and, hence, smaller is T_{cl}. The amount of deep dip varies with the machine (about 7.5mm for SLA-250 and 18mm for SLA-500).

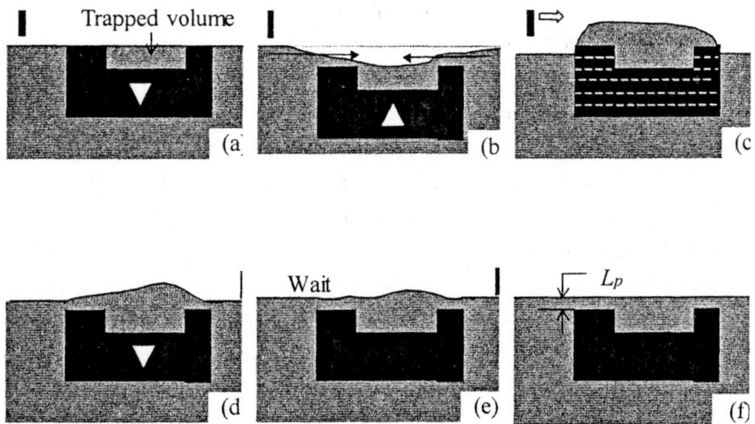

Figure6-7. A recoating cycle.

The third stage starts once the depression has closed reasonably. The platform is elevated until the top of the part is above the resin surface. As a result, there will be a layer of thickness exceeding L_p resting on part's upper surface. This excess resin needs to be removed by executing a scraping step immediately after the platform is elevated. The blade is moved over the part at a certain velocity and with the gap equal to L_p between the bottom of the blade and the top of the part. The gap is controlled by the height of the elevator. If the layer thickness created at this stage significantly deviates from L_p, the accuracy and quality of the final part will deteriorate. If the layer is too thick, adjacent layers may not even attach to each other (delamination). If the part is too high, the blade might hit the part during the scraping process. Figure 6-7c shows the configuration at the end of this scraping process. Since the part's top surface has already been elevated above the resin level, the rest of the resin is not disturbed during the scraping process. For many part geometries, the optimum scraping period is about 5 seconds.

Next, the elevator is lowered again such that the part is in the right position for scanning the next layer—Figure 6-7e. The resin surface at the beginning of this stage is still disturbed. In particular, owing to finite surface tension effects, there is usually a visible crease at the solid to liquid interface around the perimeter of the part. Experiments have shown that this crease fades away exponentially over time. Hence, a waiting period (called 'Z-Wait') is introduced for the resin surface to blend and reach the configuration shown Figure 6-7f. The Z-wait period is usually determined through a compromise between surface nonuniformity and build time. The problem of nonuniformity can be particularly problematic when thin layers are used. Therefore, one needs to choose a longer Z-wait when a large L_p is chosen. These observations might give the impression of unduly long Z-Wait. However, fortunately, with current technology, it is possible to complete computation and adjustment of resin level per layer within one second. The problem is still noteworthy because practical parts contain thousands of layers.

Owing to its importance, the scraping stage has attracted considerable experimental scrutiny and mathematical modeling. Many interesting observations have been derived from such studies. It appears that recoating behavior is quite sensitive to the presence of trapped volumes of resin in the part. For instance, for parts without trapped volumes, changing the blade velocity in the range 5 to 120mm/s does not significantly affect the final layer thickness. However, this is not the case when there are trapped resin volumes. As the part grows and the trapped volume (if any) becomes deeper, the blade velocity needs to be increased so as to ensure uniform layers of correct thickness.

Figure 6-8a schematically illustrates resin flow while scraping over a solid substrate. The blade width is W_b, and V_b is the blade velocity. If we ignore the viscous behavior of the resin, we would expect the resulting layer thickness, L, to be equal to the preset gap, g. However, depending upon the rheological properties of the resin, the fluid layer in the vicinity of the blade can assume a much more complicated configuration. Note that a bulge is formed at the leading edge of the blade and, owing to finite surface tension effects, some resin has adhered to the trailing edge of the blade. A consequence of this is that the actual layer thickness, L_p, following the scraping the process will be smaller than the gap, g.

When $W_b \gg g$, the velocity distribution within the resin layers immediately under the blade may be assumed to be triangular. In such a case, resin flow rate under the blade would be equal to $W_b g/2$. However, once the resin has flowed past the blade, the resin is no more in contact with the blade. Hence, at a location well beyond the trailing edge of the blade, the velocity distribution within the resin layer of thickness L can be expected to be uniform. Consequently, the resin flow rate there should be equal to $W_b L/2$. Equating the two flow rates, the theoretically expected magnitude of L turns out to be $g/2$ irrespective of blade width or velocity.

Experiments have shown that the above idealization works reasonably well when the substrate is large in comparison to blade width and the blade velocity is within a certain range. However, significant deviations from the above theoretical predictions can occur when the substrate is either of a size comparable to the blade width or has trapped volumes of significant size. In such situations, one may need to determine the resin layer characteristics through experiments such as those described in (Renap and Kruth 1995).

Figure 6-8b shows the experimentally observed resin flow pattern when a blade scrapes over a trapped volume. When the blade is directly above the trapped volume, the moving blade imposes a 'forced' velocity profile that is not a perfectly triangular. However, owing to the very large gap between the blade and the bottom of the trapped volume, the resin can very easily flow back towards the trailing edge. The backflow compensates for the surplus of resin removed by the forced flow. The result is a 'crater' seen beyond the trailing edge. A consequence of these phenomena is that the magnitude of L decreases with increasing V_b. It therefore becomes necessary to select the optimum blade velocity for the given resin, W_b, and g. It is found that the optimum blade velocity increases significantly with increasing height of trapped volume.

Resins with lower viscosity result in thicker layers mainly because the backflow of the resin becomes easier with decreasing resin viscosity. In contrast, forced flow is essentially independent of resin viscosity. For similar reasons, a wider blade yields thinner layers. For instance, an increase in the depth of the trapped volume from 2 to 18mm could result in an increase in the optimum blade velocity from about 5mm/s to about 220mm/s. Wider blades lead to smaller layer thickness values. For instance, increasing the blade width from 5 mm to 20mm could result in the laser thickness being decreased by about 0.3mm (Renap and Kruth 1995).

In many situations, it might not be possible to select the optimum blade velocity (assuming that we know its magnitude). For instance, the machine may not have a mechanism for adjusting blade velocity. Or, the optimum could be outside the available velocity range. In such cases, one needs to resort to re-scraping. Experiments have shown that the layer thickness decreases after each scraping pass. Scraping more than once at a certain velocity has the same effect as scraping just once at a higher velocity.

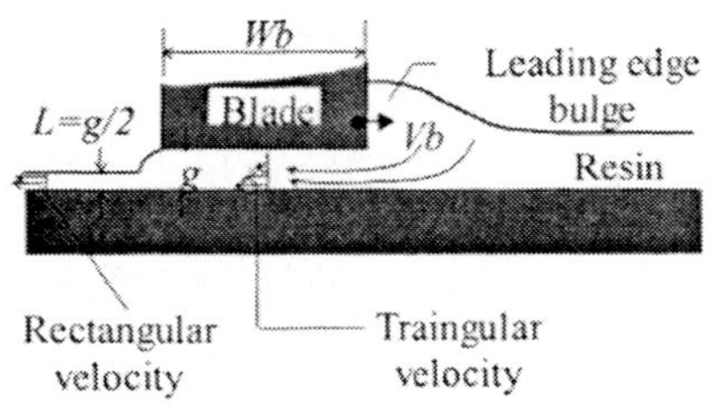

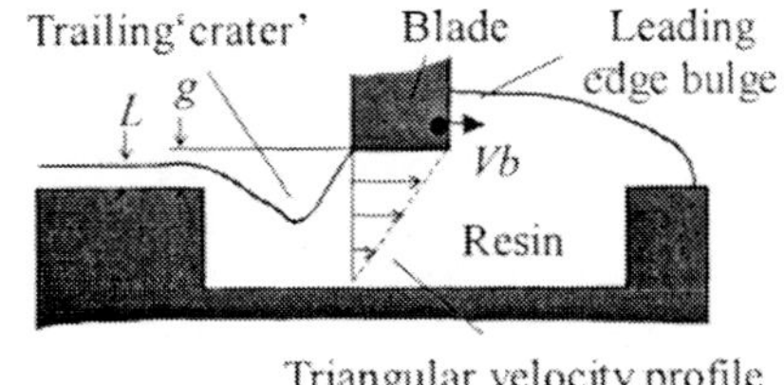

<table>
<tr><td align="center">(a) Scraping over a solid substrate</td><td align="center">(b) Scraping over a surface with a trapped volume</td></tr>
</table>

Figure 6-8. Issues in scraping.

Special problems arise when SL is used for building functional prototypes with inserts: laser shadowing, support structure complications, possibility of the recoater blade colliding with the inserts, and so on. Some solutions to these problems have been proposed recently (Kataria and Rosen 2001).

6.4.2 Resin Level Control

As noted earlier, the liquid resin is leveled at the beginning of the part-build sequence to ensure optimum laser focus. This would suffice if the density of the resin has remained unchanged during photo-polymerization. However, unfortunately, most SL resins undergo volumetric shrinkages between 3% to 5% owing to polymerization, so there will be a fall in liquid level after each layer has been built. To overcome this problem, most SL machines incorporate a level compensation module.

Some early SL machines had detected resin level changes by means of a mechanical float. Such a system is sensitive to changes in resin density, and mechanical effects such as hysteresis in bearings. Some other machines direct a helium-neon (He-Ne) laser beam on to the resin surface so that the reflected beam is received by a silicon bi-cell detector or on larger machines, by a silicon linear cell.

Linear cells consist of a linear array of silicon cells. The difference in voltage outputs from each end of the cell and where the light from the diode hits the linear cell is used as the signal. This approach greatly simplifies the level control software module is greatly simplified. A typical linear cell used in SL machines is about 5mm long and 2mm wide. In contrast, the typical laser spot size is only about 0.2mm. This configuration simplifies the production, installation, and alignment of the system.

After a layer has been built, the level sensor checks the resin level. If the level is outside a tolerance band, a computer-controlled stepper motor activates a plunger that displaces the required amount of resin. In some machines, the entire vat of resin is moved up or down.

A problem with both types of systems is that they are sensitive to discontinuities in the resin surface. Such discontinuities usually arise from small waves or bubbles created while the part is moving inside the vat. The resulting spurious signals could get interpreted as genuine signals providing information about the resin level. This could result in a 'hunting' system.

Later systems tried to solve the problem by taking advantage of developments in low-cost solid-state lasers. Several SL machines started incorporating an indium gallium aluminum phosphide (InGaAlP) visible diode lasers for resin level detection. Such lasers have a much longer operational life than He-Ne lasers. The diode laser leveling (DLL) systems have demonstrated a leveling accuracy of about ±7.5 µm. This is quite acceptable since SL layer thickness is usually in the range of 100 to 200µm.

6.4.3 Gap Control

We have already noted that layer thickness, L_p, is approximately equal to $g/2$. Hence, to control L_p, we need to control g. One way to do this is to have a separate mechanical position control system for the blade. But, this is an expensive solution. A simpler and commonly applied solution is to leave the blade at a constant vertical height and control the resin level through software. Initially the blade is set at a baseline gap. When a new part is to be built, the gap control software module accesses a file containing information on resin properties. Next, based on the accessed properties, it calculates the required up or down offset from the baseline gap. The level control system recognizes this offset while executing the level control action after each layer has been built.

Note that the offset will result in the actual size of the laser spot over the resin surface being different from the baseline spot size owing to the divergence of the laser beam. Fortunately, in the 'far field', the divergence of a laser beam is constant. Hence, the spot size is linearly related to the distance of the resin level from the location (which is constant) of the laser beam's waist. Typically, the change in spot size is of the order of 0.02% over a 508mm dimension (Jacobs 1992a). Fine-tuning the SLA equipment to compensate for this scaling error is usually left to the operator. Another way is to calculate this correction manually and add it in the shrinkage compensation factor.

An associated problem is the need to control resin volume within the vat to a value sufficient to begin the part building process. The estimation of the required resin volume must take into account the shrinkage of the part as a consequence of polymerization and the fact that the liquid level rises whenever the elevator mechanism descends further down into the vat. Further, it must be ensured that the level of resin is within the focal plane limits of the laser imaging system.

6.5 CURING AND ITS IMPLICATIONS

6.5.1 Degree of Curing and 'Green Strength'

Experimental determination of the degree of curing using DSC, DSP, and/or Raman spectroscopy (see Chapter 4) is common in studies of phenomena involving polymerization including research studies concerning SL. Since polymerization is a highly exothermic process, heat is released during an SL cycle. However, some uncured resin can still be found in an SL-built part. The uncured volume of the built part can be determined as (Fuh *et al.* 1999)

$$Degree\ of\ cure\ (\%) = (1 - H_p / H_{liq}) \times 100 \qquad (6.21)$$

where H_p and H_{liq} are the quantities of thermal energy released from the laser cured specimen and liquid resin respectively.

For polymer resins, the peak in the Raman spectrum at delta wave-number equal to $1.633 cm^{-1}$ is known to be the result of opening up of carbon double bonds, C=C, in the resin (Freeman 1974). Hence, the quantity of monomer in the resin can be determined from the intensity of the peak. The intensity may be calibrated with reference to a different material such as the quantification of the percentage of ethylideneuorborene in ethylene propylenediene (Freeman 1974) that exhibits a prominent peak at $1.444 cm^{-1}$ corresponding to methylene deformation. The ratio of the peak intensity of the carbon double bond in the test sample to the peak related to methylene deformation can be used to find the percentage cure of the sample. Alternatively, the ratio of the areas under the peaks can be used.

Investigations using DSP and Raman spectrometer have been conducted on single polymer lines produced by UV laser scanning (of SCR-310) before and after post-curing (Fuh *et al.* 1999). It is found that, in accordance with Beer-Lambert law, the degree of curing in the interior of a scanned line is smaller than that at the top surface of the line. The lowest curing degree can be as low as 62%. Post-curing can change cure degree by about 30%. The effect of post-curing is more prominent at the bottom of the cured line.

Similar studies have been performed on prismatic parts produced through multi-line scanning (Colton and Blair 1999 and Fuh *et al.* 1999). It has been reconfirmed that SL-built parts are not in a fully cured state when removed from the SL machine. Operating parameters such as layer pitch and hatch spacing significantly influence the degree of curing. Depending on the magnitudes of these parameters, there can be totally uncured regions within a green part. The degree of cure depends upon the post-cure method, the post-cure time, and the material location within the SL machine. More fully cured parts are produced when the UV post-cure duration is longer. Thermal curing yields more fully cured parts than UV curing. The hardness (and, hence, life) of a part is influenced by the degree of curing.

The strength of the part immediately after the build process and before it is subjected to any post-curing is called the 'green strength'. The strength is affected a by a host of other basic properties of the material such as its elastic modulus, hardness, and layer-to-layer adhesion. However, the quantitative characterization of green strength may be based mainly on the photo-modulus, Y, defined as the initial slope of the stress strain curve of the photo-polymerized material immediately after the build process (Jacobs 1992b).

Experimental studies on thin slabs of photo-polymeric material subjected to uniform actinic exposure, E, have suggested that the variation of Y is approximately governed by the following linear differential equation (Evans and Jacobs 1991):

$$dY / dE = K_p \{(Y_{max} - Y) / Y_{max}\} \tag{6.22}$$

where K_p is a constant and Y_{max} is the magnitude of Y when the material is fully cured.

Solving the above equation,

$$Y / Y_{max} = 1 - \exp[-\beta(E / E_c - 1)] \tag{6.23}$$

where

$$\beta = K_p E_c / Y_{max} \tag{6.24}$$

It follows that $Y=0$ when $E=E_c$. This is understandable since at the gel point, E_c, the resin is liquid, so it has negligible mechanical strength. On the other hand, when $E \to \infty$ (high exposure), $Y \to Y_{max}$ (saturation of curing).

Further, the following approximate relationship can be shown to hold when E is just above E_c (Jacobs 1992b):

$$Y = K_p(E - E_c)\big|_{(E/E_c - 1) = \text{a small number}} \tag{6.25}$$

Note that the green strength is not proportional to E but the excess exposure, $E_{ex}=E-E_c$. For many classical SL resins, K_p ranges between 1.6×10^5 and

8.7×10^5 m^{-1} whereas β (a dimensionless number) ranges between 0.007 and 0.064 (Jacobs 1992a).

Since K_p has the dimensions of that of reciprocal length, one may define $S=1/K_p$ as the characteristic length of he photopolymer. Typically, S ranges between 1 and 7µm. These values are much larger than the typical wavelength of laser irradiation, λ ($= 0.32$ to 0.36µm for some commonly used SL lasers), and the typical average length of a long-chain polymer ($= 0.001$ to 0.007µm). At the same time, S is much smaller than the typical spot diameter (150-300µm) used in SL. It is suggested that S signifies the average size of islands of local polymerization before they coalesce into a cross linked mass (Kloosterboer 1988). It is useful to note that $\lambda \ll S \ll D_p$.

If we want a part that is solidified to a sufficient green strength level by means of SL, we need to expose the part to laser energy beyond the level of E_c. Noting that deeper sections of the cured layer receive exponentially lesser energy, so the excess exposure, E_{ex}, can be expressed as (Jacobs 1992a)

$$E_{ex} = \frac{1}{C_d} \int_0^{C_d} \{E(z) - E_c\} dz = E_c \left\{ \left(\frac{D_p}{C_d} \right) \left[\exp\left(\frac{C_p}{D_p} \right) - 1 \right] - 1 \right\} \qquad (6.26)$$

Note that the E_{ex} is directly proportional to E_c. This implies that, with a view to lowering build time, if we use a resin with smaller E_c at constant D_p, the green strength will be decreased owing to a decreased E_{ex}.

It is possible to accelerate photo-polymerization by increasing the concentration of photo-initiator in the resin. However, this results in a smaller penetration depth. According to equation 6.26 this will result in a decrease in E_{ex}. It appears that, as with most production processes, one needs to find a compromise between production rate and production quality.

Further, only when $(C_d/D_p) > 2$ does E_{ex}/E_c increase rapidly and it becomes possible to obtain experimentally detectable curing. It is therefore advisable that reporting of green strength values be accompanied with the corresponding values of laser power, P_L, and cure depth, C_d. Sometimes, the layer pitch, L_p, is specified instead of C_d since the two are closely related for a given resin and exposure (Fuh *et al.* 1995).

Equation 6.26 is based on the assumption of single line curing using a Gaussian beam. However, as an SL laser ages, its beam profile can degrade. A solution is to have overlapping lines. Hence, it is difficult to compare the green strengths of two specimens processed in different ways. The best one can do is to have equal cure depths before making a comparison. These observations suggest that green strength is a function of laser drawing style in addition to resin properties. We will describe some advanced drawing styles later in this chapter. .

Although we have so far focused on the tensile photo-modulus, Y, it must be remembered that actual green parts are mainly subjected to flexural forces. Hence, particularly for the purpose of characterizing the green strength of a multi-layered part produced by an advanced drawing technique, it is advisable to measure 'the green flexure modulus (GFM)' of thin and long green part specimens. GFM values of SL parts depend strongly on build style, laser beam radius W_0, hatch spacing h_s, layer pitch L_p, and cure depth C_d. It is found that that the following phenomenological equation can be used to relate GFM to C_d, W_0, and h_s (Pang 1992):

$$\text{GFM} = k_{wv}(C_d - C_{d0})\left(\frac{2W_0}{h_s}\right)^2 \tag{6.27}$$

where k_{wv} is the resin green strength constant for the WEAVE build style, and C_{d0} is the intercept on the x-axis of the extrapolated curve.

Generally, the GFM values of epoxy-based resins have been found to be much larger than those of acrylate-based resins (Pang 1996). For instance, while the tensile modulus of parts built from urethane acrylate have tensile modulus values (values of Y) between 690 and 1100MPa, those of epoxy-based resins are between 2700 and 4200MPa.

6.5.2 Effects During Post-curing

Post-curing apparatus produced in the period 1988-89 mainly used UV emitting mercury arc lamps working at high pressures and irradiance values (Jacobs 1992a). Such lamps had wavelength bands very near the prevailing maximum photo-initiator absorption bands. As a result, D_p values were small so that only the top layers were getting cured. Since, the actinic exposures were high, the temperature in the thin top layers reached very high values (about 200°C) owing to the exothermic nature of polymerization. Interestingly, the maximum temperatures were reached quite some time after lamps were turned off. Sometimes, the high initial temperatures led to high thermal strains and, hence, to stress cracking during the cooling off period.

Subsequently, improved forms of post-curing lamps were installed, e.g., actinic fluorescent lamps. These lamps exhibit significantly lower spectral irradiance values than mercury lamps. As a result, the rates of polymerization and exothermic energy release are much lower so that the maximum temperature during post-cure is reduced to about 45°C. This means that the possibility of thermal stresses and post-cure distortion are reduced, thus improving part accuracy.

Fluorescent lamps also yield higher penetration depth values thus leading to more uniform part curing. Experimental results have shown that the cure depth advances like a propagating front moving at nearly constant post-cure velocity, V_{pc}. From these results, it is shown that the maximum post-cure time, $(t_{pc})_{max}$, is given by:

$$(t_{pc})_{max} = \frac{1}{V_{pc}} \left(\frac{3M_p}{4\pi\rho} \right)^{1/3}$$

(6.28)

where M_p is the mass of the part, and ρ is the density of the resin. Note that post-cure time is proportional to the cube root of the mass of the part. In other words, a 1000 times heavier part will only require 10 times larger post-cure time. However, as evident from the hardness data reported in (Blair 1998), post-curing time has little effect on the degree of curing once the parts have been aged for more than one week.

A problem with equation 6.28 is that it does not take into account the effect of 'oxygen inhibition.' This phenomenon mainly concerns the exterior surfaces of the part that are in contact with ambient air. These surface layers require a finite time to reach an exposure level large enough to overcome the effects of oxygen inhibition. Thus, when the part mass is below a threshold value (typically of the order of 100 grams), maximum post-cure time will be essentially constant.

A useful measure of the effectiveness of the post-cure process for a given UV wavelength, λ, is the 'volumetric efficacy', $\gamma_v(\lambda)$, defined as (Jacobs 1992a)

$$\gamma_v(\lambda) = D_p / E_c$$

(6.29)

For many SL resins, $\gamma_v(\lambda)$ peaks at a particular wavelength that is characteristic of the resin. For instance, for XB 5081-1 resin, the optimum wavelength is equal to 360nm whereas it is 375nm for XB 5139. Post-cure times can be significantly reduced by having a wavelength in the vicinity of the optimum. Some fluorescent lamps used in modern PCAs come with filters to enable optimum wavelength selection.

6.6 PART QUALITY AND POCESS PLANNING

6.6.1 Shrinkage, Swelling, Curl and Distortion

During polymerization, the resin can undergo between 5% and 7% volumetric shrinkage. Of this, between 50% and 70% is the initial shrinkage that occurs within the vat (Jacobs 1992a). The rest occurs during post-curing. Shrinkage leads to increase in part density. The degree of shrinkage is strongly influenced by the resin itself, part building style, and operational parameters (layer pitch, hatch spacing, etc.,).

Shrinkage leads to uncertain changes in the resin level. The level control mechanism is usually capable of taking care of this. A greater problem is that differential shrinkage across the part results in curling, warping, or distortion of the part (Ullett *et al.* 1994). It can also lead to the formation of cavities hence adversely affecting elongation to fracture. All these effects lead to poorer part accuracy and quality. As a result, the SL process planner needs to anticipate and compensate for the effects of shrinkage. However, the severity of the problem is getting reduced in recent times owing to the development of low shrinkage SL resins.

Part building in SL occurs through laser-induced polymerization of successive single strands of resin. Apparatus for continuously measuring linear shrinkage and shrinkage force of the strands have been developed and several kinds of resins have been tested (Chartoff 1995, Guess and Chambers 1995 and Narahara *et al.* 1999). Methods have been developed for shrinkage and distortion prediction (Flach and Chartoff 1994 and Kishinami *et al.* 1997).

Two simultaneous processes take place when the laser beam strikes a 'minute volume' along a strand. Firstly, the reaction heat raises the temperature of the minute volume. This leads to thermal dilatation (expansion). Secondly, shrinkage occurs owing to polymerization. Usually thermal dilatation dominates within the minute volume so that there is a net expansion of the volume.

However, when the laser beam leaves the minute volume, the reaction heat stops and the volume starts cooling. This stoppage occurs immediately in free radical polymerization and after some delay if it is cationic polymerization. In ether case, the rate of cooling depends upon the immediate environment, whether it is liquid resin or a previously cured and solidified strand. When the reaction heat is fully dissipated (this may take over half a minute), the thermal dilatation disappears so that only the shrinkage arising during polymerization remains.

When a strand of finite length is polymerized, the minute volumes at different locations along the strand are in different states of thermal dilatation although the shrinkage induced by polymerization might be the same. Hence, the overall shrinkage of the strand needs to be viewed as an integration of the individual dilatation/shrinkage occurring at all the minute volumes along the completed strand (Narahara *et al.* 1999). It also follows that a major reason for a green part exhibiting shape distortion lies in the uneven temperature distributions arising during the part building process.

The time $t_{s,o}$, needed for a photopolymer to exhibit significant and measurable shrinkage is usually about a fraction of a second. However, the time, $t_{s,c}$, for the completion of measurable shrinkage is of the order of 4 to 10s. Note that these times are much larger than the characteristic exposure time, t_e. On the other hand, $t_{s,c}$ is much smaller than the time, t_d, needed for drawing a single layer, which is of the order of 20 to 300 seconds.

Initial s hrinkage i s u sually s ensitive t o b uild s tyle b ecause o f t he f act that the cooling rate of a minute volume depends upon the immediate environment. For instance, the shrinkage arising in 'skip scanning' (alternating scanning lines are skipped in the first pass and are cured in a subsequent pass) is found to be smaller that in 'continuous scanning' (scanning proceeds from one line to the next adjacent line immediately) (Narahara *et al.* 1999). The presence of support structures can also influence shrinkage. Initial shrinkage is important only in the early stages of part building. It is virtually complete when the material has acquired significant green strength (Hunziker and Leyden 1992).

The n ature o f c hemical r eactions o ccurring d uring p olymerization c an significantly affect shrinkage. For instance, it is observed that the opening of vinyl bonds causes substantial volume shrinkage (about 35% for vinyl chloride liquid, and 6% for di-vinyl-bi-benzyl) that increases at increasing rate as the concentration vinyl bonds is increased.

Interestingly, while the overall part might shrink as a result of polymerization, the part can exhibit swelling over time as a result of liquid resin in the trapped volumes being absorbed into solidified lines. This absorption is more severe at solidified lines with liquid resin on both sides. Swelling is usually specified as the percentage change in a dimension resulting from a 24-hour immersion of the part in liquid resin. The 'Swell-Tower' test described in (Nguyen *et al.* 1992) is a commonly used diagnostic test for swelling.

Shrinkage resulting from post-cure can be estimated by measuring the same dimensions on the green and post-cured parts. The ratio of the difference between green and post-cured dimension to the green dimension is called the 'post-cure shrinkage, S_{pc},' in the direction of the dimension (Nguyen 1992 and Lu *et al.* 2001). Another method is to use embedded sensors (Nau *et al.* 1994).

A method of investigating post-cure shrinkage is the 'Nine-Box' method developed by 3D Systems (Nguyen *et al.* 1992). A set of nine 50mm×50mm boxes with wall thickness values varying from 2.5mm to are generated during a single build using 0.25mm layer pitch. Twelve readings of two each of x and y measurements and wall thickness are taken using a coordinate measuring machine (CMM). Next, they are post-cured under 10 actinic fluorescent lamps of 40W each for 1 hour. The measurements are repeated in the post-cured state. From these readings, the mean and standard deviation values of S_{pc} are calculated. Such tests conduced at 3D Systems have shown that shrinkage is generally not uniform and depends upon the part building style. Shrinkage is low for thin walls, increases to a maximum for medium wall thickness, and decreases slightly for large wall thickness values.

While shrinkage remains at the end of a post-curing cycle, during the cycle itself there can be the opposite trend of thermal dilatation. As result, there can be net expansion while a part is being heated during a thermal curing cycle. For instance, in an experiment on SCR-310 reported in (Fuh *et al.* 1997), thermal dilatation over a specimen of 12mm length reached about 0.18mm over a post-curing temperature interval equal to 50 to 180°C. However, after the part was cooled back to 50°C, a shrinkage value of 0.22mm was recorded.

Some further shrinkage trends during thermal post-curing reported in (Lu *et al.* 2001) are worth noting. Shrinkage was measured in the scanning as well as transverse directions. Layer pitch was varied between 140 and 300 μm, laser power between 100 and 250mW, and post-curing time between 5 and 28 hours. It was found that shrinkage increased approximately linearly with increase in layer pitch whereas, as the curing time was increased, shrinkage increased rapidly in the beginning and exhibited a saturating trend later. Laser power did not have much effect on the final shrinkage. However, at a higher laser power, shrinkage was less sensitive to variations in layer pitch. The trends were similar when the experimenters switched to UV post-curing.

A major consequence of thermal and shrinkage effects is part distortion immediately after laser irradiation and after post-curing. A single unsupported layer can shrink without distortion. However, the situation is different while we are building parts layer by layer. Upon reaching gel point, a freshly drawn strand adheres to a previously cured strand. The additional shrinkage of the new strand is transmitted to the strand below causing it to curl upwards with its free end rising slightly above the resin level. However, as further layers are cured, the supporting layer becomes stronger since it is now a composite of several layers. As a result, part curl reduces. But it does not vanish and the final part is distorted in a complex manner depending on the nature of the resin, part geometry, part building style, and build parameters. Further, owing to residual stresses (Karalekas and Rapti 2002), parts can continue t o exhibit c reep distortion long after the part has been built.

Curling is most severe over unsupported regions. The curled layers can be modeled as an arc of radius $R_c=a/s$ where a is the layer thickness and s is the linear shrinkage (Hunziker and Leyden 1992). It follows that the distortion distance, δ, over an unsupported part region of horizontal length, l, is given by

$$\delta = R_c[1 - \cos(l / R_c)] \approx sl^2 /(2a) \qquad\qquad (6.\ 30)$$

Cheah has applied a Moiré f ringe method to study surface distortions during UV polymerization (Cheah 2001 and Lu *et al.* 2001). A linear grating is bonded on to the top surface of the part. The reference grating is bonded onto a glass plate and carefully aligned with the part-grating. The assembly is then subjected to UV lighting. In-plane displacements and strains can then be studied on the basis of the resulting Moiré fringe patterns.

A series of standardized diagnostic tests for curling and distortion occurring during build and post-cure have been developed by 3D Systems, Inc. These include the 'Twin Cantilever Curl Distortion' test, the 'Vertical Wall Post cure Distortion' test, the 'Horizontal Slab Build Distortion' test, the 'Horizontal Slab Postcure Distortion' test, and the 'Horizontal Slab Final Distortion' test. A separate diagnostic test called the 'CHRISTMAS TREE[TM] METHOD' has been developed for determining linewidth compensations (LWC) and shrinkage compensation factors (SCF). Details of these tests are available in (Nguyen *et al.* 1992).

Although several experimental studies concerning part curl and distortion have been reported in literature (Marutani and Nakai 1989 and Hunziker and Leyden 1992), owing to the complex nature of the phenomena addressed, the conclusions diverge considerably. However, certain observations seem to be generally applicable (Hunziker and Leyden 1992). Curling increases with increasing laser exposure and overcure. Resins exhibiting faster polymerization also exhibit greater curl. Thus, simple acrylate monomers that polymerize 3 to 10 time faster than methacrylate monomers exhibit greater curl. While shrinkage is certainly a reason for curling, there is no direct relationship between total linear shrinkage and curl. Often, there are additional thermal effects such as shrinkage due to cooling.

6.6.2 Surface Deviation and Accuracy

Figure 6-9a illustrates the stairstepping problem associated with layered manufacturing processes in general (including SL). While the desired profile may be smoothly curved, the actual surface is analogous to a staircase with the height of the step equal to the layer pitch, L_p. Clearly, the deviation of the actual surface from the desired surface increases with increasing values of L_p and profile angle σ. Since L_p is usually of the order of a fraction of a millimeter, the stairstepped surface appears as surface roughness. Hence, one can use the centerline average value, R_a, of the surface roughness as a measure of the surface deviation. Figure 6-9b illustrates the configuration of stairstepping for a given part profile angle, σ, and layer pitch, L_p. Note that, owing to certain effects of SL process, the faces of the 'steps' are not necessarily vertical, i.e., the layer profile angle ϕ is usually greater than zero. The figure also illustrates the wavy surfaces (composition errors) usually found on the up facing ($0<\theta \leq\pi$), and down facing ($\pi<\sigma <2\pi$) surfaces. From the configurations shown in the figure, it can be shown that (Reeves 1997)

$$(R_a)_{up} == \frac{L_p(\tan\phi\sin\sigma + \cos\sigma)}{4} + K_{up}\Big|_{(\sigma-\phi)<90^o}$$

$$(R_a)_{down} = \frac{L_p(-\tan\phi'\sin\sigma' + \cos\sigma')}{4} + K_{up}\Big|_{(\sigma-\phi)\geq90^o}$$

$$(6.31a,\ b)$$

where

$$\phi' = -1.\phi, \sigma' = (\pi-\sigma) \tag{6.32a, b}$$

and K_{up} and K_{down} are the composition roughness values of the up facing and down facing surfaces respectively.

The composition roughness values are functions of the scan pattern used and subsequent material contraction. Moreover, the function for down facing surfaces needs to take into account the laser cure parabola and the effects of support structure design. Owing to these considerations, K_{up} and K_{down} cannot be assumed to be simple constants. However, it has been shown that (Reeves 1997)

$$K_{up} = \frac{K_{up}(\pi/2+\phi)-\sigma}{\pi/2+\phi}, \text{and } K_{down} = \frac{K_{down}[\pi-(\pi/2+\phi)-(\pi-\sigma)]}{\pi-(\pi/2+\phi)} \tag{6.33}$$

where the magnitudes of κ_{up} and κ_{down} need to be determined through experiments.

Experimental studies have shown that the above model yields reasonably good surface deviation estimates except when σ is in the range 100 to 150° (down facing surfaces). However, fortunately, the actual surface deviation is usually smaller than the theoretical estimate in this σ-range. This 'smoothing effect' appears to be due to the fact the additional resin at the interface of a pair of adjacent layers can get cured if the boundary of the subsequent layer exceeds that of the previous layer, thus resulting in a partial 'fillet' (Figure 6-9c). This effect is called 'the print-through' effect' (Jacobs 1992a).

The composition roughness values are functions of the scan pattern used and subsequent material contraction. Moreover, the function for down facing surfaces needs to take into account the laser cure parabola and the effects of support structure design. Owing to these considerations, K_{up} and K_{down} cannot be assumed to be simple constants. However, it has been shown that (Reeves 1997)

$$K_{up} = \frac{\kappa_{up}(\pi/2 + \phi) - \sigma}{\pi/2 + \phi}, \text{ and } K_{down} = \frac{\kappa_{down}[\pi - (\pi/2 + \phi) - (\pi - \sigma)]}{\pi - (\pi/2 + \phi)} \qquad (6.33)$$

where the magnitudes of κ_{up} and κ_{down} need to be determined through experiments.

Experimental studies have shown that the above model yields reasonably good surface deviation estimates except when σ is in the range 100 to 150° (down facing surfaces). However, fortunately, the actual surface deviation is usually smaller than the theoretical estimate in this σ-range. This 'smoothing effect' appears to be due to the fact the additional resin at the interface of a pair of adjacent layers can get cured if the boundary of the subsequent layer exceeds that of the previous layer, thus resulting in a partial 'fillet' (Figure 6-9c). This effect is called 'the print-through' effect' (Jacobs 1992a).

One obvious way of correcting stairstepping is to produce the part with thinner layers. However, this will increase the total build time. Hence, control of the step-size is usually achieved through the adaptive slicing technique described previously. Essentially, the technique aims to compensate for the effect of a larger σ by appropriately decreasing L_p.

Attempts have been made to minimize stairstepping while the part is being built. One method that takes advantage of the viscosity and surface tension properties of the resin is illustrated in Figure 6-9d (Reeves and Cobb 1997). Immediately after a layer has been drawn, it is raised out the liquid resin by a certain amount. This results in a meniscus of liquid resin being formed at the corner with the previous layer. The meniscus is then cured by subjecting it to laser scanning. Since the cured meniscus retains its concavely curved nature, the part's surface deviation is significantly reduced. The surface roughness of post-cured parts can be reduced by means of handheld sandpaper and a Dremel tool. The latter is capable of probing difficult-to-reach part features to remove supports that are not built from the top.

Other methods of post-cure finishing include barrel tumbling, centrifugal tumbling, vibratory finishing, abrasive blasting, or abrasive flow finishing (Spencer *et al.* 1993 and Williams and Melton 1998). In barrel tumbling, a horizontal barrel is loaded with abrasive media and rotated. The method is capable removing stairstepping on angled surfaces but not on vertical or horizontal surfaces. Centrifugal tumbling is similar to barrel tumbling, but is executed at much higher rotational speeds. However, being more aggressive, this method could result in parts with chipped edges and convex corners. Abrasive blasting involves propelling abrasive particles from a nozzle on to the part. The method is found to be capable of smoothing flat surfaces. Other types of surfaces remain less affected. The method is also quite aggressive, thus leading to part damage.

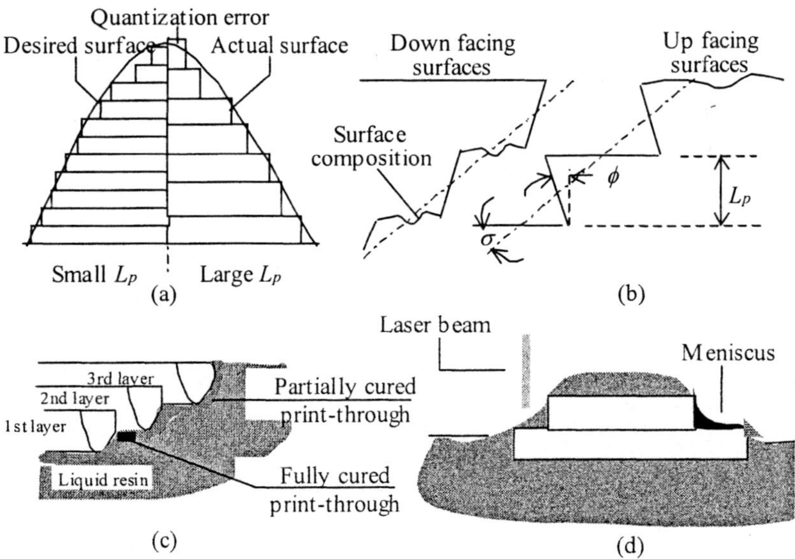

Figure 6-9. Issues concerning surface deviation.

Abrasive flow machining (AFM) is a non-traditional finishing method where an abrasive-laden viscoelastic compound is made to flow through a restricted passage formed by workpart/tooling combination so that the slug containing the tightly held abrasive particles acts as a deformable grinding stone (Williams and Melton 1998). The method has been found to remove stairstepping significantly just after one or two cycles regardless of SL parameters. However, the process is sensitive to media pressure and abrasive grit size. Likewise, metal removal rate is affected by extrusion pressure, abrasive grit size, and build orientation. Occasionally, flow lines appear on the part surfaces owing to minute brittle fractures.

Blair has reported some useful experimental results on SL-parts subjected to polishing (Blair 1998). He observed that the layer thickness and build location affected the as-built roughness. Towards the edge of the vat, out-facing surfaces were smoother than the in-facing surfaces. The roughness was relatively uniform on faces in the middle of the vat. The resin type, polishing time and applied force did not affect the final roughness. Smaller abrasives yielded smoother surfaces.

In principle, part accuracy is a function of SL machine accuracy and process accuracy. Machine accuracy and repeatability are usually studied by building an 'Acceptance Test Part' designed to suit a range of SL machines (Richter and Jacobs 1992). Part building is performed without any shrink factor or linewidth compensation. The resulting part is measured on a coordinate measuring machine. It turns out that the accuracy of contemporary SL machines is sufficiently high to permit one to ignore issues concerning machine accuracy.

SL process accuracy is usually studied by building and inspecting a benchmark part called the 'User-Part'. The part is square in shape with cross-bars and five square pockets, four at the four corners and one in the center. Several interesting conclusions have been derived from measurements conducted at 3D Systems, Inc. (Richter and Jacobs 1992). Most importantly, the 99% confidence limit error seems to increase as the square root of part dimension. Thus, when a part dimension is doubled, the process error increases only by some 41%. Further, on certain engineering plastics, the SL-produced User-parts are more accurate than the NC (numerical control) milled versions. Further views on SL accuracy can be found in (Gargiulo 1992 and Onuh and Hon 2001). Critical analyses of existing benchmark part geometries are available in (Childs and Juster 1994 and Byun *et al.* 2002).

Several phenomena such as shrinkage, swelling, curling, and distortion contribute to part errors. Since all these are sensitive to the way part interior is hatched, there has been a continual search for new and improved hatch patterns (build styles).

Another common error is 'quantization error' that arises when, as is common, a desired z-dimension on the part deviates from the summation of the layer thickness values composing that dimension (Figure 6-9a). Obviously, a smaller layer thickness setting results in a smaller quantization error.

6.6.3 Build Styles and Decisions

While building a part, the borders are drawn first, and then the interior is drawn using one of the several available hatching patterns. Next, skin filling is done in layers where there is nothing solid either immediately above or below.

One of the earliest hatching patterns was 'Tri-Hatch' (Triangular Hatch). This style used a scanned line parallel to the x-axis combined with lines at $60°$ and $120°$ to the x-axis. As a result, the internal structure consists of hollow vertical columns of the form of equilateral triangles. The columns are capable of containing liquid resin. The presence of a large amount of liquid resin (up to 5 0% in v olume) c an l ead to e xcessive p ost-cure distortion. I t also wastes the resin. The vat needs to be refilled with resin more often.

WEAVE^{TM} hatch pattern involves drawing each layer by a set of y-hatch lines followed by a set of x-hatch lines, thus forming an orthogonal grid. The two sets are separated by a small gap so that no curling is likely to occur during the y-hatch period. Scanning speed during hatch is selected such that the time required to draw each hatch line is much smaller than the characteristic time for significant polymer shrinkage. The maximum centerline exposure for the y-hatch lines is set so as to achieve a cure depth slightly smaller than the selected layer thickness. The second phase of drawing x-hatch lines (for each layer) is performed at the same exposure. It can be shown that this condition ensures that the additional increment of cure depth achieved in the x-pass will be a constant positive value irrespective of the absolute level of exposure. This feature leads to adhesion between successive layers especially at the points of intersection between the y- and x-hatch lines (the intersection points act like rivets). During the y- and x-passes, the hatch spacing is so selected as to result in a high percentage (about 96%) of the polymer within the part volume reaching the gel point and, hence, solidifying. Hence, post-cure distortion is dramatically reduced. However, the hatch lines belonging to the same set (x- or y-) continue to be vertically a ligned t o f orm v ertical c olumns. C onsequently, t he p roblem o f part distortion due to internal stresses induced during building remains.

The problem of internal stresses was solved later through the development of STAR-WEAVE^{TM} (<u>ST</u>aggered Hatch <u>A</u>lternate Sequencing <u>R</u>etracted Hatch WEAVE). 'Staggering' refers to the fact that the x- and y-lies in a layer are offset by $h_s/2$ from the corresponding lines in the previous layer. This feature prevents the problem of 'vertical columns' and reduced the possibility of micro-fissures from developing. 'Alternative sequencing' refers to reversing the sequencing of x- and y-lines, i.e., if y-lines are drawn first on odd numbered layers, x-lines would be drawn first on even numbered layers. In addition, the direction of line propagation is also altered between layers. For instance, if x-lines are drawn from the front of the vat to the back on the nth layer, they are drawn from back to front on the $(n+1)^{th}$ layer. This process helps in isolating dislocation lines. 'Retracted hatching' refers to the idea of slightly retracting one end of a hatch line from one side of the border while letting the other end attach to the other side of the border. Owing to this feature, individual cured strands do not bend and part distortion is reduced. All these complex adjustments are controlled by means of a special software module.

ACES is a more recent building style that aims to produce <u>A</u>ccurate, <u>C</u>lear, <u>E</u>poxy, <u>S</u>olid parts with highly improved dimensional stability. The method almost completely eliminates post-cure distortion and internal stresses. Curing each layer in two stages minimizes the bi-metallic strip effect. In the first stage, the layer is cured nearly to the full depth but not to the extent of bonding to the next layer. Hence, no curling occurs in this stage. In the second stage, the layer is subjected to a similar exposure to ensure uniform curing and adhesion to the previous layer. The clarity of the built part is usually affected by the presence of air bubbles. When the top layer is dipped below the resin level during the recoating phase, an air bubble forms in the center of the layer as the liquid resin converges from all sides. The expectation is that the bubble will be swept away during a single recoating sweep. However, often, this does not happen. To minimize this problem, ACES uses three sweeps. The first sweep separates the bubble from the solidified surface. The second sweep moves the detached bubble away from the build area. The third sweep is executed with the regular sweep parameters to ensure the desired layer thickness.

Next to the selection of part building style, the selection of part orientation is the most important step (see Section 5.4 for more information regarding part orientation determination and support structure design). This is also a very complex step since the way the part is oriented relative to the machine's coordinate axes can profoundly influence part equilibrium, part height and hence the number of slices, the proportion of part surface area exhibiting undesirable stairstepping, the proportion of entrapped volume and hence part warpage (Ullett *et al.* 1994), the support structures required and their height (Kirschman *et al.* 1992 and Webb and Gerdes 1994), etc. If any of these requirements is not reasonably met, part build time can increase and/or part quality suffer. Usually, consideration of these issues is left to the machine operator and the process planner who are expected to somehow make the right decisions on the basis of their experience.

Several heuristics have been identified with a view to improving the building process from the viewpoints of accuracy, surface finish, part-build time, and support removal (Chen and Sullivan 1996, McClurkin and Rosen 1998 and Pham *et al.* 1999). For instance, the following heuristics are suggested with regard to the selection of part orientation: <Place user specified critical surfaces in a horizontal and upward facing orientation>, <Orient surfaces of revolution such that their axes are vertical>, <Orient deep cuts and shafts such that their placement planes are horizontal>, and <Except when the older deep-dip recoat technique is applied, orient shells such that their concave side faces upward> (Pham *et al.* 1999). The following measures should be helpful in minimizing part-building time: <Minimize the total number of slices>, and <Minimize the height of required support structures>(Chen and Sullivan 1996).

There are ongoing attempts to computerize the analytical and decision-making aspects related to part orientation (Bablani and Bagchi 1995, Cheng *et al.* 1996 and Lau *et al.* 1997), the selection of the corresponding process variables (McClurkin and Rosen 1998), and the estimation of part build time (Chen and Sullivan 1996). Several algorithmic and non-algorithmic techniques have been attempted to address these issues: expert systems, genetic algorithms (Wodziak *et al.* 1994), response surface methodology (McClurkin and Rosen 1998), progressive multi-objective optimization (Cheng *et al.* 1995), and artificial neural networks (Lee *et al.* 2001).

A software architecture that combines empirical data, analytical methods, heuristics, and multi-objective optimization has been proposed recently for SL process planning (West *et al.* 2001) with the goal of minimizing build-time while maximizing surface finish and accuracy. The system starts by receiving a CAD model in association with user interaction. An 'orientation module' arrives at a set of candidate part orientations satisfying two constraints: support structures, and horizontal planes. Parts with the candidate orientations are sliced using a separate module. The variable considered here is layer pitch whereas the main constraint relates to horizontal planes. Next, the sets of slices are processed by a 'parametric' module that outputs a set of process plans. A separate module investigates the sets of plans and selects the most suitable process plan.

Build parameters are usually selected through interaction with the machine's computer. The machine software usually contains a database of basic machine and resin properties. These should be sufficient while executing many routine jobs. However, quite often, one comes across special jobs requiring deviations from standard practice. A process planner who understands SL fundamentals deeply and has extensive SL–planning experience should be an asset in such circumstances. However, the task of process planning for a deviant job may not always be a straightforward task. For instance, some of the machine-specific information needed may not be available in the catalogues supplied by the machine vendor. As observed in (Chen and Sullivan 1996), it is common practice to select the scanning velocity on the basis of equation 6.17. However, for some reason not revealed by the vendor, the actual scanning speeds were found to be significantly smaller than estimates from equation 6.17 by a constant magnitude.

6.6.4 Build-time and Build-cost

Since SL aims to be a *rapid* prototyping process, build time estimation and control is of paramount importance. Some commercial build-time estimators are available. The following equation may be used when an automated estimator is not available:

$$
\begin{aligned}
build\_time &= \\
(recoat\_time) &+ (suppport\_draw\_time) + \\
(border\_draw\_time) &+ (hatch\_draw\_time) = \\
\left(\frac{height \times recoat\_time\_per\_layer}{L_p} \right) &+ \left(\sum_1^n \frac{area\_supports}{L_p \times V_s\_supports} \right) + \\
\left(\frac{\sum_1^n projected\_vertical\_surface\_area}{L_p \times V_s\_border} \right) &+ \left(\frac{2\sum_1^n volume}{h_s \times L_p \times V_s\_hatches} \right)
\end{aligned}
\tag{6.34}
$$

where n is the number parts in the build.

Note that the application of the above equation requires the magnitude of part height, hatch spacing, and layer thickness to be known. Part height can be minimized through iterative selection of part orientation. Hatch spacing, h_s, is mainly decided by the part building style chosen (see the previous section). With regard to layer thickness, L_p, note that it appears in the denominators of all the four components of *build_time*. This implies that a larger L_p should lead to a smaller *build_time*. However, this is not the case in practice. When we seek a larger L_p, we would have to correspondingly increase the cure depth, C_d. However, because of the exponential term in equation (6.17), this would warrant a substantial decrease in the scanning velocity, V_s. Consequently, for a given machine and resin, there exists an optimum layer thickness at which the *build_time* is minimized.

It must be recognized that build time is quite sensitive to SL resin properties so that much of the progress in SL has occurred through the development o f i mproved materials. F or instance, S L r esins p rior t o 1 997 (e.g., SL 5180) had suffered from relatively weak resistance to high humidity and high heat. These problems were solved when Ciba Speciality Chemicals Corporation and 3D Systems Inc. jointly released a new epoxy resin called SL 5140 n 1997 (Pang *et al.* 1997). Thermo-mechanical properties also improved significantly relative to those of SL 5180 released previously. Heat deflection temperature increased from 15°C to 40°C. Glass transition temperature, T_g, increased to 88 °C when UV post-cured and further to 105°C when t hermally post-cured. Improvements in mechanical properties were achieved while further improving part accuracy and vertical surface finish. Productivity increased 2½ times over SL 5180. Also, SL 5410 required no pre-dip delay, hence cutting the overhead time. In time, the new improved resin led to new fields of applications.

Obviously, one would try to minimize the total part-building cost within time constraints. In this regard, the following approximate cost equation should be useful (Pham *et al.* 1999):

$$total\_part\_cost =$$
$$labor\_cost + software\_cost + maintenance\_cost + \quad (6.35)$$
$$laser\_cost + depreciation + resin\_cost + overheads$$

where

$$labor\_cost = (pre\_processing\_time +$$
$$post\_processing\_time\_per\_unit\_area \quad (6.36a)$$
$$\times overhanging\_area) \times labour\_cost\_rate$$

$$software\_cost = pre\_pocessing\_time \times sofyware\_cost\_rate \quad (6.36b)$$

$$maintenance\_cost =$$
$$(pre\_processing\_time + buid\_tme) \times maintenance\_rate \quad (6.36c)$$

$$laser\_cost = build\_ime \times laser\_cost\_rate \quad (3.36d)$$

$$depreciation = build\_time \times depreciation\_rate \quad (3.36e)$$

$$resin\_cost = (part\_volume + support\_volume) \times$$
$$resin\_cost\_per\_unit\_resin\_volume \quad (3.36f)$$

$$Overheads =$$
$$(labor\_cost + software\_cost + maintenace\_cost + labor\_cost \qquad (3.36g)$$
$$+ depreciation + resin\_cost) \times overheads\_rate$$

A generic cost model (Alexander *et al.* 1998) was also introduced in Section 5.4.2 for part orientation determination.

6.6.5 Functional Prototyping using SL

One of the problems with SL is that conventional resins yield polymers that are weak and brittle, thus rendering the parts useless for functional and structural applications. One way of solving this problem is to incorporate suitable reinforcing so as to form composite resins.

Reinforcement of plastics for improving the mechanical properties of parts has long been in practice. More recently, the strategy has been extended to polymer-based parts produced by layered manufacturing (LM) involving manual spreading of fibers over a photosensitive resin followed by photo-curing (Barlage *et al.* 1992 and Charan *et al.* 1994). Among the fiber structures investigated were long fibers, short fibers, and micro-spheres of glass. Experiments showed that the mechanical properties LM parts improved significantly with long and short fibers but not with micro-spheres (Ogale *et al.* 1991 and Renault and Ogale 1992).

The filling materials can be continuous carbon fibers (Greer *et al.* 1996), continuous or short glass fibers (Cheah *et al.* 1999), or glass micro-spheres. The resin, which acts as the matrix, develops sufficient green strength in the composite even though the fibers might be opaque to ultraviolet radiation. The main advantage of carbon fibers lies in its high elastic modulus. Although it may fail at the same load, a panel made of carbon fiber can be more than three times as stiff as a panel of equal weight made from fiberglass. The disadvantage of carbon is that the failure mode is brittle, with little allowable deformation or toughness. Regarding matrix materials, it appears that epoxy resins are superior for structural applications. Part shrinkage during post-curing may be decreased when short fibers are used. Green strength can be improved by using higher laser power and smaller thicknesses.

An automated approach to layered composite manufacture, called RLCM, is described in (Zak 1999). The desired resin fiber mixture is stored in an external container equipped with a continuous stirring facility. The part building process starts with the withdrawal of a precise volume of the resin mixture from the container and depositing it on to the platform or the previous layer. The liquid is then leveled using a straight edged blade. Next, the layer is cured selectively by a laser beam. The XY scanning pattern of the beam is programmed to be such as to leave a linear array of small and uncured volumes in each layer. The volumes are staggered in successive layers. Fibers protrude into the uncured volumes from the surrounding resin. When the next layer is deposited, the new resin gets cured around the fibers protruding from the uncured volume just beneath it. This strategy ensures good i nter-layer b onding. The p latform i s t hen lowered to p repare for the deposition of the next layer. This cycle is repeated until the entire part has been built. Upon part completion, the platform is raised, and the part removed. The final part is cleaned and post cured as required. A model to predict the mechanical properties of short-fibered RLCM parts is available in (Zak *et al.* 2000).

More recently, SL has been used to produce ceramic parts whose mechanical, thermal, and electrical properties are close to those obtainable by conventional ceramic processing techniques (Hinkzewski *et al.* 1998 and Chartier *et al.* 2002). In one such method (Chartier *et al.* 2002), a thickener, dispersants that act by electrostatic and steric repulsion, and a photo-initiator (Irgacure 651) are first dissolved in an acrylate monomer. Next, up to 60% by volume of alumina, zircon, or silicate powders are added. The suspension is milled for about 30 minutes to achieve homogeneity. In contrast to conventional SL where a liquid polymer contained in a vat is polymerized, the homogenized paste is delivered to the working area using a piston. An argon-ion laser (351-364nm) is used to supply coherent energy. It is found that the depth of penetration, D_p, can be expressed as:

$$D_p = \frac{2d_p}{3Q\phi_v} \tag{6.37}$$

where d_p and ϕ_v are the mean diameter and volume fraction of the powder and Q is a factor characterizing the ability of the medium to diffuse radiation. In turn, factor Q, can be estimated from the following empirical relationship (Griffith and Halloran 1996):

$$Q = \frac{h}{\lambda}(n_p - n_m)^2 \tag{6.38}$$

where h is the mean inter-particulate distance, λ is the wavelength of the radiation, and n_p and n_m are the refractive indices of the ceramic powder and monomer respectively. However, experiments have shown these equations to be unreliable. In any case, energy density needed to be optimized experimentally t o avoid s eparation (pull o ut) o f a layer f rom t he p revious layer. The process is repeated until the full part geometry is obtained. Once the green part is completed, it is cleaned using a solvent to get rid of unpolymerized paste. Finally, the green part is debinded and sintered to reach about 97% of the theoretical density.

6.7 OTHER LASER LITHOGRAPHY SYSTEMS

The success of 3D Systems' SLA technology spurred several teams across the world to engage in R&D work related to find better materials and machines for S L, a nd to improve t he a ccuracy a nd surface f inish o f p arts while minimizing the time required for building a part. Some succeeded commercially and some failed, but none has so far matched the success of 3D Systems.

Amongst the companies who have maintained commercial presence are Teijin Seiki (since 1991), and Aaroflex (since 1994). Both companies had started by licensing Soliform Solid Forming System technology (SOMOS) developed earlier at DuPont. Teijin Seiki's strategy was to gain competitive advantage t hrough s uperior t ooling r esins, s uch a s T SR-752 w ith 5 0% b y volume of inorganic filler. The resin is intended for the direct creation of injection molding tools materials. Aaroflex is marketing several models including Solid imager and AACURA 22 Solid State that employ solid-state lasers with the highest power in the SL industry. There continue to be several other European and Japanese players in the SL market. Fockele & Schwarze from Germany has developed a photopolymer-based system called the Laser Modeling System (Prinz 1997). Laser 3D of France has developed a process called Stereophotolithography. Computer Modeling and Technology from Japan, a joint venture involving Mitsubishi, is offering a line of machines called Solid Object Ultraviolet Plotter (Kochan 1993). Design-Model Engineering Center, a subsidiary of Japan Synthetic Rubber, is selling machines made by Sony called the Solid Creation System (SCS or JCS) (Kochan 1993). Other Japanese photopolymer systems include the Colamm by Mitsui Zosen (Kochan 1993), the Solid Laser Plotter by Denken Engineering (Weiss 1997)), the Uni-Rapid by Ushio, Inc. (Weiss 1997), and a s ystem f rom t he Meiko C orporation. Variants o f S L h ave b een u sed b y Japanese researchers for building small, high-resolution structures (Ikuta and Hirowatari 1993 and Takagi and Nakajima 1993).

Two examples of failed companies are Light Sculpting Inc. (Kochan 1993) and Quadrax Laser Technologies, Inc. The latter was the second company to offer stereolithography equipment (Heller 1991). Both were involved with photopolymer materials and systems (Ashley 1991) but could not capture or maintain market share for several reasons. In particular, Quadrax failed because of patent infringement and litigation—a common barrier to the creation and success of a new venture in the RP industry (Wohlers and Grimm 2001a). Ultimately, 3D Systems gained the rights to several of Quadrax's patents. One of these resulted in the Zephyr coating system used on their SLA machines. Other enhanced versions of SL are being developed in the academic arena, e.g. Shape Deposition Manufacturing (to be discussed in Chapter 8).

There are a few laser photolithography systems somewhat different from that discussed in the previous sections. For instance, Denken Engineering, Japan, builds parts in an inverted manner. The part is attached to a platform that rises as each successive layer is drawn. The liquid resin layer is deposited on a specially prepared window that is transparent to the laser. The design is such that the cured polymer adheres poorly to the plate. The platform and the part in progress are lowered into the resin such that the liquid film between the part and the plate has the correct thickness for the next layer. The new layer is cured by laser irradiation from beneath the plate. After the layer is drawn, the new structure is raised, separating the layer from the plate. The COLLAM machine manufactured by Mitsui Corporation also uses a similar idea of exposing the part from below.

A method that had raised much interest initially but had to be discontinued owing to its complexity and large size is 'Solid Ground Curing (SGC)' developed originally in 1991 by Cubital of Israel and Germany (Levi 1991). This method does not attempt to draw out each cross-section with laser photolithography. Rather, an entire cross-section is imaged in a single operation by means of an erasable mask plate produced by charging the plate via an ionographic process. The image is then developed with an electrostatic toner just as in the more commonly known Xerography process. The mask is then positioned over a uniform layer of liquid photopolymer, which is selectively cured by means of an intense pulse of UV light. A vacuum system removes the uncured polymer from the layer and replaces it with water soluble, low-melting-point wax that serves as the sacrificial support. After it has cooled, the wax layer is milled to produce a flat surface. Next, the toner on the exposed mask is wiped off before repeating the entire process. Finally, the wax is removed by melting from the completed part. With SGC, many parts can be created at once because of the large work-space and the fact that a milling step maintains vertical accuracy. Part support is ensured because liquid resin in non-part areas is replaced by wax. The various processes used to implement are performed at different stations, thus resulting in a bulky system. Further, the system consumed large amounts of resin due to contamination during the build process. Consequently, apart from being unreliable, the technology was very expensive to utilize (Burns 1993) (Wohlers 2001a). The result: Cubital no more manufactures SGC machines.

Envision Technologies of Germany has developed a technology combining the use of acrylate photopolymer and Digital Light Processing (DLP) technology of Texas Instruments (Wohlers and Grimm 2002). Visible light from the DLP light source is projected from below the build platform in such a manner that the projected image represents the cross-section that is being solidified in photopolymer.

Clearly, part-build time in SL can be reduced substantially if multiple laser beams working in parallel are used (Kamitani and Marutani 1996, Loose *et al.* 1997, Lose *et al.* 1999a and Loose *et al.* 1999b). Likewise, machine cost can be reduced by using laser diodes rather than He-Cd or Argon-ion lasers. An example of this combination is the multiple-LED photographic curing (MPC) system (Loose *et al.* 1999a) capable of producing very small parts (micro-stereolithography).

Direct Photo Shaping is a multi-layer fabrication process developed at SRI International for the fabrication of ceramic, metal, or polymer components. In this process, each layer is selectively photo-imaged by digital light projection via a Digital Micromirror Device (DMD™) array that uses more than 500,000 microscopic mirrors. The mirrors are electronically tilted at speeds greater than 1000 times per second and the light is digitally reflected on the photo-curable slurry. The technique allows the photo-curing of highly filled suspensions of ceramics with high refractive index, such as silicon nitride.

Chapter 7

SELECTIVE LASER SINTERING (SLS)

Selective Laser Sintering (SLS) uses a laser beam directed by a computer onto the surface of metallic or non-metallic powders to produce copies of solid or surface models rapidly. It is one of the few RP processes capable of producing durable and functional parts from a wide range of prototype materials that have physical properties similar to production materials. The functional prototypes are capable of withstanding high heat and chemical exposure and can be made to bend, snap or bolt together, and form flexible hinges.

SLS was first developed in 1987/88 by Carl Deckard at the University of Texas at Austin, TX, to produce plastic parts (Deckard and Beaman 1987 and Deckard 1988). The process was patented in 1989 and then commercialized by DTM Corporation, USA. Around the same time, Electro Optical Systems (EOS Gmbh) of Germany also started an SLS venture in 1989. In 2001, DTM was taken over by 3D Systems.

This chapter describes the principle and operation of SLS. Particular attention is paid to the modeling of the process with a view to enabling an appreciation of the factors affecting the capability of the process.

7.1 THE PRINCIPLE OF SLS

Figure 7-1 shows the typical configuration of an SLS machine. Note the powder-feed and part-build chambers. In some machines there could be two powder-feed cylinders, one on each side of the part-build chamber. At the beginning of each cycle of layer formation, the powder-feed piston moves up certain distance to deliver a fixed quantity of powder onto the table. The roller then sweeps across the table to spread the powder evenly over the table surface. In the EOS system, powder spreading is accomplished by a slot nozzle vibrated from side to side while the leading and trailing edges of the nozzle apply a normal force on to the surface is vibrated from side to side. An electrostatic powder deposition system replacing the roller feed system is explored in (Melvin III *et al.* 1991) with a view to reducing space requirement, achieving a more packed bed, and enabling powder storage in a different location than the workspace. The system uses compressed air to force the powder particles through an electric field so as to spray them on to a grounded plate. However, problems have been found with regard to the ability of the system to deliver powder to specific locations in the workspace. A modification of the electrostatic technique is described in (Liew *et al.* 2002a and Liew *et al.* 2002b). In this system, the powder is forced through a grounded or charged sieve on to the sintering surface before being leveled by a roller. It is claimed that the system yields up to 20% increase in final part density.

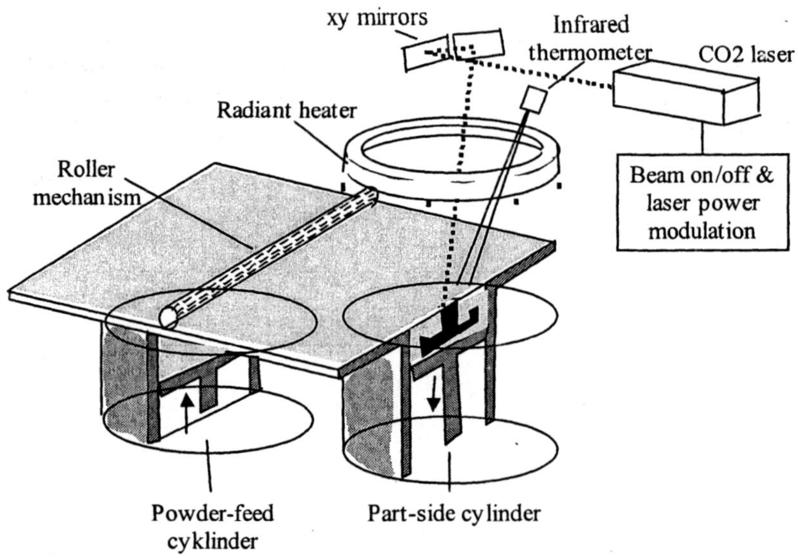

Figure 7-1. The principle of Selective Laser Sintering (Subamanian *et al.* 1995).

An infrared radiant heater heats the part of the powder layer above the part-chamber to a temperature just below the melting temperature. Such preheating of the part chamber decreases the shrinkage of the part.

Many machines use a CO_2 laser beam that moves across the heated layer under computer control in accordance with the information contained in the .STL file for that particular layer. A CO_2 laser is much more powerful than the lasers used in SLA. However, a substantial part of the laser radiation

(10.6μm) is reflected from metal surfaces (Carter and Jones 1993). Hence, in some machines, a continuous or Q-switch pulsed Nd-YAG laser of 1.06μm wavelength is used.

The power of the beam is adjusted to bring the selected powder areas to a temperature just sufficient for the particles in the powder to get attached (i.e., get sintered). Sintering may occur through partial melting or softening of powder particles themselves, or of coating on the particles. After allowing sufficient time for the sintered layer to cool down without causing significant internal stresses, the part-cylinder moves down one layer thickness to be ready for the next layer-building cycle. The part is a growing cake embedded inside the powder. Hence, no specially designed supports need be built and subsequently removed. When done, the cake (the part) is carefully removed from the powder mass with the help of putty knives and spatulas. Excess powder is brushed off from the part. The powder can be reused. Models can be sanded down and painted if required, to give a smooth finish.

There are two basic types of SLS: indirect and direct. Indirect SLS was developed by University of Texas. The first commercial machine was offered in 1995 by DTM Corporation. In this technique, metal powders are mixed with a small amount of polymer. Alternatively, metallic and ceramic powders are coated with polymer. Particle binding occurs essentially through polymer-polymer bonds. For instance, during SLS of a steel powder, the polymer softens/melts and forms necks between adjacent particles. The resulting part is called the 'green part'. Owing to melting of polymer, the green part is usually quite porous. Hence, it cannot be used without some post-processing.

Direct SLS utilizes a high-energy laser beam to directly consolidate a metal or cermet powder to densities greater than 80% of theoretical density. The advantage is that many of the pre- and post-processing steps associated with indirect SLS are eliminated.

Note that, in contrast to conventional sintering, no compaction pressure is applied in SLS. As a result, while the green parts are of lower density, they are nearly isotropic and free from internal stresses. While the latter two features are desirable, the former can lead to problems. For instance, if the part density is too low, it can distort during post-processing in a furnace or get damaged during cleaning and handling. Thus, it is desirable to aim at a green part density that is at least 50% of the theoretical value and a strength value larger than 1.7MPa (Beaman *et al.* 1997).

The key advantage of SLS is that it is capable of yielding functional parts in essentially final materials. It is possible to produce parts considerably stronger than those produced by SLA. Further, the process is simple, e.g., there are no milling or masking steps required. Living hinges are possible when thermoplastic-like materials are used. Uncured material is easily removed after a build by brushing or blowing it off.

A wide variety of materials suitable for SLS are available, e.g., wax for investment casting; and a polymer (e.g., nylon) for assembly prototypes. In principle, virtually any material that softens and has decreased viscosity upon heating can be used. Several thermoplastic materials such as nylon, glass filled nylon, and polystyrene suitable for SLS are commercially available. Also mixtures like sand with thermosetting binders, metals with thermoplastic binders or high melting point metals in combination with low melting point metals (e.g., steel/bronze) can be used. The process has also been applied successfully to powders made up of steels, Ni-based super alloys, Ti and its alloys, cermets, and refractory bronze-Ni (perhaps the most effective combination). As-built material properties can be quite close to those of the intrinsic materials.

One does not need to incorporate supports since the powder bed itself supports any overhanging sections of the part. The completed part also exhibits reduced distortion from stresses. The system is capable of producing functional parts and tooling inserts. Finally, it is possible to produce several parts simultaneously.

However, SLS is not without its share of disadvantages. Compared to most other RP equipment (e.g., SLA), an SLS system is mechanically more complex. As with parts produced by any layered manufacturing (LM) technique, SLS parts also suffer from the stair-stepping problem. Surface finish and accuracy are not quite as good as with stereolithography. The first layers may require a base anchor to reduce thermal effects such as curling. SLS parts exhibit powdery, porous surfaces that are absorbent and mark easily. The porosity and, hence, the density can vary across the part. Finally, a change in part material requires thorough cleaning of the machine.

However, part finish can be improved by applying a sealant. A sealant also strengthens parts. Likewise, infiltrating the part with some other material can solve some of the problems arising from part porosity.

When creating multiple-part models, it is advantageous if different parts could be colored differently. An inkjet technique that is capable of depositing color toner in liquid form over SLS models is described in (Ling and Gibson 1999). The system may be considered as a 3D printing subsystem incorporated into a SLS machine. After the powder has been selectively sintered over a layer, the arm of the printing head swings to position the ink cartridge to print the color toner over the selected region(s). Next, the arm is moved away and the roller mechanism spreads a new layer of powder onto the powder bed. The process is repeated until the entire multi-colored assembly has been completed.

New RP applications of powder sintering that may or may no be similar to SLS are being explored continually (Wohlers and Grimm 2002, Wohlers 2002). For instance, Concept Laser GmbH, Germany, has combined laser sintering, laser marking and laser machining in a single machine capable of making 100% dense parts from stainless steel powder. Speed Part AB, Sweden has patented a new process using infrared lamps to sinter layers of plastic powder through a mask printed on a glass plate, so the entire cycle time for a layer can be produced within 10 seconds. The company anticipates that the machine will cost about a third of current laser-based sintering machines of equivalent capacity.

7.2 INDIRECT AND DIRECT SLS

7.2.1 Powder Structures

During the early stages of the development of SLS, attempts were made to process blends of copper, lead, tin, and zinc (Hease 1989, Manriquez-Frayre and Bourell 1990 and Bourell *et al.* 1992). However, these attempts failed because of balling (formation of small spheres approximately of the size of the laser beam) as a result of the large degree of melting that occurs when the laser power was increased or the laser scanning speed was decreased Subsequently, it was found that the effect could be overcome by using a two-phase powder. This realization led to the development of indirect SLS. Figure 7-2 shows four possible two-phase structures (Klocke *et al.* 1995).

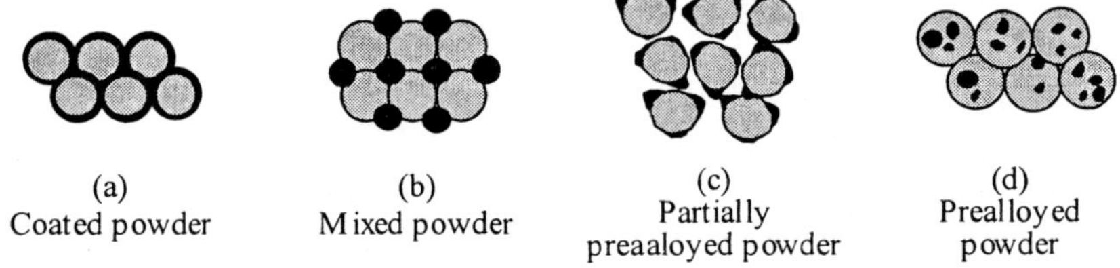

Figure 7-2. Powder structures (Klocke *et al.* 1995).

In the first of these, the primary metallic phase is coated with a metal or polymer binder with lower melting point so as to facilitate binding (7-2a). This approach is not only expensive but also brings problems arising for nonuniform coating thickness. The other extreme is mixing the two components mechanically (Figure 7-2b). This approach yields high packing density. Even higher densities can be obtained by vibrating the bed.

However, the components can get segregated while the wiper spreads it over the table. The problem of segregation does not occur when a prealloyed powder is used (Figures 7-2c and d). Prealloying can also be used to reduce oxidation of metallic powders. For instance, oxidation of copper powder during s intering i s f ound to d ecrease w hen p realloyed w ith t in (Klocke *e t al.*1995).

When the volume percentage of the melt increases, there could be oxygen pick up that can lower surface energy because metal oxides have lower surface energy than metals. The resulting oxide contaminants can mar the wetting behavior during sintering. One way to control this effect is to adopt reducing/protective atmospheres made up of nitrogen, argon, hydrogen, r nitrogen-hydrogen mixture. Another way is to adopt deoxidizer coatings or components such as $ZnCl_2$, $NaCO_3$, $KCrO_3$, Si, or H_3BO_3 (Klocke *et al.* 1995 and Lu *et al.* 2001).

7.2.2 Indirect SLS using Coated Powders

Indirect SLS can be applied to any material that exhibits a lower viscosity upon heating. Linear and branched thermoplastics satisfy this requirement. They also can be repeatedly heated to soften or melt and cooled to solidify as the temperature passes through the glass transition temperature, T_g, and the melting temperature, T_m. Both amorphous and crystalline polymers may be subjected to SLS. However, their processing characteristics differ owing to differences in the absorption of laser radiation. Amorphous polymers such as epoxies, polycarbonate, polymethyl methacrylate (PMMA), polystyrene, and polyvynil chloride (PVC) are transparent unless a second filler phase is deliberately added. In contrast, crystalline polymers such as nylon, polyethylene, polypropylene, and nylon are opaque.

When ceramic or metal parts are desired, thermoplastic polymers are used as fugitive binders in the powders Vail and Barlow 1990, 1991, Balasubramanian and Barlow 1992, Subramanian *et al.*1993 and Subramanian *et al.* 1995). The idea is to have a low melting point material adjacent to the powder substrate so that the latter is more easily compacted to produce green parts (Vail *et al.* 1996). Subsequent to green part formation, the binder is either chemically converted to a higher melting point substance (Lakshminrayanan 1992, Lee and Barlow 1993 and Jakubenas and Marcus 1995), or thermally decomposed and eliminated (Blasubramanian and Barlow 1991 and Vail 1994). When the binder elimination route is taken, the part can be sintered to full density, or the voids left can be infiltrated with a suitable m aterial (Tobin *et al.* 1993, Deckard and Claar 1993 and Barlow and Vail 1994).

Polymer binders used for coating inorganic powders used in SLS need to meet several requirements (Beaman *et al.* 1997). Firstly, the binder must be applicable to powder particles with diameters larger than 10µm since such powders have been found to spread, level, and sinter reasonably well during SLS. Secondly, they should not be soluble in water, so the flow characteristics of the polymer-coated powder are independent of ambient humidity. Further, many infiltrating (cementing) agents used subsequently are water-borne and can wash away the coating thus weakening the part. The green part will then be too fragile to handle. Hence, in practice, the cementing agent is allowed to dry and, only then, is the polymer binder burnt out. This procedure allows the linear shrinkage relative to the green part to be reduced to below 1%. Thirdly, the binder application process should be such that the resulting powder can flow freely. In practice, this is achieved through an aqueous vehicle. Note that water is used as a vehicle, not solvent. Another way is to emulsify the inorganic particles with a suitable monomer and an emulsifying agent, as is common in water-based acrylic paint systems. Typically, polymerization of the monomer into the emulsified polymer is achieved through water-borne ionic initiators. The emulsion particles are quite small (about 100nm), so the polymer gets well distributed through the aqueous vehicle. If the particle concentration in the aqueous vehicle is 40-50% (i.e., $2\text{-}5\times10^{20}$ particles per cubic meter (Bovey *et al.* 1955), gravity is unlikely to have a noticeable effect on the emulsion. This procedure is capable of producing parts with high molecular weight (MW). Finally, in contrast to its use in paints, the polymer binder must soften within the temperature range 40-100°C. This will ensure that the polymer is rigid and non-tacky at room temperature. This also facilitates storage of the binder. The upper limit of 100°C is dictated by the boiling point of water vehicle. If the upper limit is much higher, wetting of the inorganic surface, film formation, and polymer flow will not occur simultaneously. One way of adjusting the softening temperature is to take advantage of equation 4.1 characterizing the glass transition temperature of a copolymer. This equation implies that the T_g-value of the copolymer will be somewhere between those (T_{g1} and T_{g2}) of the two individual polymers. At the same time, equation 4.2 implies that the copolymer's T_g depends on its molecular weight. These two considerations can be used to guide the search for suitable copolymer constituents.

Based on the above considerations, it is found that a metal or ceramic powder applicable in SLS can be coated with copolymers of MMA (methyl methacrylate) and BMA (n-butyl methacrylate) to produce parts made of glass, alumina, and silicon/zircon mixtures (Vail and Barlow 1991), copper (Balasubramanian *et al.* 1992), silicon carbide (Vail *et al.* 1993), and calcium phosphate (Lee and Barlow 1993 and Lee and Barlow 1994). (The last material is useful for biomedical implants.) A characteristic of this binder material is that, when heated at 10°C/min, degradation occurs mainly in a relatively narrow temperature range of 275-400°C., and less than 0.5% residue is left after complete degradation. Further, PMMA-based binder emulsions in water can be spray-coated on to inorganic particles (Vail and Barlow 1991, 1992). After selectively sintering by means of an infrared laser, the PMMA-based binder can be thermally removed and the polymer-free part sintered so as to create a part made up solely of metal or ceramic (Vail and Barlow 1991, Bourell *et al.* 1992 and Barlow 1992). The average zip length of PMMA in nitrogen reduces significantly as the polymer degrades and converts to gas, or as the temperature is increased from 260 to 324°C. However, thermally induced oxidation can be practically eliminated by using a neutral or reducing environment so as to control oxygen to below 2%. When this is done (most SLS machines are designed to care of this), virtually no degradation in molecular weight is observed even after tens of hours of SLS operation.

During sintering, only the polymer layer is melted. In some cases, the polymer content is 'debinded' in a stepwise manner. Whatever the method, the green part has high porosity and cannot be used without further processing. One way to build the strength of the green part is to infiltrate it with a water soluble polymer, or a metal such as copper. This can be done either before (Vail and Barlow 1992) or after (Tobin *et al.* 1993) removing the polymer binder. However, one needs to be aware that the debinding process causes between 4 to 5% shrinkage (Naber and Breitinger 1996). Interestingly, horizontal shrinkage is often smaller than vertical shrinkage.

The degradation kinetics of the binder depends upon the surrounding gas that can be air or a less chemically reactive gas like nitrogen. The following kinetic model is suggested in (Vail *et al.* 1996):

$$\frac{d}{dt}\left(\frac{m}{m_0}\right) = -k_d\left(\frac{m}{m_0}\right)^n \tag{7.1}$$

where m_0 is the initial mass, m is the mass remaining at time t, n is the 'pseudoreaction order', and k_d is the rate constant that is assumed to follow the Arhenius relationship:

$$k_d = A\exp\left(-\frac{\Delta E}{RT}\right) \tag{7.2}$$

where A and E are the frequency factor and the activational energy for the decomposition reactions respectively, R the universal gas constant, and T the absolute temperature.

The magnitudes of the kinetic parameters can be determined by heating a known mass, m_0, of the material in the gaseous environment of interest and monitoring the remaining mass, m, over time. For instance, the following thermal degradation parameter values were reported in (Vail *et al.* 1996) for PMMA over the temperature range 290-350°C: $n=3/2$, $\Delta E=40.3\text{kcal/mol}$, and $A=1.266\times10^{12}\text{ sec}^{-1}$.

Several investigators have experimentally studied the influence of process parameters in indirect SLS. For example, Nelson *et al.* observed sintering depths in polycarbonate powders under conditions of incomplete layer-to-layer bonding (Nelson *et al.* 1993b). Their observations suggest that the sintering depth decreases significantly with increasing scan pitch and laser spot diameter. It can be taken as a thumb rule that sintering depth increases in proportion to laser power, beam radius, and bed temperature. In contrast, it can be assumed to be inversely proportional to beam speed, air velocity, and scan pitch. However, the influence of air velocity is small, thus indicating that most of the cooling at the bed surface occurs through radiation rather than convection. Likewise, the influences of beam radius and bed temperature are smaller than those of scan pitch, laser power and beam speed. The transition front between unsintered and sintered regions moves with a decreasing velocity after the laser spot has moved on and the region begins to cool down.

The temperature gradients within the bed can be quite high owing to the high laser energy input and low diffusivity of powder bed. However, the temperature gradients become smaller as sintering proceeds because of decreasing porosity and the consequent increase in thermal conductivity. For the same reasons, there is good layer-to-layer adhesion. It is possible to develop thermal models (to be discussed later in this chapter) capable of predicting sintering depths within ±0.025mm.

Badrinarayan studied the influence of energy density on the density of test bars made from 40volume% PMMA and 60volume% copper (Badrinarayan 1995). His work shows that the relative density of a sintered part increases with increasing laser energy density.

It is important to minimize binder losses during the production of green parts since the presence of binders improves the green strength of parts, so they can be handled more easily without breaking. Vail *et al.* modeled the thermal degradation of polymer binders during SLS and verified model predictions against experimental data gathered with regard to on PMMA and other binders (Vail 1994, Vail *et al.* 1996 and Beaman *et al.* 1997). (We will describe this model in some detail later in this chapter.) The experimental results suggest tat the polymer content decreases at an increasing rate with increasing energy density. The agreement between model predictions and experimental data can be within ±5%. At the same time, the bend strength of green parts increases with increasing laser energy density until a tendency to saturate is found at very high values of energy density. An interesting observation is that bend strength decreases with increasing scan vector length owing to increasing energy losses through radiation and convection. A longer scan length results in a longer time between successive scans. Hence, there is greater time for the previously heated segment to cool down. Consequently there is lower sintering or polymer degradation.

7.2.3 Direct SLS using Mixed Powders and LPS

Mixed powders enable direct SLS without the use of a polymer binder or a specially developed metal powder. The firsts Direct Metal laser Sintering (DMLS) machine was developed in 1995 by Electro Optics (EOS) of Munich, Germany. EOSINT M250 RP, a popular EOS machine, uses a 200W Infrared laser with a spot size of about 300μm. The system uses a prealloyed powder that yields a laser cure width of about 600μm at a scan speed of 3m/s with the standardized layer thickness of 50μm.

Mixed powder technology is found to be applicable in the SLS of TiB_2-Ni, ZrB_2-Cu and Fe_3C-Fe powders (Wang *et al.* 2002). The two-phase powder approach has also been successfully applied to metallic and ceramic systems such as Cu-Sn, Cu-(70Pb-30Sn) solder, bronze-Ni, Ni-Sn, alumina-ammonium dihydrogen phosphate, SiC-Si, WC-CO, and aluminum-boron oxide (Lu *et al.* 2001). More recently, Hon and Gill investigated the possibility of making SiC/Polyamide matrix composites using SLS (Hon and Gill 2003)).

The material to be sintered comprises a structural powder with a high melting point and a binder powder with a low melting point so as to promote liquid phase sintering (LPS)—see Chapter 4. For instance, while sintering Ni-Sn, Ni melts at 1455°C whereas Sn melts at 232°C and the resulting compound can be composed of Ni_2Sn, Ni_3Sn_2, and/or Ni_3Sn_4. Another example of the binary material approach is low-purity alumina (95% Al_2O_3, 3%TiO_2, 0.75% SiO_2, 0.46%Fe_2O_3, 0.25% MgO) with T_m>2000°C and ammonium dihydrogen phosphate ($NH_4H_2PO_4$) with T_m=190°C that forms a gassy phase at or below 200 °C to act as a binder (Lakshminarayan and Marcus 1991). A variation of this approach relies on partial melting of prealloyed 89Cu-11Sn involving a two-phase zone between the solidus and liquidus (Prabhu and Bourell 1993).

In a direct SLS process, the build up of part strength occurs mainly though liquid phase sintering (LPS). The nature of LPS will clearly depend upon the characteristics of the powder blend and the sintering process itself. One should therefore be cautious in generalizing. However, it is useful to obtain a basic understanding of at least one sintering case study in some detail.

With this objective, consider SLS of the bronze-nickel system that exhibits solidus temperature of 840°C and liquidus temperature of 1020°C (Agarwala *et al.* 1995a and Agarwala *et al.* 1995b). Above the solidus, bronze melts partially and above the liquidus it is completely melted. LPS in this system proceeds in three stages. In Stage I, liquid bronze flows into the pores between adjacent solid nickel particles, and rearranges the solid particles. In the process, the part rapidly shrinks with the consequent increase in density. The degree of densification depends upon the particle size and the rate of heating. In fact, most of the part densification occurs in this stage. Further, the final density of the part is independent of the starting density that, in turn, is dependent upon the laser power and scanning speed (Agarwala *et al.* 1995b). With increasing temperature, the amount of liquid phase increases whereas the viscosity decreases. Hence, if the temperature is too high, the part can distort. In Stage II, there is continued sintering through similar mechanisms as in Stage I but leading to homogenization of the nickel-bronze mix through diffusion between solid nickel and bronze particles, and/or through solution precipitation liquid bronze and solid nickel. The result is that the bronze phase disappears completely leaving behind the pores (called Kirekendall pores) and an expanded solution of copper (since the original bronze was an alloy of copper and tin) and nickel. The new Cu-Ni solution, however, has a solidus that is higher than the sintering temperature. Hence, further sintering proceeds essentially in the solid state. Hence, even after prolonged sintering, there is only a small amount of densification.

However, during SLS, the laser beam supplies energy for much less than 1s. This means that the time is insufficient for the second and third phases of LPS, i.e., solution precipitation and final densification to be completed. The green parts can therefore be quite porous and may require post-processing such as further sintering or infiltration.

In contrast to polymer binding mechanisms, melting is the key to binding in LPS. Here, neck formation is mainly due to the flow of liquid phase (Frenkel 1945). Hence, this effect is called viscous sintering. Note that viscous sintering can occur when sufficient amorphous polymer glasses are present (not to be confused with polymer binding).

Viscous sintering arises from a mechanism that equates the energy dissipated by viscous flow to that gained by the decrease of surface area during densification, so the rate of sintering can be approximated by the rate of coalescence of spherical adjacent particles given by

$$\left(\frac{x}{r}\right)^2 = \frac{3}{2}\left(\frac{\gamma t}{r\eta}\right)$$

(7.3)

where x is half the thickness of the neck formed between adjacent particles of radius r at time t, γ the surface tension, and η the viscosity (Frenkel 1945).

It is observed that the densification rate varies in inverse proportion to viscosity (Scherer 1977a, Scherer 1977b, Scherer 1977c and Swinkels 1981) and follows the shrinking cube analogy illustrated in Figure 7-3 (Sun *et al.* 1991). Based on these considerations, the densification rate can be expressed as

$$\dot{x} = \frac{3(1-\rho)\pi\gamma a^2}{24\eta_m \rho^3 r^3}\left\{r - (1-\xi)x + \left[x - \left(\xi + \frac{1}{3}\right)r\right]\right\}\frac{9(x^2 - r^2)}{18rx - 12r^2}$$

(7.4)

where $2x$ is the cube size, ρ the apparent density, γ the particle surface tension, ξ a parameter with its magnitude between 0 and 1, and η_m the viscosity of the solid-liquid mixture given by

$$\eta_m = \eta_l[1 - (V_s/V_m)]^{-2}$$

(7.5)

where η_l is the viscosity of the pure liquid phase, V_s the volume fraction of the solid phase, and V_{cr} the critical volume fraction above which the mixture exhibits essentially infinite viscosity (German 1990).

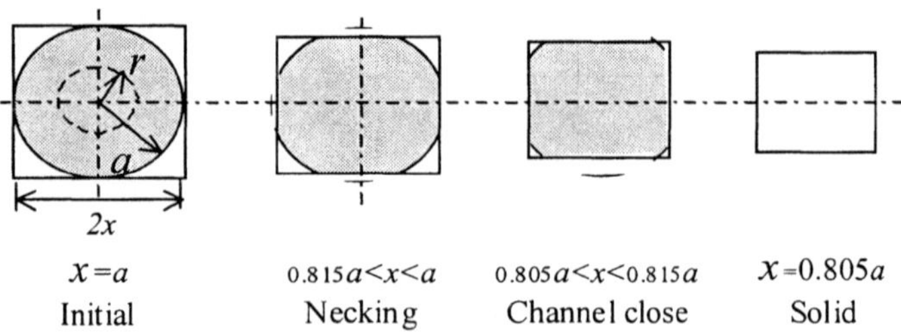

Figure 7-3. The shrinking cube analogy of powder densification.

The tendency for balling can be controlled by controlling η_m. Thus, binding can be favored over balling by having a higher volume fraction of the high melting point phase. The speed with balling occurs depends on the material. For instance, pure copper will flow to a spherical shape very quickly owing to its single melting temperatures whereas stainless steel needs up to 30 times more fusion time (Klocke and Wagner 2003). Process parameters leading to moderate heating and more evenly distributed temperatures. Are likely to reduce the tendency to balling. Hence, slow scanning velocities and low laser powers are preferable.

Two further factors influencing LPS are the liquid solubility in solid and the solid solubility in liquid. The latter is typically 1 to 20% in volume terms. When liquid solubility is high, the transient liquid phase can lead to heightened process sensitivity and swelling of the compact during heating (Lu *et al.* 2001). In contrast, solid solubility promotes densification because it results in initial grain shrinkage thus helping the rearrangement phase. Thus, a combination of low solid solubility with high liquid solubility can promote swelling. This effect can be reduced by using a more homogenous powder. Further, since a higher laser energy density leads to increased solid solubility,, higher laser powers are usually preferred in SLS based on LPS. However, there could be some densification error before the formation of the liquid phase (Lu *et al.* 2001).

Another way is to add small particles, or to use coated particles. For instance, in the EOSINT M250 SLS process, copper-phosphide is used as the binder for the purpose of enabling sintering of bronze/Ni powder. When the laser beam scans the powder, the binder melts at about 660°C. The melt penetrates the surrounding cavities and wets the bronze/Ni particles.

Several investigators have experimentally studied the influence of process parameters in direct SLS. For example, Das *et al.* investigated the effects of combining direct SLS with hot isostatic pressing (HIP) in the prototyping of high performance metal components made from Inconel 625, etc. (Das *et al.* 1998). Parts so processed were found to be comparable to those produced by conventional processes.

Khaing *et al.* assessed the process capability of DMLS of EOSINT M Cu 3201 powder from the viewpoint of rapid tooling requirements (Khaing *et al.* 2001). The powder consisted of nickel, bronze, and copper phosphide with average grain size equal to 30μm. The following results were found: dimensional errors between 0.003 and 0.082 mm, hardness of sintered part ranging from 26 to 33 Rockwell B which increased to 65-69 HR B after epoxy infiltration, roughness on untreated parts of 12-16μm, and porosity without infiltration from 30 to 45%.

Wang *et al.* compared the SLS (using an Nd-YAG laser) performance of WC-Co powder mixture against those of Fe-Cu powders (Wang *et al.* 2002). WC-Co powder had 91wt% of W (T_m=2687°C) and 9wt% of Co (T_m=1496°C). Copper has a much lower melting temperature (T_m=1083°C) than that of cobalt. Co has a higher absorptivity than the more reflective Cu. Interestingly, while energy absorption generally followed Beer-Lambert law (exponential decay with increasing depth) at higher depths, the maximum absorption did not occur at the surface but slightly below (about 15μm) the surface. This shift depends on the laser source and powder bed characteristics. It is important that the shift is recognized in simulations of sintering depth. The accumulated energy is concentrated in a small layer (of a fraction of a mm thickness). Hence, quite large thermal gradients develop within the top layers. The thickness and width of the sintered zone decrease with increasing scan speed. At lower scan speeds, the molten layer can be maintained in a liquid state for a longer time. Very large sintering zones can be obtained when a high laser power combined with a low scan speed is used. Thin-walled parts with overhanging features are easily built from WC-Co powders owing to the material's good flowability and supporting characteristics.

Kruth *et al.* have recently developed a ferro-powder that promotes SLM rather than LPS, so quasi-dense parts can be produced without any post-processing like furnace sintering (Kruth *et al.* 2003). The material is a mixture of spherical particles of Fe, Cu,, Ni, and Fe_3P. The presence of P considerably lowers the melting temperatures of Fe and Cu. At the same time, it lowers the surface tension of molten Fe, so balling is prevented. Another advantageous property of P is lowering of the oxidation rates of Fe ad Cu particles. Ni is added for its strengthening effect. A substantial quantity of liquid is produced when the powder is exposed to a pulsed Nd-YAG laser. The liquid fills the pores in the previously sintered layers through the action of gravity and capillary forces. Parts are preheated to promote better wetting. The first sintered layers are anchored to a base plate to avoid warping of parts. Process parameters are controlled such that the melt pool resolidifies rapidly behind the laser spot (short melt length).

7.3 MODELING OF SLS

7.3.1 Modeling of Material Properties

SLS involves several physical phenomena. These include energy input and absorption, heating of the powder bed, binder melting, and sintering and cooling of the sintered sample. Another complication is that its thermal modeling requires the application of unsteady state heat transfer equations (Carslaw and Jaeger 1959). A complete solution of the SLS modeling problem requires the application of a range of numerical and computational techniques (Vail *et al.* 1996, Williams and Deckard 1998, Bugeda *et al.* 1999, Boillat and Glardon 2000, Childs *et al.* 2000, Childs and Tontowi 2001, Katz and Smith 2001 and Matsumoto *et al.* 2001). The description of these techniques is beyond the scope of this book. In the following, we will confine our discussion to certain important features of selected models with a view to obtaining some physical insights into the process of SLS.

The size, distribution, and shape of particles can be of significance in most powder-based processes. Particle size is usually measured by sending the particles through a series of graded sieves. Other methods of particle size measurement include sedimentation, microscopic analysis, light scattering, optical sensing, and electric sensing of particles suspended in a liquid. Particle distribution is usually analyzed by plotting the frequency distribution, and evaluating the associated statistical measures. Shape is usually specified in terms of the mean values of aspect ratio (ratio of length to thickness) and the shape index (ratio of surface area to volume).

The sintering effect in SLS is a result of the heat produced by the interaction between the laser beam and the powder bed. Further, SLS is a dynamic process since a region in the powder bed gets heated as the laser beam approaches and then cools down as the beam passes beyond. In view of these features, modeling of SLS necessarily involves unsteady heat transfer equations (Carslaw and Jaeger 1959). A recurring parameter in such equations is thermal diffusivity, $\alpha = k/(\rho C_p)$, where k is the thermal conductivity, ρ is the density, and C_p is the specific heat capacity of the conducting medium. Hence, it is important to have accurate estimates of the magnitudes of k, ρ, and C_p, for the given powder bed in an SLS operation.

The density of a powder bed depends not only on the density of the powder particles but also on the degree of packing. If the powder has just been poured, the particles will be randomly oriented. If the powder has been 'tapped' subsequently, some directionality is imposed and the density increased in the vicinity of the surface areas tapped. Many powders, including ceramic powders, tend to agglomerate due to Van der Waals surface forces. Agglomerated powders do not fill space efficiently. Clearly the determination of the degree of packing of a powder bed is quite an uncertain process.

The degree of packing of a powder bed may be characterized by the relative density, ρ_R, defined as

$$\rho_R = \frac{\rho}{\rho_s} = \frac{\rho}{\sum_{1}^{n} \phi_i \rho_i} \tag{7.6}$$

where ρ is the density of the bed, ρ_s is the theoretical solid density of the aggregate powder consisting of n types, and ϕ_i is the volume fraction of particle type i of density ρ_i. The density, ρ, of the powder can be measured by the cup method described in (Vail *et al.* 1993). The method involves the preparation of cups of precise geometry using SLS and measuring the mass of the non-sintered powder contained in the cup.

Alternatively, one can use the porosity parameter defined as the ratio, ε, of the difference between the bed volume and solid particle volume to the bed volume. It is easy to show that

$$\varepsilon = 1 - \rho_R = 1 - \left(\rho / \sum_1^n \phi_i \rho_i \right) \tag{7.7}$$

While the determination of actual powder density is an uncertain process, it is useful to keep in mind certain idealized extremes that are deterministic in nature. One such idealization involves the assumption that all the solid particles in the powder are spheres of equal size and density arranged in a cubic array. In such a case, the bed density is given by

$$\rho = \pi \rho_s / 6 \tag{7.8}$$

where ρ_s is the density of the solid particle Thus, in this extreme case, the relative density, ρ_R, is 52.3%, so the porosity, ε, is equal to 47.7%.

However, actual powder beds usually do not consist of spherical particles of equal size. Hence their relative densities can be higher or lower than the idealization described above. Actual beds used in SLS exhibit relative densities between 40 and 60%.

The specific heat capacity of a powder bed consisting of a range of solid particle components can be estimated by mass averaging

$$C_p = \sum \omega_i C_{pi} \tag{7.9}$$

where ω_i and C_{pi} are the mass fraction and specific heat capacity respectively of component i. This equation is found to be fairly reliable (German 1990 and Nelson *et al.* 1993a). It follows from the equation that, when solid particles are of the same material and the voids are filled with a gas (air), the specific heat capacity of the bed is practically equal to that of the solid particles (because the density of air is insignificant when compared to that of the solid particles).

The specific heat of powdered materials can be measured by a variety of methods including differential scanning calorimetry (see Chapter 2). In some RP studies, sapphire is used as the standard and DSC readings are subjected to polynomial fitting (Sih 1996). A comparison of the specific heats of some commonly used metallic, ceramic and organic powders is available in Xue and Barlow (1990). It is reported that the specific heat capacities of nylon (0.40-0.55cal/g-°C), ABS (0.30-0.3855cal/g-°C), and PVC (0.30-0.3255cal/g-°C) powders are higher than that of glass (0.20-0.225cal/g-°C). In turn, the figure for glass is higher than that for tin (0.46cal/g-°C). Within the range for a given powder, specific heat increased with increasing temperature in the range 10 to 100°C.

Heat transfer through powders mainly occurs through conduction and radiation because inter-particular distances are too small to permit convection heat transfer. The powdered bed mainly consists of solid particles separated by a gas. Since gases have much smaller thermal conductivities (0.0065 W/m-°K at room temperature to 0.049 W/m-°K at 325°C for air) than those of good metallic conductors (400 W/m-°K) or, even, plastics (0.2 W /m-°K), the thermal conductivity of a powder bed is mainly dictated by the thermal conductivity of the gas embedded within the voids (Incropera and DeWitt 1985 and Nelson *et al.* 1993b).

The thermal conductivity of powder beds can be measured by using any one of the generally applicable steady (ASTM standards C177-63, C420-62T, and C518-67) or nonsteady (Bird *et al.* 1960) heat transfer based methods. Two nonsteady methods specifically applicable to the study of selective laser sintering have been reported and modeled in (Xue and Barlow 1991 and Sih and Barlow 1994, 1995).

The thermal conductivity of a powdered bed can vary across the bed depending on the local temperature during sintering, porosity, and particle contact conditions. According to Yagui and Kunni, the effective conductivity of a of a packed bed including radiation, convection, and conduction effects can be written as

$$k_{ef} = \rho_R k_s /(1 + \Phi k_s / k_g) \tag{7.10}$$

where ρ_R is the relative density of the bed, k_s the solid thermal conductivity, k_g the thermal conductivity of the surrounding gaseous environment, and empirical coefficient taken as equal to $0.02 \times 10^2 (0.7 - \rho_R)$ (Yagui and Kunni 1957). This equation enables the conductivity at each point in the powder bed as a function of the local relative density.

In (Zehner and Schlunder 1970), another thermal conductivity model was presented by including the radiation term defied in (Demkohler 1937), k_R, to take into account inter-particle radiation heat transfer that is likely to be significant at the high temperatures prevailing SLS (Vail *et al.* 1996):

$$k = k_g \left(1 - \sqrt{1-\varepsilon}\right)\left(1 + \frac{\varepsilon k_R}{k_g}\right)\left[\frac{2}{1 - \frac{k_g}{k_s}}\left(\frac{\ln\frac{k_s}{k_g}}{1 - \frac{k_g}{k_s}} - 1\right) + \frac{k_R}{k_g}\right] \qquad (7.11)$$

where k is the thermal conductivity of the powder bed, k_g is the as thermal conductivity—usually a function of temperature (Perry and Chilton 1973), k_s is the solid thermal conductivity, and

$$k_R = 1.2\varepsilon_R \sigma D_{par} T_b^3 \qquad (7.12)$$

where ε_R is the material emissivity, σ the Stephen-Boltzmann constant, D_s is the particle diameter, T_b and the powder bed temperature (Perry and Chilton 1973). This model is found to yield better results than the Yagui-Kunni packed bed thermal conductivity model (Yagui and Kunni 1957) utilized in some earlier SLS models.

The estimation of k_s can be problematic while modeling heat transfer associated with polymer-coated particles since such particles usually exhibit a tendency to agglomerate. A model for k_g recognizing particle agglomeration is available in (Balasubramanian and Barlow 1990). The effective solid conductivity of a mixture of uncoated and polymer coated particles can be estimated by the following series resistance model (Vail *et al.* 1996):

$$k_{s,eff} = \phi_{sc}/k_{sc} + (1 - \phi_{sc})k_{su} \qquad (7.13)$$

where ϕ_{sc} is the volume fraction of polymer-coated particles in a hypothetical perfectly solid mixture, and k_{sc} and k_{su} are the thermal conductivities of the coated and uncoated solid particles respectively.

Another complication is that the SLS heat transfer problem becomes nonlinear when the thermal properties of the powder bed are temperature dependent. In such a case, numerical solution of the problem is not feasible unless the temperature dependence can be expressed as known functions. The following empirical functions were identified in (Nelson *et al.* 1993a) for bisphenol A-polycarbonate:

$$C_p(T) = 932 + 2.28T \qquad (7.14)$$

$$k_s(T) = 0.0251 + 0.005T \qquad (7.15)$$

where T is expressed in degrees K, C_p in Jkg^{-1} K^{-1}, and k_s in Wm^{-1}K^{-1}.

7.3.2 Energy Input Sub-model

Figure 7-4 shows the factors needing consideration while modeling energy input to an SLS process. We have already noted in Chapter 5 how radiation distribution, H_r, and laser power, P_L, are related (see equation 5.5). Thus, once the power setting used by the operator is known, one can estimate the radiation intensity distribution at any radial location, r, from the instantaneous center of the laser spot.

Since, the laser beam is moving at speed, V_s, in the scanning direction, the duration, τ, for which any point on the powder bed is irradiated during a single scan is given by

$$\tau = 2R_L / V_s \qquad (7.16)$$

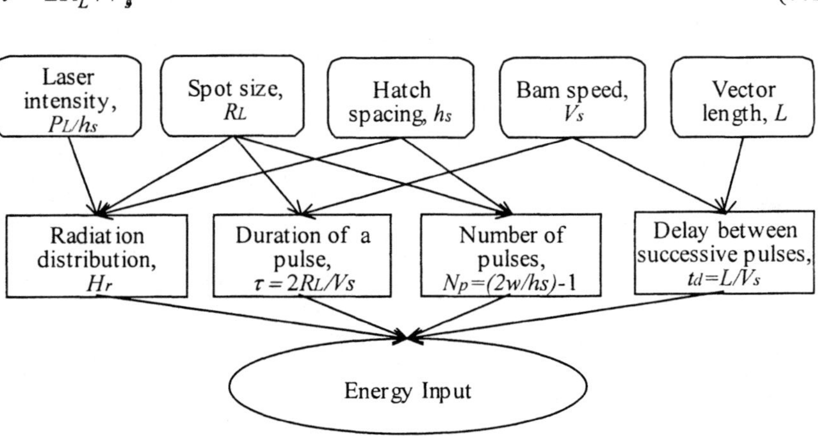

Figure 7-4. Factors affecting energy input in SLS (Williams and Melton 1998).

Usually, the hatch spacing, h_s, is smaller than the spot diameter, $2R_L$ (see Figure 7-5), so there is an overlap between successive raster scans. Clearly, the overlap is equal to $(2R_L\text{-}h_s)$ so that fractional overlap,

$$O = 1 - h_s / R_l \qquad (7.17)$$

Owing to the overlapping, a given point on the powder bed is exposed to multiple scans. The number of effective exposures can therefore be expressed as

$$N_p = \mathrm{int}[(2R_L / h_s) - 1] \qquad\qquad (7.18)$$

Note that each scan lags from the previous scan by duration

$$t_d = L / V_s \qquad\qquad (7.19)$$

where L is the scan vector length. During these lag-periods, the surface cools down owing to convective and radiation heat losses (Figure 7-6). That L has an influence on the sintering process has been experimentally verified in (Subramanian *et al.* 1995, Nelson *et al.* 1993b and Vail 1994).

Studies o n p owder a bsorptance h ave b een c onducted t o d etermine t he process window suitable for sintering without significant evaporation (Bourell *e t a l.* 1 992 a nd Tolochko *e t a l.* 2 000). A laser-based m ethod for measuring absorptance is described in Tolochko *et al.* (2000).

When radiation is incident on any material, proportion, R*, is reflected and the rest, A*=1-R*, A* range from 0 to 0.95 for powdered materials. In general, the absorbed energy depends upon many factors such as the laser wavelength (λ), the nature of materials being sintered, the ambient gas (air or a nonreactive gas), the surface geometry, and the temperature field (Duley 1983). The absorptance of metals and carbide powders is found to decrease with increasing wavelength whereas the opposite is found to occur with oxide and polymer powders (Tolochko *et al.* 2000). Absorptance of metals is higher than that of polymers when λ=1.06μm, where as the reverse happens when λ=10.6μm. Additional powder phases can also be added to promote absorptance of a powder bed. For instance, the addition of graphite powder is found to increase the absorptance of polycarbonate powder under a CO_2 laser beam (Ho *et al.* 2002).

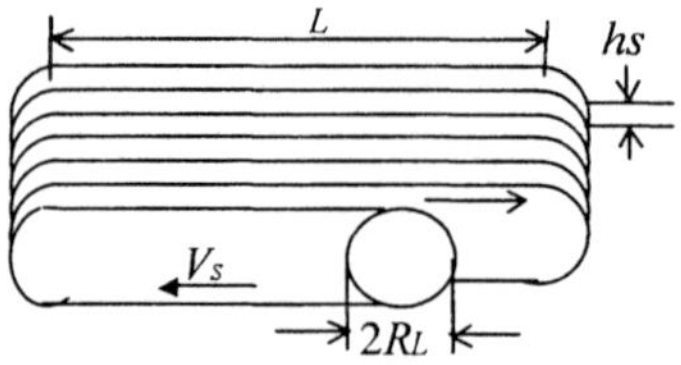

Figure 7-5. Laser scan parameters
(Williams and Melton 1998).

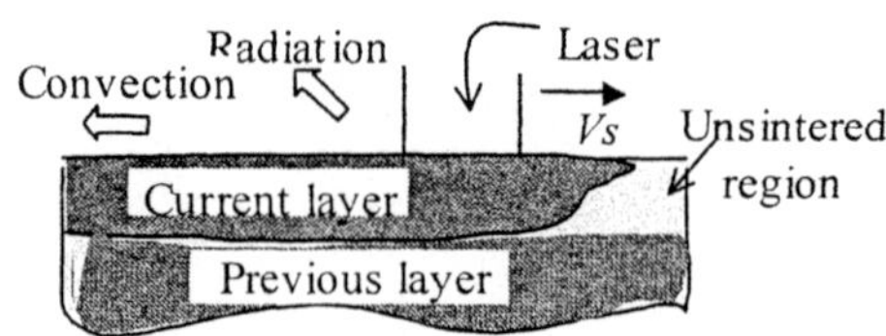

Figure 7-6. Laser material interaction
(Williams and Melton 1998).

In contrast to the situation with solid materials, radiation can pass through the pores in a loose powder, thus penetrating deeper. Absorption by pores approaches that of a black body. Since the pore structure changes radically during sintering, absorptance of a powder region also changes as sintering proceeds. For these reasons, it is more appropriate to use the term 'energy penetration' rather than 'energy absorption' while modeling powders (Wang *et al.* 2002). A simulation of energy penetration based on the combined effects of multiple specular reflections of a bundle of rays of laser radiation is described in (Wang *et al.* 2002). The particles are assumed to be spherical and follow certain size distribution. The amount of energy absorption occurring whenever a ray strikes a particle surface is dictated solely by the material of the particle. Thus, after each reflection, the energy in the ray gets diminished. Each ray is followed until it disappears by escaping from the powder or its energy is reduced to a negligible level. It is observed that transmitted radiation over depths of the order of 0.1-0.5mm in powder beds reduces exponentially with depth. However, it is possible to obtain reasonable results by approximating the distribution to be linear thus resulting in the following empirical function (Childs and Tontowi 2001):

$$\Delta z_p = \frac{a_p \varepsilon}{1 - \varepsilon} \tag{7.20}$$

where Δz_p is the penetration depth, a_p is a suitable adjustment factor, and ε is the porosity of the powder bed.

The heat loss rate is governed by combined convective and radiation heat loss rates and h_r respectively, so that

$$h = h_c + h_r \tag{7.21}$$

The convective coefficient, h_c, may be estimated from the equation given in (McAdams 1954) for heat transfer from heated plates to air:

$$\frac{h_c L}{k_f} = a(\text{GrPr})^b \tag{7.22a}$$

where Gr and Pr are the Grashof and Prandtl numbers respectively,

$$a = 0.54, b = \frac{1}{4} \text{ for } 10^5 < (\text{GrPr}) < 2 \times 10^7$$

$$a = 0.14, b = \frac{1}{3} \text{ for } 2 \times 10^7 < (\text{GrPr}) < 3 \times 10^{10}$$

$$(7.22\text{b, c})$$

The radiation coefficient may be estimated as

$$h_r = \alpha_g \sigma T^4 / ((T - T_\infty)) \tag{7.23}$$

where σ is the Stephen-Boltzmann constant and α_g is the absorptivity. Note that, under the usual gray body approximation, absorptivity is equal to the emissivity, ε_R.

A numerical procedure for the estimation of the laser energy incident on each elemental area on the scanned regions of the powder bed can now be easily implemented in terms of H_r, τ, O, N_p, and t_d. Experimental work has shown that the density and bending strength of the green part increases with increasing energy density until polymer degradation begins to occur (Vail *et al.* 1993, Subramanian *et al.* 1995, Nelson *et al.* 1995 and Balasubramanian and Barlow 1995).

Many of the commercially available SLS machines actively control the temperatures of the powder bed and feed surface so as to keep them just below the softening or melting temperatures of the material. If the temperature is too high, the support structure can fuse and it becomes difficult to break away the part. The temperature window depends on the powder material, e.g., 5-10°C for amorphous polymers (Beaman *et al.* 1997). In practice, the peak temperature can vary considerably between as well within layers depending on the scan lengths. Such temperature variations can be party controlled by indirectly controlling the energy density. When the scan lengths are very short as in small isolated part regions, a set of additional bars are built outside the actual part so as to increase the scan lengths. The bars are removed after part building has been completed.

7.3.3 Heat Transfer Sub-model

The goal of this sub-model is to determine the time dependent temperature distribution, *T*, resulting from the moving heat source resulting from the raster scan of the laser beam. This unsteady state heat transfer behavior can be described by the following classical conduction equation with respect to the Cartesian coordinate system with axis *x* directed along scan velocity, V_s, and axis z directed normal to the powder bed surface (Williams 1998, Bugeda *et al.* 1999):

$$\rho C_p \frac{\partial T}{\partial t} = \nabla(k\nabla T) + g(x,y,z,t) \qquad (7.24)$$

The term on the left represents energy storage, the first term on the right the three-dimensional heat conduction, and the second term on the right the volumetric heat generation (internal heat source). The following boundary conditions are applicable in the case of SLS:

$$-K \frac{\partial T}{\partial z}\bigg|_{z=0} = \varepsilon_0 \sigma(T_{z=0}^4 - T_{sur}^4) + h(T_{z=0} - T_{env}) \qquad (7.25)$$

where ε_0 is emissivity, h is convection coefficient, and T_{sur} is the temperature of the surrounding and T_{env} that of the environment. Further, no heat is expected to be lost at the bottom of the powder bed (of thickness z_p), so

$$-K \frac{\partial T}{\partial z}\bigg|_{z=z_p} = 0 \qquad (7.26)$$

The following initial condition may also be applied to recognize the existence of uniform temperature, $T_{0'}$, throughout the powder bed prior to the SLS process:

$$T(x,y,z,0) = T_{0'} \qquad (7.27)$$

Equations 7.24 to 7.27 may be simultaneously solved through appropriate numerical procedures to yield the time dependent temperature distribution throughout the powder bed.

Depending on the particular SLS application, the SLS heat transfer problem described above can be solved in different ways. For instance, while addressing polymer coated ceramic powders, the beam speed is likely to be quite large and the thermal diffusivity of the powder bed quite low (Vail *et al.* 1996). Hence, the thermal gradients in direction z are likely to be much larger than those in the xy plane. Consequently, it is possible to reduce the problem to its one-dimensional version expressed solely in terms of dimension z without significant error (Vail *et al.* 1996). In particular, it appears that the one-dimensional approximation is applicable when the parameter $(V_s R_L / \alpha_p)$ is greater than 3.9 (Festa *et al.* 1988)—where R_L is the beam radius and is the α_p thermal diffusivity of the powder bed. The value of this parameter is of the order of 1000 when the powder bed is made up of polymer coated ceramic particles (Vail *et al.*1996).

When applied to ABS powder, no significant difference was found between the 3D and 1D solutions (Sun 1992). Further, the second term on the right hand side of equation 7.24 can be negative when the SLS process includes depolymerization. This is because depolymerization, being the opposite of polymerization (which is exothermic), is an endothermic process. Further, the negative heat is quite large at about 13.8 kcal/mol (Brandrup and Immergut 1989). A method for the heat sink term, $G(t,z)$ arising from this endothermic action is described in (Vail *et al.* 1996). However, it is found that, this term is effectively equal to zero during laser heating, so the second term on the right hand side of equation 7.24 can be ignored.

On the other hand, while modeling heat transfer during SLS of a crystalline and a glass-filled crystalline polymer, Childs and Tontowi found it convenient to adopt a two-dimensional model (Childs and Tontowi 2001). In this model, the heat source is idealized as a blade source with Gaussian mean width equal to $2w_0$, length equal to L, mean power (q) equal to $P_L/(2w_0L)$, and the energy s ource m oving i n d irection y a t s peed e qual to $V_s h_s/L$.

7.3.4 Sintering Sub-model and Solution

The sintering sub-model aims to describe the rate of change of porosity, ε, of a powder with temperature. It is difficult to measure sintering kinetics arising due to exposure to a laser beam. However, it is possible to make such measurement in a separate experiment for amorphous glasses and extrapolate the results to laser processing (Beaman *et al.* 1997).

Frenkel's equation (equation 7.3) provides qualitative appreciation of how the main process parameters control sintering rates in amorphous systems. However, the equation is not sufficiently global to provide a quantitative understanding of the behavior of a powder bed.

Bugeda *et al.* have implemented a combination of two different sintering models reported earlier (Bugeda *et al.* 1999). The first of these models was developed by Scherer to arrive at an explicit relationship between porosity, ε, and strain rate, $\partial e/\partial t$ (Scherer 1977a, b, c). This model is applicable when $\rho \leq 0.94\rho_s$ where ρ is the density of the powder bed and ρ_s that of solid particles. The model approximates the powder bed as a network of cubically arranged cylinders. The cylinder is set equal to the particle diameter and the cylinder length proportional to the pore diameter. As sintering proceeds, the cylinder heights collapse so as to reduce the void fraction. The process stops when the cylinder walls touch each other. At this stage, Bugeda *et al.* move over to the explicit relationship reported by Mackenzie and Shuttleworth (Mackenzie and Shuttleworth 1949) that is applicable when $0.94\rho_s < \rho \leq \rho_s$. This model describes the powder bed in terms of contiguous spheres with closed pores. The following are two explicit relationships developed in (Scherer 1977a, Scherer 1977b and Scherer 1977cc) and (Mackenzie and Shuttleworth 1949) respectively:

$$\frac{\partial e}{\partial t} = \frac{(\gamma' N')^{1/3}}{\eta} \frac{(3\pi)^{1/3}}{6} \frac{\left(2 - 3C' A_{cyl}\right)}{\sqrt[3]{A_{cyl}(1 - C' A_{cyl})^2}} \quad \text{for } \rho \leq 0.94\rho_s \qquad (7.28)$$

$$\frac{\partial e}{\partial t} = \frac{(\gamma' N')^{1/3}}{\eta} \frac{1}{2} \frac{(4\pi)^{1/3}}{3} \left(\frac{\rho_s}{\rho} - 1\right)^{2/3} \quad \text{for } 0.94\rho_s < \rho \leq \rho_s \qquad (7.29)$$

where γ' is the surface energy, η the viscosity of the material at the given temperature, A_{cyl} the aspect ratio of the cylinders idealized in by Scherer, N' is the number of particles per unit volume of fully densified material, and C' a constant equal to $8\sqrt{2}/(3\pi)$.

The viscosity of the material can be expressed as an Arhenius-type function of temperature, T:

$$\eta = \eta_0 \exp\left(\frac{\Delta E_\eta}{RT}\right) \qquad (7.30)$$

where η_0 is the viscosity coefficient and ΔE_η is the activation energy.

As sintering proceeds, the void fraction, ε, would decrease. In the Scherer model, ε is a function of A_{cyl} such that

$$\varepsilon = 1 - 3\pi A_{cyl}^2 + 8\sqrt{2} A_{cyl}^3 = 1 - \frac{\rho}{\rho_s} = 1 - \frac{\rho_0}{\rho_s} \exp(-3e) \qquad (7.31)$$

Hence, for a given ρ, we can use equation 7.31 to find the corresponding magnitude of A_{cyl} and substitute this value in equation 7.28 to determine the corresponding free strain rate, $\partial e / \partial t$.

The final solution of the sintering problem is obtained in (Bugeda *et al.* 1999) by applying an iterative finite element routine that couples the effective thermal conductivity (equation 7.10), the heat transfer sub-model (equations 7.24 to 7.27) with the sintering sub-model (equations 7.28 to 7.31). In particular an SLS exercise involving sintering of polycarbonate is considered. The input to the routine consists of the array $\{H_r, R_L, V_s, k_s, k_g,$

initial ρ_R, ΔE_η, η_0, γ'} in addition to the positions of the heat source and the dimensions of the powder bed. The outputs include the distributions of T and ρ_R across the powder bed. It is an easy step to deduce the depth of sintering from the latter.

Procedures analogous to the above have been implemented by several other researchers (Vail *et al.* 1996, Williams 1998, Childs and Tontowi 2001 and Matsumoto *et al.* 2001). The differences mainly stem from the powder materials b eing c onsidered, a nd t he idealizations u sed w hile m odeling t he heat source and the properties of the powder bed. A brief summary of a selection of such variations is given below.

For amorphous polymers, the sintering law is expressible as

$$d\varepsilon / dt = C\varepsilon \tag{7.32}$$

where C is a constant depending on the temperature, powder material, and size (Nelson 1993 and Childs *et al.* 1999). The powder's surface tension, γ, drives the sintering rate whereas viscosity, η, retards it. For spherical powder particles of radius r, the constant can be expressed as

$$C \propto \frac{\gamma}{\eta r} = A_s \exp\left(-\frac{E_s}{RT}\right) \tag{7.33}$$

where A_s is an Arhenius constant, and E_s the sintering activation energy (Frenkel 1945 and Nelson *et al.* 1993b).

However, crystalline polymers do not s inter by viscous f low, hence a different approach is required (Nelson 1993). It is suggested in (Childs and Tontowi 2001) that one can follow the approach used while modeling melt spinning of crystalline polymers (Ziabicki *et al.* 1998), so

$$C = A_s \exp\left(-\frac{E_s}{RT}\right)\exp(-\beta X^{n'}) \tag{7.34}$$

where the first part on the right side represents the viscous sintering of the melt, and the second the retardation caused by the solid or crystalline fraction, X.

It is not trivial to determine the values of A_s and E_s for a given polymer powder. To overcome this problem, it is suggested in (Childs and Tontowi 2001) that these could be determined from the known values of some other polymers and measuring the values of surface tension, viscosity, and size of the polymer of interest as well as of the reference polymer. Following this procedure, it can be assumed that

$$\left[\frac{rA_s \exp\left(-\dfrac{E_s}{RT}\right)\exp\left(+\dfrac{E_\eta}{RT}\right)}{A_s \exp\left(+\dfrac{E_\gamma}{RT}\right)} \right] = \text{constant} \qquad (7.35)$$

where E_η and E_γ are the activation energies for viscosity and surface tension respectively. Some useful viscosity and surface tension data are available in (Nelson *et al.* 1993b) and (Brandrup and Immergut 1989) respectively.

The viscosity, η_{0fil}, of a filled polymer may be related to the viscosity, η_0, of the corresponding unfilled polymer as

$$\eta_{0fil} = (1 + C'V_{fil})\eta_0 \qquad (7.36)$$

where the constant C' is between 2.0 and 2.5, and V_{fil} is the volume fraction of the filler. Following the above procedures, it is found that $n'\approx 1$, and $\beta\approx 10$ for nylon-12 and glass-filled nylon-11 powders (Childs and Tontowi 2001).

Shiomi *et al.* performed a finite element simulation of the meting and solidification processes occurring during direct SLS of metal powders under the influence of a pulsed laser (Shiomi *et al.* 1999). Observations from simulations were verified against data collected from specially constructed experimental setup. They assumed that, when a laser pulse strikes the powder surface, a small volume of the powder in the vicinity of the beam melts and takes up a spherical form owing to surface tension effects. The ball cools and solidifies by the time the next pulse strikes in its vicinity, when the ball and a part of the nearby powder melts and combines, so resulting in a larger molten ball. This melting-combining-solidifying sequence continues until a stable configuration is reached. Note that this approach does not recognize the capillary effects associated with liquid phase sintering that is characteristic of many direct SLS operations performed on metal powders. Notwithstanding this limitation, it is useful to note that the maximum temperature of the powder in a heating/cooling cycle and the corresponding solidified volume is more sensitive to the peak power than to the duration of laser irradiation, so there exists an appropriate peak power for a given metal powder.

Subsequently, the melting-combining-solidifying using a pulsed Nd-YAG was implemented to yield an RP method called selective laser melting (SLM) laser (Abe *et al.* 2001). Tests showed that melt-balls were not formed always. Frequently, linear shaped pools leading to groove appearance on the parts were observed. To overcome this problem, a continuous CO_2 laser was added to the system. The focal points of the pulsed and continuous beams were aligned in the scan direction with an adjustable offset in the range –2 to +2mm. When the offset was negative, the continuous beam was ahead of the pulsed, so the continuous laser preheated the zone that would subsequently be melted by the pulsed laser. In contrast, a positive offset meant reheating. Experiments showed that the hardness and bend strength of the solidified part varied in a systematic manner by about 20%. This observation was helpful in optimizing the beam offset. Tests using aluminum, stainless steel, and nickel-base alloys yielded some encouraging results with the resulting parts or dies occasionally exhibiting distortion and cracking.

Melting as a route to RP has recently started to be pursued on a commercial basis. For instance, F&S GmbH of Germany is commercializing a SLM. Likewise, Arcam AB, Sweden, is developing an approach called Electron Beam Melting (EBM) that uses an electron beam gun to melt steel powders to deliver up to 99.5% dense part (Wohlers and Grimm 2002).

7.4 POST-PROCESSING

Since SLS processes (indirect as well as direct) yield green parts with significant porosity, they need to be post-processed so as to build part strength. Post-processing may involve some or all of the following four steps: binder removal, sintering of the remaining metal or powder, infiltration, and surface finishing.

Removal of binder, if any, should occur before sintering so as to avoid part distortion owing to the gaseous substances released during its thermal dissociation. When heated, the polymer binder can flow out of the green part under the action of gravity combined with wicking. Polymer binders decompose 2-3 times faster in air than in the neutral atmospheres, such as nitrogen, used in SLS systems. Some ceramic powders tolerate oxidation binding better and may not need to have a neutral atmosphere.

Parts to be debinded are usually packed in alumina and then placed in a cold oven. Next, the oven is heated to $200°C$ and held there for 3-4 hours to let the polymer drain into the packing material. Little depolymerization occurs at this temperature. Next, the part is heated at a rate of $100°C$/hr to a higher holding temperature.

A model of the debinding processes occurring in such a part is presented in (Beaman *et al.* 1997). The model starts with several assumptions. Firstly, it assumes that none of the binder flows out though gravity or wicking. Secondly, a nitrogen atmosphere is assumed. Thirdly, the part is assumed to be surrounded by 2cm thick layer of alumina. Likewise, the oven walls surrounding the part are assumed to be made of alumina. The following unsteady state, one-dimensional heat transfer equation system is applicable in this situation:

$$\rho C_p \frac{\partial T}{\partial t} = \frac{\partial}{\partial z}\left(k \frac{\partial T}{\partial z}\right) + G(z,t)\Big|_{(z_{max}-2)>z>2cm}$$

$$\rho_{pac} C_{pac} \frac{\partial T}{\partial z} = \frac{\partial}{\partial z}\left(k_{pac} \frac{\partial T}{\partial z}\right)\Big|_{0<z<2,(z_{max}-2)<z_{max}\ cm} \tag{7.37a, b}$$

subject to the following boundary conditions:

$$\rho_{pac} C_{pac} \frac{\partial T}{\partial z} = -(h_c + h_r)(T - T_o)\Big|_{z=0,\ z_{max}\ (cm),t>0} \tag{7.38}$$

$$k_{pac} \frac{\partial T}{\partial z} = k \frac{\partial T}{\partial z}\Big|_{z=2cm,\ z_{max}-2\ (cm),\ t>0} \tag{7.39}$$

$$\frac{\partial T}{\partial z} = 0\Big|_{z=z_{max}/2,\ t>0} \tag{7.40}$$

where ρ, C_p, and k are the density, specific heat capacity, and thermal conductivity of the powder respectively; ρ_{pac}, C_{pac}, and k_{pac} those of the packing material respectively; G the heat sink due to depolymerization, T_o the temperature of the oven wall; h_c and h_r are parameters representing convention and radiation effects. Calculations have shown that the heat sink term can be ignored in the case of PMMA. The convection parameter can be calculated as (McAdams 1954)

$$h_c = \rho_{nit} C_{nit} V_{gas} (Pr)^{-2/3} (Re)^{-1/2} \tag{7.41}$$

where ρ_{nit} and C_{nit} are the density and specific heat of nitrogen (atmosphere); V_{gas} the gas velocity (typically, 50cm/s); (Pr) the Prandtl number; and (Re) the Reynold number (with the characteristic length set at 6cm).

The radiation parameter can be calculated as (Pitts 1977)

$$h_r = \frac{\sigma(T_o^4 - T^4)}{(T_o - T)\left[\dfrac{A_{pac}}{A_o \varepsilon_o} + \left(\dfrac{1}{\varepsilon_{pac}} - 1\right)\right]} \tag{7.42}$$

where σ is the Stephen-Boltzmann constant; A_{pac} and A_o the surface areas of the packing material and oven wall respectively; and ε_{pac} and ε_o the emissivities of the packing material and oven walls respectively. Calculations have shown that the magnitudes of h_c and h_r are of the order of 0.9 cal/cm^2-hr-$^\circ$C and 9.5-8.0 cal/cm^2-hr-$^\circ$C respectively.

Application of the above model to debinding of PMMA from SiC compact has shown that, for a part of 6mm thickness, the time to 99.8% debinding increases linearly in the range 1-10s with increasing value of $(1/T)$. Debinding time is also sensitive to part and bed thickness values. Relative density of a debinded part is usually between 40 and 80%.

During the sintering stage, microscopic welds form at particle contacts. The necks so formed lead to atomic diffusion even in the absence of isostatic pressing. Until the neck structure has formed, the parts are too weak to be handled. However, there remains sufficient porosity that could be filled through subsequent infiltration. If the relative density exceeds 90%, interconnections between pores becomes unstable, so the pores become isolated and mass transport mechanisms such as evaporation, condensation and surface diffusion stop. Further densification leads to spherodization of isolated pores with the latter acting as pinning points for the boundaries of previously existing particles. Even further densification leads to shrinking of the spherical pores, so they are unable to pin the particles. However, since diffusive processes have been stopped, densification becomes even slower.

In the early stages of sintering, there is relatively little shrinkage. Hence it is desirable to aim at modest sintering. However, the disadvantages of full densification can be overcome by external pressure resulting in external tractions. Parts may be subjected to hot isostatic pressing (HIP) where the part is encapsulated in a metal foil or glass to close the pores and placed in a high pressure gaseous inert gas environment. Alternatively, one may use the Ceracon Process ('CERAmic CONsolidation') where the part is unidirectionally pressed (Ferguson and Smith 1984). For parts with simple shapes, unidirectional pressure suffices.

Infiltration can be done from the top or the side. If done from the top, infiltration should be done in a separate furnace cycle where the lower melting infiltrating medium is placed on the top of the part. If done from the side, the debinding, sintering, and infiltration steps can be performed in the same furnace cycle. The infiltration front penetration distance, r_{fr}, increases with time, t, as

$$r_{fr} = \left[(t\gamma_{LM})/(\eta_l d_s) \right]^{n_i} \qquad\qquad (7.43)$$

where γ_{LM} is the liquid-metal surface energy, η_l the liquid viscosity, d_s the solid particle size, and n_s a constant of the order of 0.5 (Bunnell *et al.* 1995).

A customer survey conducted in 1995 indicated that there are as many ways of finishing SLS parts as there are customers (Anon. 1995b). The reason for this variety is the fact that the range of materials that can be subjected to SLS is much wider than that for other processes such as SLS. Bead blasting is used to remove excess powders embedded in part surfaces. Abrasive tumbling is used to improve finish on all accessible surfaces. However, these two processes can damage small features such as sharp corners on the part. Pores and stair steps can be covered by means of surface coating. Hand sanding is used to remove excess material as well as to obtain improved finish. More recently, robotic finishing has been attempted (Shi and Gibson 2000). Robot end effecter paths are derived from the CAD model of the part after suitable compensation for shrinkage, etc. Different finishing strategies can be applied to different parts and different part areas. The process is capable of eliminating the stair-stepping effect. However, it is not applicable to inaccessible surfaces such as those of internal cavities.

7.5 PROCESS ACCURACY

Parts inaccuracy in SLS can arise from several sources: stairstepping, quantization, galvanometer errors, temperature variations within the powder bed, trapped volumes, and shrinkage/curling. However, error sources such as those resulting from external supports are not pertinent in the case of SLS. Issues and solutions related to galvanometer errors, stairstepping and quantization are similar to those already discussed with respect SLA. Hence, they will not be discussed here.

An overall appreciation of the accuracy of parts producible on a given machine can be obtained by fabricating and inspecting 'benchmark' parts. Specifications of a benchmark part for assessing SLS process accuracy while producing parts made up of nylon, glass filled nylon, polycarbonate, Protoform composite, and acrylic based TrueForm PM materials and the corresponding accuracy testing protocol are available in (Beaman *et al.* 1997). The part consists a core and cavity, and has 64 dimensions ranging in size from 1.27 t 152.4mm. Tests have led to several useful conclusions. Machines equipped with M3 galvanometers exhibit 30-40% smaller dimensional errors than those with G325 galvanometers. Nylon and Protoform arts exhibit similar levels of inaccuracy. With the exception of polycarbonate, there exists a strong correlation between shrinkage and root mean square error in dimensions. The correlation is much stronger when 'first time build' results rather than 'corrected' results are compared. These results highlight the importance of controlling shrinkage.

Temperature control may also be important during cooling. In some cases, the cooling time can reach several hours during which significant stresses and 'post-build warpage' can develop. For instance, regions near the piston or sidewalls are usually cooler. One way of homogenizing temperature during cooling is to adopt the 'downdraft' technique involving gas flow via the pores in the bed. Another method, successfully used while building nylon parts, is to control the piston temperature (Forderbase *et al.* 1995). This method is found to be capable of significantly reducing warpage in r egions w ithin a f ew c entimeters o f t he p iston. H owever, o wing t o t he insulating properties of the powders, this method has little influence on warpage in regions farther from the piston.

Part orientation can be of importance when the part has openings. A trapped volume is created during part building if the part is so oriented as to have the opening facing downwards. This results in uneven build up of heat in side the trapped volume, hence leading to uneven temperature distribution and warping (Suh and Wozny 1995).

Most materials shrink during the SLS process. Further, it is possible for fusing outside if the laser spot is of finite diameter (0.4mm). It is therefore necessary to compensate .STL file surfaces for shrinkage and offset. Fortunately, most systems have software procedures to effect such compensations. However, the determination of the parameters to be used during software control will usually require some trial and error. It is possible to reduce the trial and error significantly by performing some systematic experiments on a suitable benchmark part. Specifications of a benchmark part with dimensions ranging from 5 to 178mm are available in (Nelson *et al.* 1995). The part is built twice, once with part axis orientation along the x-axis of the machine, and once along the y-axis. No shrinkage or offset compensation is applied. Experiments have shown that the dimensional errors decrease linearly with increasing nominal dimension. Hence, one can obtain fairly reliable intercept and slope value following linear regression. The intercept indicates the shrink value, whereas the slope indicates the offset value. Most SLS machines require a scale factor to be set up. In such cases, the scale value is calculated as $1/(1 - \text{shrink value})$.

However, often, the shrinkage/offset compensation problem is not as easy to solve as described above because of nonuniform shrinkage. Shrinkage of a new layer is constrained by the substrate. The substrate characteristics can be different at different levels. Layer positioning accuracy can vary with z-depths owing to increasing weight of layers above as new layers are added (Lee *et al.* 1995). Modeling combined with experimental measurement has shown that this error increases in a parabolic fashion with increasing depth, i.e., the error is highest in the middle of powder bed thickness. This effect can be significant (up to 0.5mm) when, as is usual in SLS, the powder is not pre-compacted.

More recently, Katz and Smith presented an SLS process model applicable to simple parts made up of a few layers (Katz and Smith 2001). In particular, the net thermal shrinkage in volume percentage is written as

$$v = [1 - \alpha(\bar{T} - T_0)]^3 \qquad (7.44)$$

where α is the thermal expansion coefficient of the material, $\bar{T}$ the average working temperature, T_0 and the initial temperature. The initial temperature could be a result of preheating. However, the more recently built layers (the upper layers) experience greater thermal shrinkage, thus leading to a upward curl in the upper few layers.

The resulting layer curl radius, R_{cn}, on the nth layer can be expressed as

$$R_c = [1 - \alpha(T_n - T_0)]/[\alpha(dT/dz)_{sur}] \qquad (7.45)$$

where T_n and $(dT/dz)_n$ are the temperature and vertical temperature gradient respectively at the nth layer. The temperature gradient will depend upon the layer build interval, Δt, and the surface temperature. This relation can be determined empirically. Assuming that the previous layer (layer number n-1) cools exponentially, we can write the temperature gradient at he surface as

$$(dT/dz)_{sur} = \frac{(T_n - T_0)[1 - \exp(-\text{const}.\Delta t)]}{L_p} \qquad (7.47)$$

Chapter 8

OTHER RP SYSTEMS

The previous two chapters have described two very widely used laser-based RP processes: stereolithography (SL), and selective laser sintering (SLS). This chapter moves on to other processes of major interest to RP practitioners as well as researchers. Lasers play a critical role in some of the processes to be described. These are discussed first. The rest of the chapter is devoted to RP processes not necessarily dependent on the use of a laser.

8.1 SELECTIVE LASER CLADDING (SLC)

Laser cladding was originally developed for surface treating parts with a view to improving their wear or corrosion properties. The process aimed to clad a new material on to part surface with minimum dilution of the substrate (Steen 1997). Usually, a single layer was clad with overlapping tracks (Soares and Perez-Amor 1987). However, our interest here is in the laser processes that can selectively clad multiple layers {SLC) to create parts rapidly. This process also known as direct laser deposition (DLD), directly produces metallurgically sound components with properties equaling or, even, exceeding those of the wrought material (Lewis and Schlienger 2000).

Laser engineered net shaping (LENS^{TM}) is an SLC process capable of producing metal parts (Giffith *et al.* 1996). The technology was developed in 1993 through cooperative research between Sandia and United Technologies Pratt & Whitney (UTPW). The system uses a high power Nd-YAG laser to melt metal powder in a deposition head (Figure 8-1). Laser radiation travels through the center of the head. One or more lenses focus the laser to a small spot. The head design is such that metal powders are delivered and distributed around the circumference of the head by gravity, or a pressurized gas. A v ariety o f m aterials i ncluding a luminum, c opper, I nconel, stainless steel, tool steel, and titanium (a reactive metal), plastic coated or otherwise, can be continuously fed to produce linear beads of material with a designated overlap. An inert shroud gas such as argon, helium, or nitrogen shields the melt from atmospheric oxygen. This feature also improves layer-to-layer adhesion by providing better surface wetting. The part is built on an XY table placed below the deposition so that the laser/powder head can move on it in a raster fashion to selectively deposit the melt. After a layer is built, the head is moved up by the desired layer thickness. Files controlling machine motions are written in stereolithography format (.STL). While fabricating features or assembly components requiring deposition in planar layers at angled orientations to the parent or base feature, the laser beam-powder delivery head can be tilted so that the beam axis is normal to the deposition plane. This facility is useful while producing overhanging features without building underlying support structure typically required in other plastic rapid prototyping processes.

An attractive feature of LENS is t hat the powder composition can be changed dynamically and continuously, so it is possible to build multi-material and functionally graded parts (Griffith *et al.* 1996 and Griffith *et al.* 1997). Further, the parts can be fabricated to full density at reasonable speeds while ensuring good grain structure and other metallurgical properties. The parts can be near net shape. However, in general, finish machining is required. They have good grain structure, and have properties similar to, or even better than the intrinsic materials. In comparison to SLS—the other RP process capable of producing metal parts directly—LENS has fewer material limitations while not requiring secondary firing operations. Further, material can be selectively applied to existing metal parts without compromising the integrity of the parent material. Owing to fast localized cooling of the molten pool, the process is capable of fabricating thin walled parts with holes as small as 0.35mm in diameter and depth-to-diameter aspect ratios of more than 70 to 1. In contrast, hole aspect ratios of no greater than 10:1 are achievable using conventional CNC machining. These features make LENS an attractive choice for the direct fabrication of injection molds. Recently, the use of powder as a support material has been explored (Schlienger *et al.* 1998)

A technology very similar to LENS is Direct Light Fabrication (DLF) offered by Los Alamos National Laboratory, Los Alamos, New Mexico (Thoma *et al.* 1996). The main difference is that DLF permits the use of up to four different powder compositions. This is made possible by separating the laser head from the powder feed head. The latter is tilted towards the laser's focal region. Different laser power settings can be used while building the same part to suit different powder compositions. Laser cladding is typically performed in argon or other inert gas with oxygen impurity below 5 ppm. The part is placed on a rotary table so that the beam axis is normal to the deposition plane while fabricating features or assembly components r equiring d eposition i n p lanar l ayers a t angled o rientations t o the parent or base feature. Motion control of the five axes is performed by means of files formatted as in conventional CNC machines. As with LENS, functionally graded parts can be produced (Lewis *et al.* 1997 and Lewis and Schlienger 2000).

Another technology similar to LENS is Controlled Metal Buildup (CMB) developed in 1992 at the Fraunhofer Institute for Production Technology (IPT) (Klocke *et al.* 1996). The main difference is that a 2-1/2D milling cutter placed within the work chamber is available to finish the walls of metal parts. CMB is capable of direct fabrication of metal prototypes and prototype tools.

There have been a few other attempts to develop hybrid technologies by combining selective laser cladding (SLC) with conventional CNC milling. One version, similar to the LENS system in many respects, uses a laser/powder feed head with an integral water cooling jacket supplied by two or three independent DC motor controlled hoppers (Jeng and Lin 2001). The technology has been used to produce injection molds from a mixture of Fe, Ni, and Cr in the weight ratio 72:10:18. The process is found to be applicable to the direct fabrication of new injection molds as well as repairing existing molds.

SLC allows complex shapes to be manufactured from a variety of metal powders including titanium alloys, Inconel and H13 tool steel. The powders are produced by different atomization techniques: gas, water, centrifugal, vacuum and ultrasonic. The most commonly used materials are gas-atomized, spherical powders. However, more recently, it has been shown that, despite their lower costs, water-atomized irregular powders exhibit superior deposition quality in terms of microstructures, inter-layer bonding, deposition uniformity, and surface finish (Pinkerton and Li 2003). The main reasons for the superior performance lie in differences in particle shapes and oxygen content.

8.2 LAMINATED OBJECT MANUFACTURING (LOM)

LOM was originally developed by Helisys Corporation in 1985, but the first machines were shipped much later in 1991. In this technology, objects are built by gluing profiled pieces of paper, plastic, or another web material in a layered manner (Feygin and Hsieh 1991). The paper used has one of its surfaces coated with a plastic coating that melts when heated. While the original system had used Kraft paper, other materials were introduced progressively, e.g., polymeric material (Pak 1997), polymer matrix composite (Klosterman *et al.* 1997b), and composites for tooling (Ogg 1998).

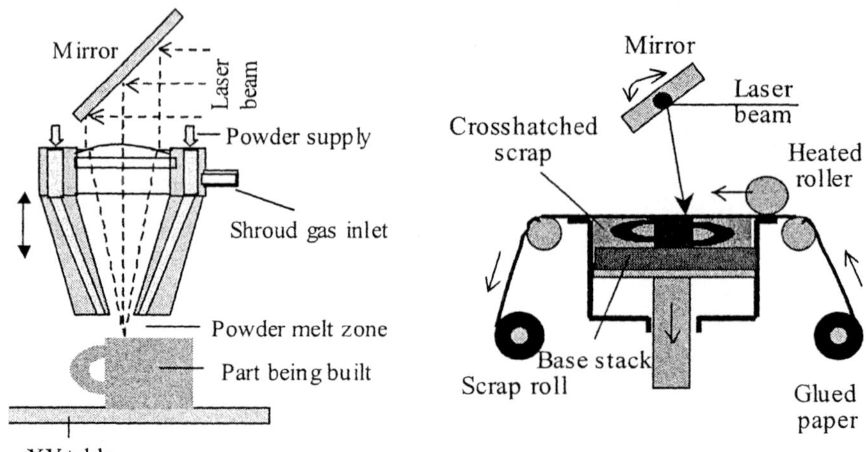

Figure 8-1. The LENS process.   *Figure 8-2.* The LOM process
                                        (Helysis Corp.).

Part building sequence begins with the lamination of 20 layers of paper to the retractable platform (Figure 8-2). The paper is unwound from a paper supply roll such that the coated surface is at the bottom when the paper is placed in the workplace. After the paper segment for a layer has been so fixed, a heated the glue. Then a CO_2 laser beam scans over the surface to cut out the outline of that layer as dictated by the desired part geometry. The laser intensity is set at just the level needed to cut through a single layer of material. Next, the rest of the paper is crosshatched heavily (diced or 'cubed') to make it easier to break later. Once a layer has been processed thus, the base plate moves down, and the whole process is repeated till the complete part has been built. The sheet of material is made significantly wider than the base plate, so that a neat rectangular hole is left behind when the base plate moves down (Chartoff *et al.* 1996). The scrap material is wound onto a take-up roll, pulling a new section across the base plate as it goes. At the end of the build process, the crosshatched columns are broken away to free the object. Objects made from paper have the look and feel of wood and can be finished just as wood. However, the finish, accuracy and stability of paper objects are inferior to objects made of other materials using other RP methods.

Variations of the LOM process have been developed by several companies and research groups. Of these, Paper Lamination Technology

(PLT) of Kira Corp., Japan, is particularly noteworthy (Inui 2000). The main differences are that a knife is used instead of a laser to cut each layer, and the adhesive is applied to bond layers using the xerographic process. The computerized blade can cut fine detail and can last for up to 10,000 sheets of paper. The use of a knife entirely eliminates the combustion issues associated with laser cutting. Part building on a PLT machine starts with the automated placement of new paper layer on the block. A hot plate then descends to press the paper block. Next, the hot plate withdraws and a computerized knife cuts the layer to shape in accordance to the .STL file of the part. The three steps are repeated until the part is removed. Finally, waste paper is removed and the part is extracted. It is claimed that the high lamination pressure used makes the model 25% harder than wood and that the process induces minimal strain and curling. On the other hand, independent benchmarking tests have shown that breaking the part out can be very time consuming and parts can get glued back together and break (Hopkinson and Hague 2001). There are also variations of the LOM process that achieve higher speeds and greater material versatility by cutting the edges of thick layers diagonally to avoid stair stepping.

LOM technology can be used to produce full sized inexpensive visualization models, shell models for styling, and large architectural models. It is highly suited for thick-walled applications like patterns for sand and investment casting, and cavity molds used in injection molding (Mueller and Kochan 1999 and Wang *et al.* 1999).

Whatever the application, several problems with the LOM process need to be recognized. Among the major problems are delamination, significant dimensional error in the z-direction, and the labor-intensive nature of the waste removal process. Other problems include wiggling of the laser beam due to smoke particle build up on the laser's tracking mechanism, mirror, and lens(es).

Park *et al.* (2000) conducted differential scanning calorimetry tests on LPH 042 paper (0.107mm thickness) with the expected bond strength of the adhesive equal to 5MPa. The experiments showed the melting temperature of the adhesive and the degradation temperature of the paper to be 65.4°C and 270°C respectively. The weight of the paper decreased by about 5% owing to moisture loss during after processing. This was noted to be the main reason for the z-error of up to 2.5%. The key parameters affecting dimensional accuracy were crosshatch size, left and right heater roller margins, heater speed and temperature, and laser power and speed.

Sonmez and Hahn conducted a quasi-steady state finite element-based thermomechanical analysis of the LOM process (Sonmez and Hahn 1998). The analysis assumed Coulomb friction between the roller and the laminate, and recognized the insulating effect of the air gaps between laminates and the temperature dependency of material properties. Contact length between the roller and the laminate were calculated through stress analysis. Several useful conclusions emerged from this work. It is important to ensure that the pressure experienced by an element of the laminate through contact with the heated roller is maintained long enough to reach the glass transition temperature of the bond. This may be achieved by using a larger roller or a secondary roller. Thinner layers experience higher stresses. Uniform pressure between laminates may be achieved by controlling roller indentation pressure. There is an upper limit to roller speed since higher roller speeds imply smaller contact time. Interestingly, notwithstanding the several attractive features of LOM, Helisys Corp. faced the problem of changing market demand in favor of functional parts. The company did try to respond by changing its technological and marketing focus, but it was too late (Wohlers and Grimm 2001). As result, in 2000, the company stopped manufacturing LOM equipment. However, market interest in LOM continued because several other LOM equipment manufacturers had already emerged—through licensing relevant technology from Helisys, or otherwise. Further LOM development led to applications in durable models for assembly and functional testing, and solid models for fit verification and packaging studies. LOM applications have also emerged in thermal forming and prototype stamping.

LOM machines usually provide default values for crosshatching layer regions to be discarded. The laser beam scans the regions in two perpendicular directions (x and y) with penetration depth approximately equal to the layer thickness. Typically, crosshatch spacing is kept uniform in both directions and across all 'waste' regions, so the waste regions get 'cubed' in effect. Thus, at the end of a part-building cycle, the part would be embedded within one or several piles of 'cubed' regions, so the part needs to be extracted out of the cubed pile(s) by breaking out the cubes. However, as it turns out often, the scrap material is not only as strong as the part itself, but also adheres to the part. Therefore, the clean up needs to be handled with care using special tools (dental tools usually come in handy) such that only the waste material is removed and delicate sections of the part are not fractured. All these considerations usually make the waste removal process a difficult, time-consuming and labor-intensive task (Cohen 1991 and Klosterman *et al.* 1997a, b). For instance, in one case study it was found that the time for part contour machining was only 25% of the total cycle time whereas the proportions of cycle time for crosshatch machining and decubing were 35% and 40% respectively (Liao and Chiu 2001b). An algorithmic method for maximizing the proportion of cutting time spent on part contour machining is presented in (Karunakaran *et al.* 2001).

Clearly, decubing time can be reduced by choosing a smaller crosshatch spacing. However, this would correspondingly increase crosshatch machining time. One way of solving this problem is to adopt a fine/medium/coarse hatching scheme depending upon the x-y of the sliced part only (Feygin and Pak 1997 and Feygin and Pak 1999). A further variation of this adaptive approach takes into account variation of part geometry in the z-direction too. The latter approach involves creating an offset region around each part region in a layer (Figure 8-3) with the offset value being determined by a set of heuristics tat takes into account part geometry characteristics in the x-, y-, as well as z-directions (Liao and Chiu 2001b). The offset regions are then fine-hatched while the outer regions of waste material are coarse-hatched. It has been demonstrated that this adaptive crosshatching approach is capable of reducing the total cycle time significantly (in one case study, the reduction was around 45%) while significantly reducing laser power consumption as well as the possibility of part damage.

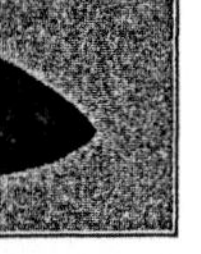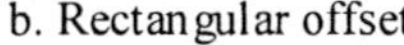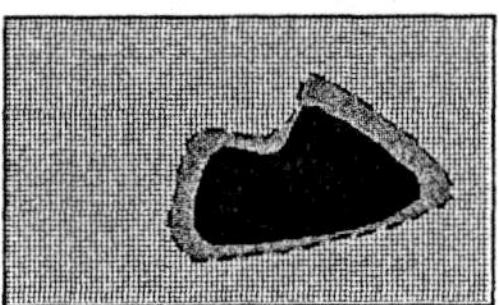

a. Conventional   b. Rectangular offset   c. Contour offset

Figure 8-3. Crosshatching strategies in LOM (Liao and Chiu 2001b).

Several variants have been proposed for eliminating the time-consuming and tedious decubing process associated with conventional LOM. One of the better known variant is 'computer-aided manufacturing of laminated engineering materials (CAM-LEM) developed by Newman in collaboration with CAM-LEM Inc. in collaboration with CAM-LEM Inc. CAM-LEM applies cut-then-stack strategy which is exactly the opposite of the stack-then-cut strategy adopted in LOM (Newman *et al.* 1996). After the laser has cut out the required layer outline on the build material (ceramic was used), a separate process creates a mask that is punctured with a hole pattern bounded by an outline identical to the part outline. Next, a robotic vacuum gripper lifts up the mask and positions itself over the sliced layer of material, so a gripper can lift the material by suction through the punctured mask. The robot then releases the material over the partially built stack. Once all the layers have been completed, the assembly is subjected to warm isostatic pressing so as to achieve intimate interlayer contact. Finally, the laminated 'green' object is fired to create a monolithic structure by fusing the layers and particles within the layers. Problems with this process include the possibility of sucking up some waste material when the part has thin cross-sections, and the possibility of sagging on some layers owing to the absence waste material within the stack as in the case of LOM.

Another interesting LOM variant using the cut-then-stack approach is the Genie fabricator produced by Ennex Fabrication Technologies (Anon. 1996c). This process uses a vinyl sheet with a paper backing. A knife is used cut out layer outlines on the vinyl part while leaving the backing in tact. The sheet is next picked up by a rolling drum, transferred to the part-building platform, and rolled out on top of the stack with the paper-backing side up. Next, the drum rolls back across the part, peeling away the backing while leaving the sticky side of the current layer ready for the next layer to be laid down. This approach avoids leaving the waste material within the stack to be decubed l ater as in L OM However, a s w ith C AM-LEM, t his p rocess also suffers from the problem of lack of supporting material.

Another interesting sheet deposition system is that described by Cho *et al.* from South Korea (Cho *et al.* 2000). This process starts with the cutting out of the perimeter of the void area with the aid of a knife. Next, the sheet with holes is fed to the object being constructed. In the process, the shielding paper gets peeled off along with the void area. Next, the platform is lifted so the new layer contacts the previously built stack on its adhesive side. A roller then sweeps over the new layer to bond it to the stack. A second cutting operation is then performed to cut the remaining edges. Finally, the platform is lowered and the entire cycle repeated till the part has been fully built. The advantage of the process is that no post-processing for waste removal is needed. The disadvantage is that support structures need to be built along with the part.

There have been continual efforts to extend the application of the LOM process to the fabrication of more durable and functional objects. For instance, Klosterman *et al.* have explored the manufacture of structural polymer matrix composites (PMC) by taking advantage of LOM's inherent capability to handle flat sheet materials (Klosterman *et al.* 1997b). In particular, objects are built from two commercially available thermosetting preimpregnated fiber performs (prepegs): chopped glass fiber/epoxy mats with a high degree of pre-cure ('B-staged'), and continuous fiber/epoxy sheets (52-55% unidirectional glass fiber). Both unidirectional and 0/9° layups have been tested. Since the continuous fiber prepeg is not supplied as a continuous roll, the sheets need to be cut into pieces of appropriate size. The backing paper is removed from each piece before placing it on the LOM machine. Next, the edges of the piece are taped to the platform so as to maintain the tension during lamination. Further, a Teflon release ply is placed over the layer to avoid sticking to the roller. Experiments on the chopped fiber mats specimens have shown the degree of B-staging to be of particular importance. Too little B-staging leads to excessive resin flow during lamination. Too much B-staging results in weak lamination. Interestingly, no curing advancement is observed during lamination although the roller exhibited a temperature of 240°C. Thermocouple measurements have indicated this to be due to the fact that the temperature of the stack had not exceeded 60°C although a momentary temperature spike of 160-180°C was experienced by the top few layers of the stack. These observations point to the advisability of a separate pressure cycle being carried after completing the LOM process, so the PMCs could be fully cured and consolidated. Parts prepared following this strategy have been found to be comparable to those achievable by traditional processes.

Rodrigues *et al.* have used a modified LOM machine to fabricate functional parts from tapes of green silicon nitride ceramic tapes (Rodrigues 2000). The tapes a re f abricated u sing Si_3N_4 p owder along w ith p olymeric additives and plasticizers. After part building, the binder is burnt to remove organic materials. The burnt part is then either sintered to higher density in a conventional furnace, or infiltrated with a preceramic polymer. The strength and fracture toughness values of the parts thus produced have been found to be in the ranges 700-900MPa and 5.5-7.5 $MPa.m^{1/2}$ respectively. These values are comparable to those attainable by conventional ceramic processing.

More recently, Zhang *et al.* have explored the fabrication of Al2O3 parts by LOM Harbin Institute of Technology, Harbin, PRC (Zhang *et al.* 2001). The input material consists of coated tapes of Al_2O3 formed by a roll-forming process. The organic vehicle is polyvynil butyral (PVB). Since the tapes are not stiff enough, they need to be placed manually on the machine platform. Tape segments are combined using polyvynil acetate paste. Relative densities of the green parts have been found to be of the order of 65%. Post-build binder removal parameters have determined through a heat analysis. During debinding, the organic vehicle gets oxidized and disassociated, so the weight of the part decreases. Scanning electron microscopy (SEM) has revealed mass transportation of ceramic grains, thus suggesting the presence of sintering effects. This effect increases part relative density to around 97%. However, a side effect is part shrinkage, particularly in the z-direction. Part accuracy therefore depends on laminate thickness. However, rates of heating and cooling need to be controlled particularly on large parts to prevent cracks from appearing. Further, the mechanical strength values of parts so finished are inferior to those attainable with similar material by traditional processes. Overall, it appears that s mall b atch p roduction o f A l2O3 p arts b y L OM i s feasible especially when one recognizes the low cost nature of LOM.

8.3 FUSED DEPOSITION MODELING (FDM)

Fused deposition modeling is perhaps the second most widely used rapid prototyping technology, after stereolithography. The technology was first developed in 1988, patented in 1992, and commercially offered as FDM^{TM} by Stratasys, Inc. in the same year. The technology is capable of producing, strong, durable, and fully functional prototypes. For instance, Adda Rapid Product Development Service in Hong Kong created parts for a prototype tricycle by means of FDM. Figure 8-4 shows the principle of FDM as implemented on the machines offered by Stratasys Inc. (Stratasys 1991 and Stratasys 1997). A linear or branched thermoplastic plastic filament is unwound from a coil under the action of a pair of drive wheels and supplied via a resistively heated zone (liquefier) to the extrusion nozzle. The original filament materials were polyolefin and polyamide (Comb *et al.* 1994). More recent materials are ABS and a modified, high-impact ABS material (Stratasys 1997). The material is heated to slightly above its flow point, so it solidifies quickly. The incoming cold filament acts as a piston and pushes the liquefied material forward towards the FDM tip. The tip is a nozzle with a mechanism to allow the flow of the melted plastic to be turned on and off. The nozzle is mounted on a mechanical stage that can be moved in both horizontal and vertical directions. The location of the nozzle over the work area is controlled in accordance with the .STL file of the part supplied to the machine.

The method is office-friendly and quiet. It I also fairly fast for small parts of the order of a few cubic inches, or for those that have tall, thin form-factors. However, it can be very slow for parts with wide cross-sections. The finish of parts produced with the method have been greatly improved over the years, but may not be as good as those built by means of stereolithography. The closest competitor to the FDM process is probably 3D printing. However, FDM offers greater strength and a wider range of materials.

The nozzle squeezes out the fused filament in a manner analogous to a toothpaste tube and deposits the material at the desired location on the partially constructed part. Polymer flow through the FDM tip is not by simple shear (as in SL and SLS) but a combination of shear and elongation, so it becomes necessary to recognize the viscoelastic nature of the polymer (Tadmore and Gogos 1979). In particular, the resistance to stretching needs to be modeled by using the notion of 'elongational viscosity' (Bird *et al.* 1977). Further, part properties can be affected by the flow orientation of filament placement during layer buildup (raster direction) relative to the part loading direction (Fodran *et al.* 1996 and Bertoldi *et al.* 1998). It has been found that the tensile strengths of FDM-built specimens aligned diagonal to the testing direction are generally lower (by about 18%) than those obtained by aligned specimens, thus suggesting that interbead bonding may be a limitation. Further the actual strength values may fall significantly short of the values claimed by the manufacturers.

Since the entire system is contained within a chamber held at a temperature slightly above its flow point but below the melting point, the plastic hardens immediately upon being squirted and bonds to the layer below. As a result, it is possible to form short overhanging features. However, in general, explicit supports need to be planned and included in the .STL file as in the case of SL. In particular, down-facing surfaces at an angle smaller than 45° from the horizontal plane need to be explicitly supported. Three support generation strategies embodied within the QuickSlice software of Stratasys are noteworthy (Chalasani *et al.* 1995). In the first strategy, called containment, the shadow of the part with respect to the build direction is computed by merging the contours of all layers such that a minimal containing outline is produced. Next, the outline is expanded slightly to contain the entire part. Thus there are supports both below (where some of these might be needed) and above (where they are not needed) the part. The second strategy, called regional support, allows the user to interactively specify regions requiring support. The third strategy generates the shadow of the current layer over the previous layer and applies a set of heuristics in the context of the given part geometry.

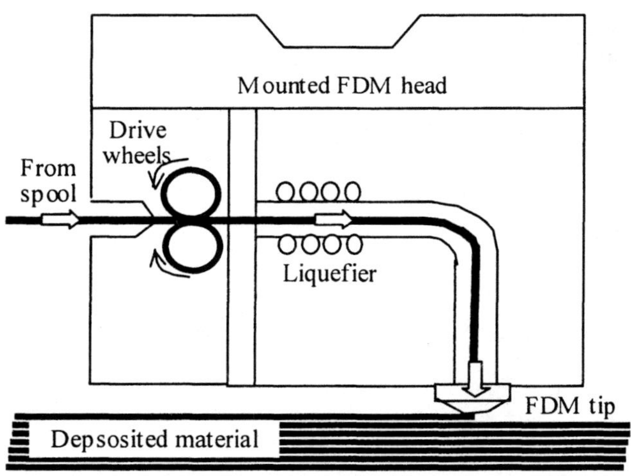

Figure 8-4. The principle of FDM (Stratasys 1997).

Once built, the supports need to be removed after part building by breaking them away from the object. In a newer Stratasys machine, two modeling materials can be dispensed through a dual tip mechanism (Gasdaska *et al.* 1998). More recently, a system called FDMC with four independent filament feed mechanisms has been developed at Rutgers, the Sate University of New Jersey, NJ, for the purpose of achieving fused deposition of multiple ceramics (Jafari *et al.* 2000). Subsequently, the system has led to the fused deposition of multiple materials (FDMM) (Pilleux *et al.* 2002). When a dual or multiple head FDM machine is available, the supports can be built from a softer material, so they can be broken away easily. Alternatively, they can be built from a water-soluble material, so that they can be simply wished away after part building has been completed.

Material shrinkage associated with temperature and phase changes can mar the accuracy of FDM parts. However, the resulting warpage and curl are somewhat attenuated by the ability of polymers to stress relax. To reduce the warpage, Stratasys deposits the material onto a removable foam base that is firmly anchored to the bed of the machine. The foam lowers the cooling rate, thus allowing the viscoelastic polymer to relax out some of the stresses generated during crystallization. Further control of the cooling rate is achieved by controlling the surrounding air temperature.

FDM p arts c an exhibit internal d efects o f several forms: v oids, pores, delaminations, cracks, etc. These can arise from several reasons, e.g., FDM system hardware and software limitations, poor material characteristics, and poor process optimization.

The main process variables that an FDM process planner needs to consider are build orientation, layer thickness, road width, air-gap, build temperature, and interior fill strategy (Ziemian and Crawn III 2001 and Ahn *et al.* 2002). Build orientation decisions need to be made within the CAD environment. The important considerations are part z-height, and the angular orientation of each facet (tessellation) with respect to the xy plane (table plane). As in the case of SL and SLS, different part orientations are evaluated for their impact on total build time including data preparation time (including support generation time), material deposition time, and processing times.

Layer thickness, L_p, refers to the travel in the z-direction (vertical direction) of the table. Usually, different layer thickness values are associated w ith d ifferent F DM t ips (e.g., t ips T10, T12 a nd T16 o n F DM 2000 machine of Stratasys). As in any layered manufacturing process, the stair-stepping effect increases with i ncreasing L_p. Owing to stair-stepping, the volume of the actual part will be smaller than the volume of the part as specified i n the CAD model (volumetric error). Since a larger volumetric error implies larger dimensional errors, it is important to control the former. A method for calculating the volumetric error as a function of part orientation is described in (Masood *et al.* 2000 and Masood and Rattanawong 2002). Using this approach, the optimum orientation for minimum error can be determined. Taguchi experiments (a popular design of experiments approach) on FDM have shown layer thickness to be the most important build parameter affecting surface roughness (Anitha *et al.* 2001).

In contrast to SL and SLS where L_p is controlled by the z-movement of the worktable (under computer control), layer thickness in FDM is determined by the orifice diameter in the FDM head. In conventional FDM, this diameter is constant for a given head. This implies that one cannot combat the stairstepping problem through adaptive slicing (Suh and Wozny 1994) when a conventional FDM head is used. A solution to this problem via the development of a laboratory scale variable diameter nozzle called 'adjustable filament deposition (AFD)' has recently been reported (Tseng and Tanaka 2001). The same report also describes a second version, called 'planer layer deposition (PLD)', that lays layers of adjustable width with a view to avoiding the need for raster scanning.

Road width, W_b, refers to the width of the deposition path. Different tips and machines have different road widths (e.g., 0.3-1mm for the FDM 1650 machine). Usually, $W_b > L_p$. Hence, the bead cross-sections can be assumed to be elliptical.

The material flow rate required for complete filling can be obtained as

$$Q = L_p W_b V_h \tag{8.1}$$

where V_h is the head speed (Agarwala *et al.* 1996a). The corresponding filament feed rate is given by

$$V_f = \mu_f Q / A_f \tag{8.2}$$

where is the coefficient of friction between the drive wheels and the filament, and is the cross-sectional area of the filament. If $\mu_f > 1$, the filament would be crushed by the rollers whereas $\mu_f < 1$ implies slipping between the rollers and the filament.

Air-gap, g, is the space between two horizontally adjacent denser structure. Positive air gap leads to a loosely packed structure. Larger gaps lead to longer build times. Build temperature refers to the temperature of the heating element that controls the viscosity of the extruded plastic. If the temperature is too low, the filament can buckle while emerging from the tip. A thermal model of FDM is described in (Yardimci and Guceri 1996).

Qucikslice 2.0 software offers three build strategies. 'raster fill' refers to motion of the FDM tip back and forth inside the defined perimeter regions to fill the entire defined area. When 'contour fill' is selected, the tip makes several c losed-loop c ontour m otions inside the d efined p erimeter until t he defined area is filled. The third alternative, called 'contour and raster fill' may be used to fill part of the defined region by contouring and the rest by raster. Based on these, it is possible to implement three basic fill strategies: 'solid', ' part-fast', a nd ' crosshatch'. I n t he c rosshatch s trategy, b eads i n a layer are aligned in a direction perpendicular to those in the previous layer.

Further, beads in successive layers can be aligned in the vertical direction or skewed (staggered) by $(W+L_p)/2$. When the skewed configuration is used, there is a decrease in the base height on which new beads are laid. Hence, one needs to apply the correction factor, δ_z, calculated as (Rodriguez *et al.* 2000).

$$\delta_z = L_p - \frac{h}{2W_b}\sqrt{4W_b^2 - (W_b + g)^2} \qquad (8.3)$$

A negative gap combined with the skewed configuration has been found to yield the greatest bond strength and the lowest void density.

FDM is capable of creating matrix-like part structures exhibiting 3D interconnectivity of pores (Too *et al.* 2002). In particular, crosshatching leads to parts with non-random porous structures having no clear and closed enveloping boundaries.

The density, ρ, of a green part with aligned bead configuration can be estimated by measuring the part's weight and volume. The overall porosity, P, of the part can then be calculated from the density of the bead material, ρ_m, as $P=1-\rho/\rho_m$. The fact that $P<1$ implies that there is less material to support external load in the direction of fiber deposition. Hence, assuming that the properties of the bead material have not changed during the FDM process, one expects the Young's modulus and yield strength of the part to be smaller than the corresponding values for the bead material by factor $(1-P)$. However, in practice, the material properties would change during FDM. The degree of change depends mainly on the extrusion temperature and extrusion rate. The extrusion process leads to molecular orientation and strengthens the p art i n the d irection o f f iber alignment. I n f act, m olecular alignment is the reason for part shrinkage in FDM. Larger the degree of molecular orientation in a direction, greater is the shrinkage in that direction. Experimental work on ABS has generally confirmed these observations (Rodriguez *et al.* 2001).

Pore diameter can be measured by means of a porosimeter. The porosimeter fills the penetrometer and sample chamber with mercury under vacuum and takes a volume reading. However, initially, the sample is not intruded with mercury because of high surface tension. Hence, a gradually increasing pressure is applied. For each incremental increase in pressure, the change in intrusion volume can be expected to be equal to the volume of the pores whose diameters fall within an interval that corresponds to the particular pressure interval. Usually the pore diameter increases with increasing gap setting.

FDM experiments on ABS (Too *et al.* 2002) have revealed that, with increasing raster gap, pore diameter increases linearly whereas fractional porosity increases nonlinearly with a saturating trend at higher gaps. Consequently, the compressive strength of the part decreases approximately exponentially with increasing porosity.

Several materials are available for the FDM process: investment casting wax, nylon (Comb and Priedeman 1993), polypropylene homopolymer (Rovick 1994), polycarbonate, polyphenylsulfone, ABS (Rodriguez *et al.* 2000 and Ahn *et al.* 2002), and polymer bound advanced ceramics including electronic bio-ceramics, and polymer bound metals (Agarwala *et al.* 1996b). The materials may come in different colors. For instance, FDM P400 ABS material is available in white, blue, black, yellow, green and red. FDM has been used successfully to produce prosthetic sockets made of nylon P301 (Tay *et al.* 2002).

Ceramic parts (including piezoelectric parts made from lead-zirconate-titanate—PZT) are produced by means of a variation of FDM called 'fused deposition of ceramics (FDC)' (Agarwala *et al.* 1995c and Bandyopadhyay *et al.* 1997). During FDC, a thermoplastic polymer binder included in the ceramic filament melts inside the liquefier and carries the ceramic particles with it. Hence, selecting suitable binder chemistry is of particular importance in this process. The polymer binder needs to be burnt out subsequently through a post-processing step. A method similar to FDC, called 'fused deposition of metals (FDMet)', has recently been developed to enable the fabrication of hard tooling from stainless steel, etc.(Wu *et al.* 2002).

Apart from stair stepping, parts produced by FDM, FDC or FDMet often exhibit a variety of surface defects (Agarwala *et al.* 1996a). When the part is built with supports there would be burrs where the supports have been broken off. Since the foam surface has significant surface roughness of it own, the bottom surface of the part will have a negative image of this roughness. The top surface will have a ridge like appearance owing to the rounded top surfaces of the beads. There can be errors on side surfaces owing to the fact that the drive rollers have to stop before the FDM tip has reached the end of a raster segment. This is necessary because the backpressure in the liquefier continues to feed the material even when the drive rollers have stopped. This delay needs to be recognized and optimized. If the rollers stop prematurely, there is the danger of voids appearing at the perimeter. If the rollers continue to rotate too long, there is the danger of over-deposition at the perimeter. A reverse situation exists when a raster segment is begun. However, fortunately, most of these surface defects can be machined away at least in the case of FDC and FDMet. The machining can be performed after green part formation, or after binder removal and partial/complete sintering of the part.

A system that takes into account many of the FDM features discussed above for the purpose of optimum tool path generation has been reported recently (Qiu and Langrana 2002). The system is based on a specially developed closed loop CAD system that includes modules for solid model design, slicing, single and multi-material tool path generation, and virtual simulation. The system can generate void-free tool paths with the help of an intelligent and adaptive algorithm that computes void sizes and their locations. The path is validated with the help of the virtual simulation module. While the system has only been implemented so far on a FDM machine, it is applicable to any extrusion-based multi-material layered manufacturing process.

8.4 3D PRINTING AND DESKTOP PROCESSES

RP processes based on inkjet and 3D printing technologies have found increasing application in office environments in recent years. Owing to the emphasis on office-friendliness, the technology is also referred to as Desktop RP. These technologies have been developed mainly through company-based research. This is evident from the fact that fewer than ten academic articles could be found on the subject in 1997 although at least five companies were marketing 3D printing machines (Carrión 1997). The major 3D printing equipment available at that time were DSPC of Soligen, Genysis of Stratasys (Anon. 1996b), Personal Modeler of BPM, Actua of 3D Systems (Anon. 1996a), and Model Maker of Sanders. A benchmarking report comparing these and other systems from the viewpoint of industrial application is available in (Hopkinson and Hague 2001).

3D Printing (3DP) was originally developed by professors Emmanuel Sachs and Michael Cima, and their students at Massachusetts Institute of Technology (MIT), Cambridge, MA (Sachs *et al.* 1990a, Sachs *et al.* 1990b, Sachs *et al.* 1992, Sachs *et al.* 1993a and Sachs *et al.* 1993b). The principle of 3DP is illustrated in Figure 8-5. The machine generates the part bottom up by successively depositing thin layers of ceramic powder from a printing head. After a layer the roller has spread a layer, a printing head squirts liquid binder (e.g., colloidal silica) on to selected regions of the powder layer to define i n a ccordance w ith t he g iven c ross-section o f t he m old. E ach d rop binds the underlying powder to create voxel cell. The voxels are knit together by overlapping. Bounded voxels form the part while unbounded ones provide the supporting material. Once all the layers are so built, the part is removed from the machine and fired at a low temperature to get rid off all moisture and harden the part, so a rigid part surrounded by unbound powder is produced. The finished parts may be infiltrated with a suitable material to increase strength, e.g., copper alloy into metal (Sachs *et al.* 1992), and glass into ceramic (Cima *et al.* 1995).

Inkjets come in two basic types: drop-on-demand and continuous-jet (Heinzl and Hinz 1984). Drop-on-demand inkjets rely on a pressure pulse created to cause a small drop of ink to be ejected from the nozzle. Refilling the cavity is achieved through capillary action. In some systems, the pressure pulse is produced by a small piezoelectric element (Kyser 1981). Other systems pulse a small r esistive hater, s o a n i nk bubble i s formed through evaporation.

In continuous-jet technology, droplet formation is driven by the reduction in surface energy resulting from a transition of a cylindrical stream to set of droplets of equal volume. Rayleigh had long back observed that this transition occurs at a characteristic frequency (Rayleigh 1878). A piezo-induced vibration of the nozzle tip regularizes the droplet formation. If the ink is slightly conductive, it is possible to electrically charge the droplets, so they can be electrostatically deflected under binary control either onto the target or into a catcher.

The accuracy and precision of the 3DP process mainly depend on the powder material used, the machine axis responsible for the particular dimension, and the magnitude of the nominal dimension (Dimitrov *et al.* 2003). The achievable accuracy lies in the range IT9 to IT16. Part accuracy can be improved by applying appropriate scaling constants to the three axes of the machine.

Soligen (founded in 1992) licensed the 3DP patents from MIT to create the direct shell production casting (DSPC) products and services by building upon 3DP. DSPC produces ceramic molds for metal castings directly from 3D CAD designs without requiring any tooling or patterns. An advantage of the direct approach over conventional pattern-based casting is that there is no need to worry about drafts, parting lines, core prints, and core assemblies. Further, one can make parts hollow to reduce weight. One can also reduce parts count by combining several simple parts to a single complex geometry. This approach has the potential of reducing inventory, assembly, and tolerance requirements while increasing strength and reliability.

Z Corp is another licensee of the 3DP technology developed at MIT, but limited to the use of the process to building appearance models. Consequently, the materials used consist of a powder combining starch, cellulose, and a water-based binder, so the parts are relatively porous immediately upon completion by the machine. However, the parts can be heated and infiltrated with wax, or a cyanoacrylate. The former is usable within an office environment whereas the latter produces stronger parts. The machines are known for their high production speeds.

The DSPC process starts with the design of a virtual pattern for net shape casting including the required gating chills and risers on the basis of the CAD model of the part received from the customer. A digital model of the mold including integral cores is then created and transferred to a DSPC machine. A physical model of the mold is created next using a process similar to 3DP. Next, the desired molten metal (e.g., aluminum, iron, and tool steel) is poured into the shell. Upon solidification of the metal, the part is finished and inspected. The method is capable of creating fully functional parts.

The technology utilized in Genisys was originally developed at the IBM Thomas J. Watson Research Center. However, IBM later joined hands with Stratasys because of similarities between their technologies and Stratasys' FDM technology. In 1995, Stratasys became the majority shareholder in the new technology. In contrast to conventional FDM where the raw material is in the form of a filament, the technology involved melting of solid wafers of a polyester compound and depositing the melt through a heated extrusion tip.

In 2002 Stratasys further extended FDM technology to market a low-cost 3D printer called Dimension. The new machine is easily networked and is capable of notifying clients about part completion via e-mail, pager, or the Web. Material cartridges plug in much like toner cartridges without involving chemicals, powders, or fumes. Each cartridge contains an EPROM chip that is programmed to track used and unused material. Cartridges may be switched midway through part building to produce multi-colored parts. The materials in a molten state are held in the respective reservoirs and fed to individual jetting heads. The temperature of the droplets decreases rapidly upon contact with the partially constructed part, so they solidify quickly. After ach layer has been built, the operation of the nozzles is checked by depositing a line of each material on a narrow strip of paper and reading the result optically. If the check is positive, the elevator table moves down a layer thickness and the next layer is begun. If a clog is detected, a jetting head cleaning cycle is carried out. Upon completion of a layer, a milling head passes over the layer to achieve a uniform layer thickness. Particles removed by the milling head are vacuumed and captured in a filter. The process is repeated to form the entire object. The system is capable of producing parts with resolution and surface finish equivalent to those produced by CNC machining. Further, the system's software automatically builds breakable supports. The supports are later physically broken, melted, or dissolved away. The size of the machine and materials are office-friendly. It is capable of producing highly precise (error of the order of a thousandth of a dimension) prototypes and patterns with fine details. However, the technique is very slow and the use of a milling head creates noise at levels that could be considered objectionable in an office environment. Further, the technology is feasible with only a limited variety of materials and the properties of the feasible materials are such that the parts are quite fragile. The applications include concept models and precise casting patterns for arts, jewelry, and industry in general.

A technology similar to Sratatasys's Genisys was the ballistic particle technology used in BPM Technology's Personal Modeler (PM) (Masters 1990 and Richardson 1991). Having issued a patent in 1987, the company had begun operations in 1988. PM was unveiled in 1989 as the first ever concept modeler. This machine used an inkjet nozzle to spray a wax-like thermoplastic material in layers of 0.075mm thickness at an object under construction. The same material was used for support structures designed such that they could break away easily from surfaces of the completed part. The material was strong enough to create investment casting patterns or for soft tooling used in making functional plastic parts (Anon. 1995). The machine had five motion axes so as to reduce stairstepping. An advantage of the machine was its relatively small size and weight (about 120 kg). However, many PMs were returned by customers citing dissatisfaction with the performance and reliability of the machine (Wohlers and Grimm 2001). In particular PM's speed and materials fell short of its immediate competition (machines of ZCorp and 3D Systems). As a result, BPM had to close shop after a few years

3D Systems, a company well known for its stereolithography-based RP equipment, made its presence felt in the 3D printer market through its Actua 2100 that utilized MultiJet Modeling (MJM) technology (Anon. 1996a). The machine has 96 inkjets positioned in a linear array to deposit layers of approximately 0.1mm thickness of a proprietary thermoplastic resembling a hard wax. Since the layer thickness is only of the order of 0.1mm, the technology is capable of producing parts with impressive detail and surface finish. However, the parts are not as strong as those made by Stratasys's Genisys

Subsequently 3D Systems developed ThermoJet Modeler™ systems incorporating several hundred nozzles in a wide head configuration. The machines use a special ThermoJet brand material available in three colors. A hair-like matrix of build material is used so as to provide support for overhangs. Subsequent to part building, the supports can be brushed off easily. Owing to the large number of heads, the process is much faster than Solidscape's inkjet process. However, the surface finish achievable is not as good.

The fourth major 3D printing class of machines includes Model Maker^{TM} and PatternMaster^{TM} developed by Sanders Prototype, Inc. in 1994. (In 2000, Sanders Prototype, Inc. changed its name to Solidscape, Inc.) Weighing under 40kg, ModelMaker uses ink droplets of a thermoplastic (T_m=90 to 113°C) of 0.013mm size and can produce features as small as 0.25mm. The accuracy is adequate for large parts to be built in segments. Two nozzles are used, one for part material and the other for the support material. Support material is so designed as to permit removal by a heated solvent without causing part damage. Because of its ability to lay down precise layers, the technology is well suited for producing small but highly accurate wax patterns for secondary processes, such as investment casting (Stoddart 1997).

PatternMaster^{TM} supports variable layer thickness values. It is intended mainly for rapid tooling including core and cavity inserts made of RTV silicone, epoxy-aluminum, investment casting, powdered metals, Kirksite^{TM} tooling or spray metal. Supports are made with easy-to-dissolve wax (Anon. 1999). The machine operates with a standard IBM-compatible PC running Microsoft Windows 95/98 or NT 4.0. Solidscape systems are, however, limited in terms of speed, range of materials available, and part size.

Figure 8-6 illustrates the principle of Solidscape's inkjet machines. As noted earlier, inkjet technologies create 3D models by depositing, in a layer-by-layer manner, tiny droplets of hot liquid thermoplastic (obtained by heating micro-particles) in the desired pattern over a XY table. Depending on the machine, there are one or two inkjets. If one jet is available, the part and supports are generated using the same proprietary material. If two jets are available, the support material can be another proprietary wax-like material. In the case of PatternMaster, part building time is reduced by means of smaller jets with vector print for profile fabrication and larger jets with scan print for interior filling. A milling cutter planes off each layer to the required layer thickness. However, this step also removes material from the interior regions unnecessarily. A variation of ModelMaker involving 'flexible layers' described in (Jeng *et al.* 2000a) provides a way out of this problem. In this variation, the build/mill steps for the profile and interior regions are performed independently with a strategy different from the conventional being used for the inner region. In particular, while building layer 1, build/cut sequence is performed only on the profile region, i.e., the interior region is skipped. However, the loss of the interior region is compensated for on the next layer by doubling the layer thickness over the interior region (followed by milling as usual). This alternating skip procedure is repeated till the full part is built. Experiments have shown that this flexible layer approach can decrease total part building time by about 40%. Further reduction in interior filling time is possible by substituting interior scanning by spraying material from a spraying nozzle (Jeng *et al.* 2000b). As a result, the total part build time can decrease to as low as 15% of the original build time without affecting part accuracy.

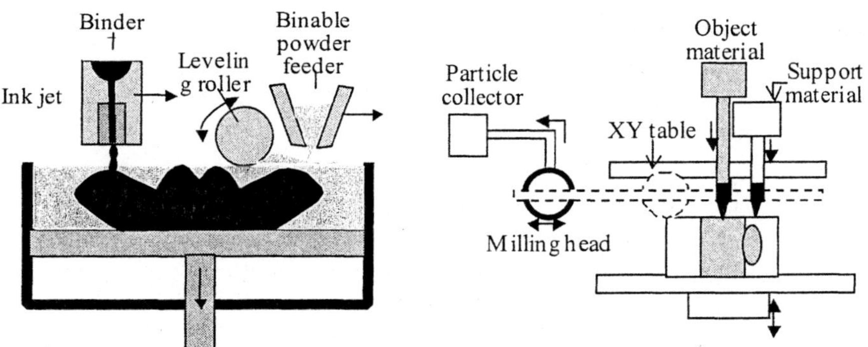

Figure 8-5. 3D Printing.   *Figure 8-6.* Solidscape inkjet system.

A wide area photopolymer-based phase change inkjet system called Quadra™ was unveiled in 2000 by Objet Geometries Ltd. of Israel. In this system, a UV flood lamp mounted on the print head is used to cure each layer after deposition. The support material made from another material can be washed away after part building. Owing to its low price and the ability to use photopolymers, the system holds promise as a substitute for stereolithography.

A similar system called InVision™ was launched by 3D Systems, Inc. in July 2002. The system is of the size of a copying machine and is capable of producing durable concept models, prototypes and patterns with high-quality finish for form and some fit from a translucent acrylic photopolymer based blend. The support material is a wax-based blend.

A range of 3D printers under the brand name of ZPrinter based on Hewlett-Packard's inkjet technology was unveiled by ZCorp in early 2003. The system utilizes a range of proprietary powders including a talcum powder, a four like substance, and a specially formulated powder called ZCast for making molds for metal casting. Figure 8-7 schematically illustrates powder-based 3D printing as implemented in ZCorp 42 3DP machine. A precisely controlled thin layer of the selected powder is first applied to the building platform. Next, the head traverses over the layer squirting a thin glue-like binding material so as to solidify the powder in the selected areas. When one layer has been traversed and solidified, the build platform descends by one layer thickness, and the process repeated bottom to top until full part has been built. The partially fused powder is contained by an enclosure. The enclosure also helps in supporting overhanging features. The system is especially good at making parts that have hard-to-reach cavities, as the un-fused can be poured and vacuumed. The saved powder can be reused. However, taking the powder out can make the system a little messy, so one needs a small cleanup area. Further, odors can be felt while penetrating and sealing models with a material similar to Superglue.

After slicing the soft module, the system organizes the area covered by each 2D slice into a set of shells (closed boundaries of part regions) and cores (shell interiors) on the basis of the binder setting value selected by the user. A higher value results in superior part quality, but at the expense of longer build time. A separate parameter called binder setting saturation level determines the degree of binder infiltration (maximum 100%). Layer thickness values typically range from 0.09 to 0.23mm. Taguchi experiments have shown that part quality and process performance depend on binder setting saturation value and level, layer thickness, location of the part relative to the position of the head, and shrinkage (Yao and Tseng 2002).

8.5 SHAPE DEPOSITION MANUFACTURING (SDM)

Shape Deposition Manufacturing (SDM) is a layer manufacturing technology in which the layers are shaped by CNC. Thus, it is an additive-cum-subtractive process. A temporary material is used for building support layers needed for overhanging, undercut, and disjoint features. The technology is capable of creating multi-material objects, functionally graded materials, and objects with embedded sensors, advance tooling, and other electronic devices (Weiss *et al.* 1997 and Kietzman 1999).

SDM technology was originally conceived and developed in the early 1990's at Carnegie Mellon University (Merz 1994 and Merz *et al.* 1994a). It was an improvement upon their previously patented 'recursive mask and deposit (MD)' process in which parts are built by spraying zinc, steel, or nylon layers of material through disposable masks (Weiss *et al.* 1991, 1992 and Prinz *et al.* 1993). Each mask is about 0.1mm thick and has the shape of the current cross-section. However, parts produced by the MD process exhibit several drawbacks such as stair-stepped surfaces, porosity, low mechanical strength compared to cast or welded parts, and warpage and delamination caused by the build-up of residual stresses. Initially, these problems were addressed by combining weld deposition of material to 'near net' but rough shapes with computer controlled milling for accurately shaping the deposited rough shapes. This strategy has the advantage of decoupling material properties affecting deposition from those affecting shape generation. Subsequently, a variety of material deposition methods and other improvements were tested at Carnegie Mellon University, and Stanford University.

Among the deposition methods explored are laser welding, gelcasting, UV curable systems, and 2-part polyurethane and epoxy mixtures. The choice of deposition system depends upon the material to be deposited. The original system used at Carnegie Mellon University was the 'microcaster'. This system melts material in wire form by feeding it into an arc, so the falling melt droplets could be used for deposition (Merz *et al.* 1994b). Another material deposition system uses a laser to create a melt-pool on a substrate and a powder feeder deposits a powder into the melt-pool By using a set of powder feeders with different materials, it is possible to make multi-material parts, or functionally graded materials including ceramics. More recently, a process combining sputtering and etching has been used to make thin features for embedded sensors. Deposition techniques for polymeric materials include extrusion and casting.

The first step in planning the SDM process is to subtract the solid model of the object/assembly from a block of solid material to produce a model of the complementary support material required to construct the part. The part and support models are then searched for the locations of silhouette loops' that divide the composite solid model into undercut and non-undercut segments. The non-undercut segments are called 'manufacturable compacts' (Ramswami 1997). Next, the compacts are sequenced such that all compacts needed for supporting a given compact would have been built previously. Sometimes, one comes across cyclic dependencies, so a specific sequence does not exist. In this case, some of the compacts are further divided to produce a manufacturable sequence. The final process-planning step is to determine the end-milling tool path required to shape each manufacturable compact in a layered manner.

Figure 8-8 illustrates the typical configuration of SDM. Material deposition is done through a nontransferred welding process called 'microcasting' that leads to the accumulation of discrete, super-heated molten metal droplets in order to build metallurgically bonded, fully dense structures (Amon *et al.* 1997, Kietzman 1999 and Binnard 2001).

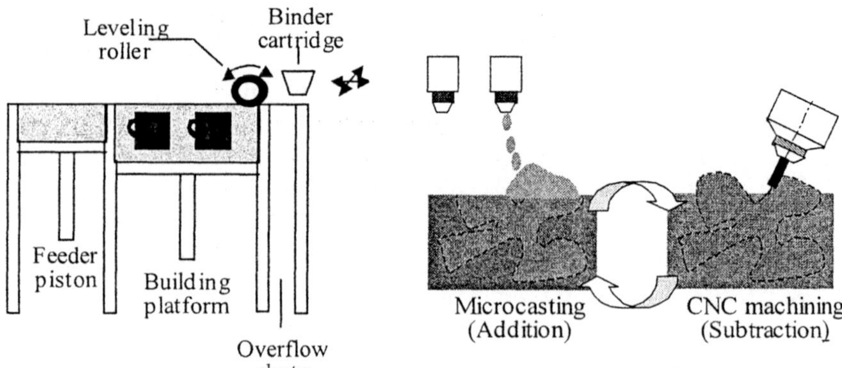

Figure 8-7. Powder-based
3D printing (ZCorp).

Figure 8-8. SDM.

SDM process begins with the construction of a support material cavity shaped like the underside of the bottom of the desired object. Part material is deposited into this cavity layer by layer. The CAD model of the object/assembly including supports is first sliced into 3D layered structures as in most other RP processes. Many SDM machines are equipped with alternative deposition heads, e.g., the Multiple Material Machining (Richardson 1998). This enables the deposition of multiple materials in the desired sequence within a layer. Thus, for example, stainless steel may be deposited as the primary material and copper as the sacrificial support material. Each segment is deposited as a near-net shape using the appropriate material. The segment surfaces visible are then machined to net shape by means of 5-acis CNC milling facility. The shaping process eliminates stairstepping, thus helping reducing fabrication time through the use of thicker layers. The deposit-remove cycle is decided on the basis of the local geometry such that undercut features need not be machined. Only up-facing surfaces are milled (Figure 8-9). If necessary, additional upward-facing features are milled into the exposed part material so additional support material can be deposited to surround the bottom layer of part material. This procedure is repeated until the part is complete. An analysis of the thermal stresses induced during SDM of metal parts is available in (Nickel 1999).

The deposit-remove cycle can be implemented in many ways. If the steps are implemented on the same machine, problems arising from nonrepeatability of part positioning can be eliminated. If different machines are used, one needs to utilize pallets with repeatable positioning systems. Pallet transfer between machines can be carried out manually or by robots (Hartmann 1994 and Prinz and Weiss 1994b).

A major advantage of SDM is that it is quite easy to create heterogeneous structures (Weiss *et al.* 1997). Since, the deposit-remove cycle is repeated many times, it is easy to interrupt the process at suitable moments to embed external objects. The feasibility of embedding sensors in metallic structures by using Nd-YAG laser assisted SDM has been investigated (Prinz *et al.* 1994c). Embedded sensors can be used to monitor parameters at critical locations not accessible to ordinary sensors. For instance, embedding of thin film thermo-mechanical sensors and fiber optic sensors has been found to be feasible. For instance, embedded thin film strain gauges have exhibited good linearity. Likewise, embedded fiber optic sensors for the measurement of temperature and strain have shown good accuracy, linearity, and high temperature capability. Also a non-contact fiber optic sensing system has demonstrated acceptable performance. Thus, SDM has enabled the creation many smart structures such as wearable computer (Weiss *et al.* 1996).

A variation of SDM called 'Mold SDM' implemented on a commercial CNC milling machine uses has been investigated for the fabrication of complex shaped fugitive wax molds (Cooper *et al.* 1998). The method is capable of producing alumina, silicon nitride, polyurethane and epoxy parts with feature sizes ranging from 0.5 to 30mm.

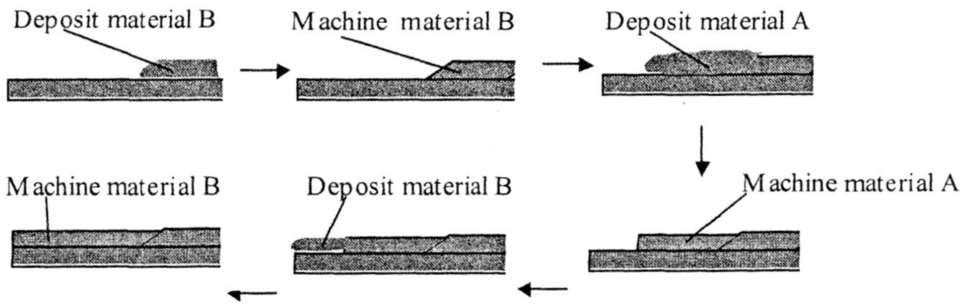

Figure 8-9. Deposit-remove sequence in SDM (Kietzman 1999).

8.6 VACUUM CASTING

There are several variations of this technology. One variation starts with a master modelwith defined parting planes. Typically, the model is made by stereolithography or selective laser sintering. Next, silicone (a flexible material) is cast around the master. In order to avoid air bubbles being trapped between the master and silicone, the casting process is conducted in vacuum. The mold is then cured and cut according to the parting planes so as to enable the removal of the master. This leaves a cavity from which multiple copies of the part can be cast. Again, casting is performed in vacuum in order to avoid air bubbles. Typically, two-component polyurethane is used as the copying material. A broad range of polyurethanes with different physical properties is available. The process is capable of fast production of high quality parts.

8.7 ELECTROFORMING

According to ASTM B-832-93, electroforming is the production or reproduction of articles by electrodeposition upon a mandrel or mold that is subsequently separated from the deposit. The process analogous to electroplating, but the 'plate' is much thicker. Generally, plating is performed on a model/pattern/mandrel produced by some other process such as stereolithography. It is important to ensure that the model material is capable of being wetted by the electrolyte. For instance, if the model is made of wax, i t can be metallized by applying a conductive metal-loaded paint through brushing, spraying, or dipping. Just as electroplating, electroforming uses aqueous solutions of a metal salt as the electrolyte. When dissolved in water, the salt disassociates into positively charged metal ions and negatively charged ions (e.g., when copper sulphate is used as the electrolyte, sulphate ions are negatively charged whereas copper ions carry a positive charge), so a current can be made to flow by applying a suitable DC potential difference between two electrodes dipped into the electrolyte. As a result, the positive metal ions 'electromigrate' towards the negatively charged electrode (cathode) while the negatively charged ions are carried towards the positively charged electrode (anode). Upon arriving at the cathode, the positively charged metal ions get neutralized by the electrons furnished by the cathode. Electroformed metal is extremely pure and exhibits superior properties over wrought metal due to its refined crystal structure. Multiple layers of electroformed metal can be molecularly bonded together to produce complex structures with "grown-on" flanges and bosses. The thickness can reach several millimeters in some applications. The distribution of metal layer deposited is a function of the part geometry. Recessed areas will attract a thinner deposit owing to lower field strength while the deposit is thicker in prominent regions such as outer corners and edges of the part. However, a smoother deposit can be obtained by incorporating certain proprietary organic additives in the electrolyte. An application of this technology for the rapid production of electrodes used in electric discharge machining (EDM) is described in (Rennie *et al.* 2001). In this application, mandrels fabricated from stereolithography were electroformed i n a n aqueous c opper sulphate e lectrolyte t o d eposit c opper layers of 0.2 to 0.3mm thickness.

By depositing a sufficiently thick layer on a mother die, it is possible to flake off the deposited layer, so the flaked off part has reverse shape of the mother die. The dimensional error of the reproduced goods can be within ±1μm over 100mm, and the distortion of the surface within 50μm. The nickel-shell electroforming process deposits a 1-2mm shell of nickel backed up by 3-4mm of copper. This process may not really be considered to be 'rapid' in that it produces a tool in about the same time as conventional machining. However, it can lead to shorter molding cycles by virtue of five to ten times higher thermal conductivity. Further, nickel provides a surface hardness of 20-40Rc, so durability is sufficient. Electroformed tools can be produced to the same precision as molds made with EDM electrodes Press tools for sheet metal forming have been successfully produced by using a variation combining stereolithography with nickel electroforming (Yarlagadda *et al.* 2001).

Recently commercial interest is growing towards the application of electroplating for the production of metal parts from the micron scale to the mesoscale of a few millimeters (Wohlers and Grimm 2002). For instance, MemGen, CA, is attempting to commercialize a technology called Electrochemical Fabrication (EFAB) developed at the University of Southern California.

8.8 FREEZE CASTING

The Freeze Cast Process (FCP) was f irst patented by Yodice in 1991 (U.S. patent no. 5,072.770). The process starts with a solid master and silicone mold. The latter is used to make an ice pattern. The pattern is repeatedly dipped into refrigerated ethyl silicate slurry and stuccoed until a ceramic shell of sufficient thickness is built. The shell is then brought to room temperature, so the ice pattern melts away. After draining and drying, the shell is ready for investment casting. However, the original FCP was plagued with problems arising from anisotropic expansion during ice formation: air bubble removal, ice-pattern demolding, and poor part complexity. However, some of these problems could be addressed through the development of Rapid Freeze Prototyping (RFP) (Zhang *et al.* 1999, Sui *et al.* 2000 and Liu *et al.* 2002). RFP disposes of the need for the starting solid matter and silicone mold by building the ice pattern layer-by-layer with water supplied by a nozzle traversing over the build platform. The nozzle can supply water continuously water supply, or deposit water drops on demand. The water used can be pure or colored. The build environment is kept below the freezing temperature of water. Any support structure needed is made of brine. Since the freezing point of brine is lower than that of pure water, the support can be removed after pattern building by utilizing the melting temperature difference. An analysis of the RFP process leading to expressions for the smallest feature sizes achievable is available in (Leu *et al.* 2003).

The advantages of RFP include clean and cheap material, good surface finish, potentially good accuracy, easy pattern removal and joining, easy expansion compensation, and high building speed. The major disadvantage is that it needs a low temperature environment.

8.9 CONTOUR CRAFTING

Contour Crafting (CC) is an additive fabrication process developed at the University of Southern California, CA, to enable rapid prototyping of large sized, smooth surfaced, complex shapes from a wide variety of materials (Khoshnevis 1998 and Khoshnevis *et al.* 2001). In a CC machine, raw material housed in a cylinder is extruded through a nozzle by means of a piston connected to a threaded feed rod. The vertical extrusion head is capable of linear motion along three coordinate axes. The part is built on a flat rotary worktable. A separate computer controlled stepper motor is used to control each of the five motions. A top and a side trowel attached to the cylinder head enable shaping and smoothing-out of the material on the top and outer surfaces of the part respectively. The actions of the trowels are similar to the way a potter manipulates clay with his hands. As with many other rapid prototyping processes, parts are fabricated layer-wise. A heat gun attached to the machine can be moved inside the hollow part to accelerate drying. It has been claimed that the system should be applicable in the fabrication of producing parts such as turbine blades which have complex surface profiles devoid of small and detailed features, large tooling for automotive and aerospace industries, and civil structures such as house walls (Khoshnevis 1998 and Khoshnevis *et al.* 2001). it is claimed that the system should be applicable in the fabrication of producing parts such as turbine blades which have complex surface profiles devoid of small and detailed features, large tooling for automotive and aerospace industries, large, and civil structures such as house walls (Khoshnevis *et al.* 2001). However, so far, only hollow rotational parts made of clay and spackling plaster have been produced successfully.

8.10 3D WELDING

During the 1960, a process called 'Shape Welding' was explored in Germany. The technology was used later by many companies such as Krupp, Thyssen, and Shulzer for fabricating large components of simple geometry. In 1991, a process called 'Shape Melting' was developed by Babcock and Wilcox for building large components made of austenitic materials (Doyle 1991). Rolls-Royce, UK, successfully used 3D welding to produce some aircraft parts made of nickel- and titanium- based alloys. Research on selective 3D welding has also been in progress at the University of Nottingham (UK) (Dickens *et al.* 1993), Cranfield University (UK), University of Wollongong (Australia), and Southern Methodist University (USA) (Kovacevic and Beardsley 1998). In particular gas metal arc welding (GMAW) has received much attention.

A continuing hurdle to the extensive use of GMAW for RP is the need to control temperature gradients caused by fusing molten droplets onto previously deposited layers. Thermal spraying creates small molten droplets of about $50\mu m$ diameter, so they do not possess enough heat to re-melt the underlying surface. On the other hand, conventional GMAW and plasma welding are associated with large heat build up that could affect the shape of the underlying material. In classical welding-based deposition approaches such as GMAW and plasma welding, metallurgical bonds will be formed, but the large heat transfer will affect the shape of the underlying material. Consequently, if welding were to succeed in the arena of RP, it is necessary to tightly control heat build up. This, in turn, requires control of the size and detachment rate of weld droplets (Beardsley and Kovacevic 1998).

Conventional GMAW is not applicable in RP for two reasons: uncertain droplet detachment instant, and inconsistent droplet size. The method relies on the 'one-drop-per-pulse' approach by properly selecting the duration of the peak current. Further, the peak current is adjusted to be larger than the transition current (250-300A). Such a high current generates excessive fumes due to the superheating of the droplet and a very high impact speed of the droplet. Consequently, the underlying geometry is deformed and the droplet volume becomes irregular. This problem has been solved by equipping the system with a high frame rate digital camera assisted with a He-Ne laser and a real time image-processing algorithm to monitor the droplet formation (Beardsley and Kovacevic 1998). Experiments have shown that switching the current from the peak level (about 200A) to the base level (about 50A) initiates an oscillation of droplets near the tip of the electrode. The oscillation is monitored by the vision system. When the droplet moves downwards, a signal is sent to the power source controller to raise the current to the peak level, so the electromagnetic force is increased.

The current pulse is designed to yield the desired depth of bead penetration as well as the allowed level of induced heat. The downward motion of the droplet in combination with the increased electromagnetic force generates a large enough detachment force to detach the droplet from the electrode.

8.11 CNC MACHINING AND HYBRID SYSTEMS

Although capable of producing free form surfaces, c onventional CNC (computer numerical control) machining is usually not considered a RP process because of the need for human planning and operation, and custom fixturing/tooling. Further, it can only produce features where a tool can reach. Reentrant surfaces will have to be machined by electrical discharge machining (EDM). However, as we have already seen machining plays a secondary or, even, integral role in many RP technologies. In particular, in the rapid t ooling industry, CNC machining remains side by side with RP technologies.

A sculpturing robot (SR) system called the SR system has been developed at Delft University, Netherlands (Tangelder and Vergeest 1994). In this system, blanks mounted on a turntable are machined by a milling cutter mounted on the end-effector of a six-degree-of-freedom robot. Minkowski operations are applied to generate an interference-free milling tool path whereas an interactive simulation of the milling process is also implemented (Walstra *et al.* 1994).

Amongst CNC machining technologies, high-speed milling (HSM) is most attractive from the viewpoint of RP. Products can be made rapidly by increasing the rotation speed of the milling cutter from 5,000rpm to 50,000rpm, and by direct control from CAD data. With HSM, neither surface quality nor accuracy pose significant problems. However, straight corners are more difficult owing to the finite diameter of cutter, tool path generation is time consuming, and undercuts are not possible. A major RT application of HSM so far has been in the fabrication of electrodes for electrical discharge machining (EDM).

Some laser based machining techniques seem to be particularly suitable for R P w ork. O ne s uch p rocess is L asercarving o f B avarian L aser C enter (BLZ), Erlangen, Germany, has been used for cutting cavities in tools and texturing tooling surfaces (Anon. 1997). In this process, cavities are cut into metal or ceramic stock in a downward, layer-by-layer fashion by simultaneously directing a 750W CO_2 laser and an oxygen source at the surface to be cut and sweeping across. The laser beam heats the metal and the oxygen source oxidizes it. The oxidized chips break away due to differential thermal expansion between the underlying unoxidized material and the oxide fragment. The overall machining times can be reduced by first roughing out material at a higher rate (up to 1,000 mm³/min) using the laser in the melting mode that removes material. An attractive feature of the technology is that it can work directly on steel hardened to 60Rc and even ceramics. It does not have the diameter problem for inner corners as in milling. However, further research is required to improve accuracy and surface quality.

Layered manufacturing has the advantages of being high-speed, single set-up, and full automation. On the other hand, CNC machining is capable of providing high accuracy, precision and surface finish and is applicable to a wide variety of materials. This complementary relationship has naturally inspired many exploratory and commercially applied hybrid systems. For instance, CAM-LEM and SDM and described already (see Sections 8.2 and 8.5 respectively) are hybrid technologies. Another commercially applied system is the SWIFT process based on solvent welding and CNC contour machining (Taylor and Cormier 2001). Each layer is built by applying a thin high-density polyethylene (HDPE) and laser printing the required image that acts as a solvent mask protecting the downward-facing surfaces from being welded to the previous layer. Next, acetone solvent is applied to the bottom side of the sheet and a pressing force is applied, so the current sheet gets welded to the fabricated part. The current sheet is next shaped by 3-axis CNC machining is then performed to shape the current sheet. The procedure is repeated till all the layers have been built. The advantages of the systems include relatively low cost and high speed. However, a large geometric error can arise due to the uniform layer thickness.

Another interesting hybrid system is ECLIPSE-RP where 3-30mm chemical wood sheets are stacked together to build a part (Hur *et al.* 2002). The layer building cycle includes a sheet inversion step so that both sides of the sheet can be machined. Machining is carried out on a machining center based on a six-axis parallel mechanism. Four types of machining features are semi-automatically extracted, machined, and glued together.

A rising trend in hybrid systems is to utilize a robot for machining tasks. For example, in the sculpturing robot (SR) system developed at Delft University in Netherlands, the milling cutter is mounted on the end-effector of a six-degree-of-freedom industrial robot while the raw material stock is mounted on a turntable (Tangelder and Vergeest 1998). Minkowski operations are applied to generate an interference-free milling tool path (Tangeleder *et al.* 1998). Another robot-based system is laser-based machining (LBM) system developed at the University of Hong Kong (Yang *et al.* 2002c).

In fact hybrid systems such as those described above only the beginnings of a larger trend towards integrated systems (Onuh 2001). For instance, the

Consortium for Advanced Manufacturing International (CAM-I) has made integration of RP with other manufacturing processes as one of the key objectives in its Next Generation Manufacturing Systems (NGMS) programs.

Chapter 9

RAPID TOOLING

The current tool, die, and injection mold industry worldwide is estimated to involve annual revenues between 4 and 65 billion dollars depending upon the estimation source and methodology. Of this the present share of rapid tooling (RT) is around $400 million. The amount RT will capture in future is still unknown. However, it must be recognized that tooling lead-time can be reduced by as much as half with the aid of RT (Denton 1995). Hence, RT is particularly important in niche markets thriving on product variety rather on production volume. Time is of the essence in such industries.

At the minimum, rapid tooling (RT) can be seen as a means of transferring non-functional models constructed from the range of RP techniques discussed in previous chapters to functional prototype parts. On the other hand, it can also been seen as the second wave in rapid prototyping because, with rapid tooling, the production process can be prototyped instead of the final product.

Unlike models produced for concept verification or communication, tools need to be manufactured to specification and must be durable enough to last the desired number of shots. Unfortunately, no single RP process is able to satisfy this requirement in all cases. Thus, most RT approaches utilize multi-step procedures involving a mixture of RP processes and conventional tool fabrication methods. Many of these multi-step processes offer advantages beyond cost and time reduction. For instance, some enable conformal cooling channels (CCC) that are not possible by CNC machining and other conventional tooling fabrication methods. Conformal channels allow coolant to pass through the mold in a pattern that closely follows the shape of the mold cavity. Compared to the conventional straight channels, conformal channels significantly reduce the possibility of hot spots, so the injection molding cycle time is dramatically reduced by as much as 10 times. Use of conformal channels can decrease tooling cost by 25 to 70% (Jacobs

1999). Some RT methods improve thermal management by enabling functionally graded tooling materials. For instance, steel can be used in sections subjected to severe mechanical loads and copper in those subjected to high heat input.

It is estimated that more than twenty multi-step RT procedures are presently in use (Wohlers 2002). This chapter summarizes the major features of a selection of these procedures. It will be assumed that the reader is already aware of conventional tooling and tool fabrication methods. Readers without such a background may refer to books such as (Kalpakjian 1995), (Wick *et al.* 1994), and (Mitchell 1996). Readers interested in further details of RT and case studies may refer to (Hilton and Jacobs 2000).

9.1 CLASSIFICATION OF RT ROUTES

Rapid tooling techniques have been classified in various ways, e.g., prototype/bridge/production, soft/bridge/hard tooling; and nondirect/direct. While these groupings do not usually have clear definitions or borders, they do help in clarifying a given RT scenario (Rosochowski and Matuszak 2000).

Soft tooling refers to tools required for small quantity production. Hard tooling is the opposite, whereas bridge tooling is somewhere in between. In view of the small production quantities, tooling cost reduction is of prime importance in soft tooling. It is difficult to amortize expensive tooling over the production of a few products. In contrast to hard tool steels used in hard tooling, soft tooling is usually produced from 'soft' materials such as silicone, rubber, epoxies, low melting point alloys, zinc alloys, aluminum, etc. Much hard tooling in industry is produced by means of computer numerical control (CNC) machining and electrical discharge machining (EDM). The methods employed for soft tooling depend on the material used. Soft tooling can be used in processes such as injection and compression molding, b low m olding, vacuum c asting a nd f orming, e xtrusion a nd g lass reinforced plastics lay-up. In the production of metal components, soft tooling can be used for the fabrication of patterns, matchplates, core boxes, etc. for sand casting; wax patterns for investment casting; and to meet some requirements of sheet metal forming, die casting, and forging. Soft tooling is also applicable in processes such as slip casting, powder compacting in presses, and isostatic forming used in ceramic components industry. Conventional methods of soft tooling fabrication include resin casting, ceramic casting, metal spraying, electroforming, silicone rubber molding, spin casting, the Keltool process, and orbital abrasion. Many RP technologies, including stereolithography (SL), are applicable in the fabrication of soft as well hard tooling.

Many RT routes utilize a master pattern (e.g. a SLA part) with a view to fabricating the final production tool by more classical methods such as investment casting and resin tooling (Styger 1994). These routes may be labeled as 'indirect' since they include at least one intermediate step in the tooling process. In contrast, direct tooling involves no intermediate steps in the manufacture of tools. Just as SLA directly provides the prototype, these techniques can directly supply the production tool(s). Direct RT aims to produce durable and functional tooling by the shortest possible path while not compromising the notion of rapidity.

Many RT companies use hybrid approaches combining two or mother RT processes. For instance, while RP is used for complex-shaped tooling components, machining might be used more economically for simple shaped metal components. The use of metal has many advantages over the materials preferred in RT methods: easy welding, repair, retexturing and/or polishing.

9.2 RP OF PATTERNS

A critical element of any casting technology, conventional or otherwise, is the master pattern. Many times the pattern is of complex geometry with fine detail. Hence, usually, pattern-making time constitutes a substantial proportion of the overall lead-time required for the fabrication of the final part or a tooling component by means of sand casting, investment casting, etc. Indeed, since the advent of RP processes capable of creating patterns, many industries have applied investment casting (INC-RPT) and precision casting for the fabrication of rapid production tooling (RPT). The RP approaches used to pattern making include stereolithography (Jacobs 1993, Schaer 1995), selective laser sintering, laminated object manufacturing LOM (Mueller and Kochan 1999, Wang *et al.* 1999), fused deposition modeling, 3D printing (Stoddart 1997) and rapid freeze prototyping (Liu *et al.* 2002). Among these, SL is most used.

In the early days of SL (around 1989), SL parts were not accurate enough to serve as master patterns for tooling. However, by 1995, RMS error on the benchmark USERPart had improved to below 50µm, thus making the application of SL in rapid tooling feasible. In particular, SL gained ground in the area of investment casting.

Conventional investment castings are produced by the 'solid mold' method or the 'shell casting' method. With very few exceptions, the ceramic shell has replaced the solid mold method. Shell molding utilizes patterns made of wax, so they can be melted away. The patterns are produced from female cavity dies and assembled on a common gating system. The assemblage is called a 'cluster' or a 'tree'. The cluster is dipped into ceramic slurry followed immediately by a coating (stucco) of dry refractory grain. The coated assembly is then allowed to dry in a controlled environment. The dip, stucco and dry steps are repeated until a shell of about 10mm thickness is obtained. When the shell is complete, the wax invested within is removed by placing the shell in a steam autoclave or directly in a preheated furnace. The melted wax exits the shell through a runner or sprue system. Prior to casting, the shell is fired to develop strength. Molten metal is then poured through the runner or sprue system under gravity, pressure, and/or vacuum. Because of metal shrinkage during solidification much of the ceramic shell literally falls off the castings. Materials that can be so cast include aluminum alloys, bronzes, tool steels, stainless steels, Stellite, Hastelloys, and precious metals. Since the mold is formed around a pattern that does not have to be pulled out as in traditional sand casting, very intricate parts and undercuts can be realized. Tolerances of the order of 0.5 % of length are routinely possible. Castings can weigh from a few grams to tens of kilograms. Normal minimum wall thickness values are in the range 0.5 and 1mm.

A major problem with conventional shell investment casting arises from the time and up-front cost of producing the patterns. SL can be used to solve these problems. Since SL is a CAD model based rapid prototyping process, it is possible to try out a range of part design solutions during the product development stage without incurring excessive costs. Knowledge so gained can be transferred to enhance the quality of the tooling in subsequent production runs.

SL patterns are typically made from epoxy resin. However, when the pattern is built using one of the conventional part build styles described in Chapter 6, there usually is a gross mismatch between the thermal expansion of the pattern and the ceramic shell used in investment casting. Since epoxy resins are thermosets, they do not melt and continue to expand (Jacobs 1993). To solve this problem, DTM developed a porous substance called Trueform with low thermal expansion. Another solution developed by 3D Systems involves allowing some room for the model to expand and subsequently collapse during autoclaving. The company realized that the problem can be solved if the pattern has a solid outer skin and mostly hollow inner structure, so the part can collapse inwards when heated. Several build styles have been developed to achieve this. The first version was QuickCast 1.0 that adopted an internal open lattice structure of closed triangles. A subsequent version, QuickCast 1.1, used closed squares (Jacobs 1995). In both styles, the structure is a combination of two identical levels that are offset from each other to allow for the drainage of uncured resin under gravity. However, despite these developments, the investment casting shells continued to crack. The hollow structures were not collapsing reliably. Further, they were not draining resin effectively. These problems were eventually resolved through the development of QuickCast 2.0. This style is constructed in three levels where each level comprises individual sides of a hexagon. Connection between the levels is through point contact between the individual sides. Because of the small degree of connectivity, the structure is quite delicate and collapses easily during investment casting. The more open hexagonal shape and the point contact features also help in reducing drainage problems caused by meniscus formation at the corners of hexagons (Hague *et al.* 2001).

Several design guidelines specifically associated with SL-based pattern fabrication are worthy of note. The most important of these is that the nature of RP patterns must be kept in mind while preparing the shrinkage-compensated solid CAD model. For instance, in conventional investment casting, it is usual to use different shrinkage factor for different groups of part features. This is difficult when SL is used. It is therefore important that the single shrinkage factor is selected with care and after consulting with all the experts available. Fillets on all corners should be as generous as possible. Since cored passage features are difficult to build using QuickCast, they should be minimized or redesigned to allow more shell material to enter. For the same reasons, nearby vertical walls should be placed far enough. The necessity for machining certain surfaces on the cast part must be anticipated. Such surfaces must be offset to ensure machining allowance. Not all mold components can be cast. Alternative fabrication processes must be found for such components while keeping cost, quality and time constraints in mind. However, all the components must be included in the CAD model.

As with any tooling, molds need to be significantly more accurate than the parts produced from them. Hence, accuracy is of particular importance while making RP patterns. In addition, the patterns should have high surface finish and it should be possible to produce fine details. Basic RP processes such as SL, SLS, and LOM are usually unable to meet such stringent requirements. In such a case a basic RP process may be used to address geometric complexity. Requirements for accuracy and finish can be met by supplementing the basic RP process with milling. Examples of such 'compound' processes are PatternMaster of Soligen, and Shape Deposition Manufacturing (SDM). Some RP processes such as electroforming are inherently capable of superior accuracy and finish.

9.3 INDIRECT RT

In the previous section, we studied rapid pattern-making approaches while leaving open the question of whether the pattern is intended to make a product or a tooling component (e.g., a mold). An indirect RT approach refers to the latter. It is said to be 'indirect' because the intermediate step of preparing a pattern or master is required. Over the last decade, several pattern-based RT approaches have been developed (Hilton and Jacobs 2000, Wohlers 2002). We will now discuss a selection of them.

9.3.1 Indirect Methods for Soft and Bridge Tooling

Room Temperature Vulcanizing (RTV) Silicone Rubber Molds (SRM) is a widely used indirect RT approach for the fabrication of soft tooling. RTV rubber molding provides the least expensive and fastest way to create about a dozen prototype parts. The use of a highly flexible material such as silicone rubber facilitates the extraction of parts or tooling components with highly complex geometry including undercuts from the molds.

It is useful to start the discussion of this topic from natural rubber. Natural rubber is formed from the sap from rubber trees. Sap comprises a dispersion of polyisoprene particles in water. Polyisoprene is a chain of molecules each of which is made up of a carbon atom linked through a single bond each to a CH_3 and a CH_2 molecule and through a double bond to a $CH-CH_2$ chain. The remainder after water removal is natural rubber, a solid that readily creeps because there are no cross-links. The material can be brought to a useful hardened state through vulcanization, a process in which rubber is heated with 3-5wt% sulfur. Some of the sulfur atoms form cross-links by attacking the CH_2 groups neighboring the double bond. Vulcanization of most rubbers leads to a material with glass transition temperature below room temperature while exhibiting high elastic deformation and elongation.

Most polymers including natural rubber have carbon-based backbones. In contrast, silicone polymers have silicon-based backbones. (Note that silicon is in the same group as carbon in the periodic table.) Many silanes are analogous to the hydrocarbons but with Si-Si bonds. For instance, the backbone of silicone rubber is composed of Si-O-Si bonds. Various organic groups such as methyl or the benzene ring may be bonded to the silicon. The Si-Si basic structural unit is found in many rocks and minerals in nature including common sand. This bond is very strong and stable with bond energy equal to 445 kJ/Mole. Silicone rubber can withstand 300°C continuously and much higher temperatures intermittently without thermal decomposition. Thus it is one of the most stable and inert polymeric materials available.

The manufacture of silicone rubber involves extensive cross-linking between chains of the polymer. As the reaction proceeds, the sites on a chain react with each other, forming a highly cross-linked network of polymer. Silicone rubber is made by letting cross-linking to proceed until all the reactive sites are linked. Most importantly, this cross-linking can be made to proceed at room temperature, thus leading to the term 'room temperature vulcanization (RTV).'

The first step in RTV-SRM is to create a pattern using a RP method as described in Section 9.2. The pattern might include the required sprues and gates. If not, the pattern is fitted with appropriate gating and sprue system with the aid of 'superglue.' Next, the pattern is surrounded by a parting surface so as to establish the parting line for the mold. Sometimes, this step is omitted and, in a subsequent step, the cured mold is carefully cut to form the parting line. The assembly is then fixed in a vat and a RTV liquid is poured over it. RTV air-cures at room temperature, so the cure time depends on the geometry, the RTV type, and the environment. Cure time can range from 0.5 to 40h. Once the RTV liquid has cured, it is removed from the vat and separated from the pattern and parting line surface. This step yields two halves of the mold. These are next aged for up to three days to improve mold life. Upon aging, thermoset liquid-silicone RTV material is poured or injected into the mold. Prior to pouring, the material is mixed under vacuum to remove air bubbles. A popular RTV material is urethane. Urethane has properties similar to elastomers or ABS. Further it can be machined, sanded, glued, and painted easily. However, urethane materials have the disadvantage that their properties are different from the thermoplastic materials used in production.

Next the part is removed from the mold by distorting it appropriately if the part has undercuts. As long as the distortion is not severe, a silicone rubber mold can spring back to its original shape. Finally, the part is post-processed by trimming any flash and, possibly, sanding. The gate and sprue are also removed at this stage.

The main advantage RTV-SRM soft tooling is that it is very fast. It is possible to produce the first part from an existing CAD file within five days. RTV soft tooling is also substantially less expensive than CNC-machined aluminum tooling. RTV is intrinsically capable of replicating extremely fine detail (including scratches on the pattern's surface). However, RTV mold durability is marginal, so the process is not applicable for more than 10-40 parts. However, in recent years, the number of parts producible from a single RTV mold has been increasing owing to the development of new materials. For instance, Santin Engineering, Inc. (MA) is using a urethane material called RP 6453 R/H to quickly and inexpensively handle runs of up to 10,000 photocopier housing components (Wohlers 2002).

Another emerging approach to soft tooling fabrication is the Rapid Mold (RM) process developed by the University of Texas at Austin, TX. This process utilizes selective laser sintering, but in a manner that avoids the high tooling c ost a rising f rom t he n eed f or a m etal-grade f urnace (Tobin *et a l.* 1993) (Badrinarayan 1995) (Barlow and Vail 1994, Barlow *et al.* 1996). RM uses epoxy for infiltration with a view to avoiding the need for a metal-grade furnace. Relative to filled-epoxy tooling, RM offers the advantages of ease and speed of manufacture owing to good, hardness, thermal conductivity, and thermal expansivity. However, since epoxy is weaker than most metals, the strength of RM parts will be much lower than the steel used in conventional injection mold tooling. Hence RM is not recommended for the construction of the entire mold. Instead, the mold is redesigned with a composite structure where the cavity geometry is created by means of RM while mold strength is ensured by surrounding the cavity with a steel ring (Barlow *et al.* Barlow 1996). Details concerning the design of such steel rings are available in several publications (Tobin *et al.* 1993) (Apelskog-Killander 1996) (Beaman *et al.* 1997).

'Bridge tooling' is intended to fill the gap between soft tooling and production tooling. This gap is generally considered to be referring to the production of 20 to 500 parts in the desired material quickly and inexpensively (Hilton and Jacobs 2000).

Composite aluminum-filled epoxy (CAFÉ) tooling is a commonly used approach to bridge tooling. This process uses premixed finely ground aluminum powder and thermoset epoxy in the ratio 1:2. Before use, the mixture (CAFÉ material) i s degassed i n vacuum to eliminate air bubbles. When cured, the material is sufficiently strong for the production of thermoplastic parts. The method is inexpensive and fast if the parts are simple in shape. The process starts with a master pattern produced by means of a suitable RP process. If necessary, the pattern is sanded to eliminate stairstepping. This is followed with coating of a commercially available mold-release agent. Next a suitable parting line is selected on the part. If the line is plane, a simple wooden parting board will suffice in the next step. If it is complicated, one needs to machine a parting board. The master, the parting board, and any conformal cooling tubes needed are then accurately registered in a chase box. Next the CAFÉ material is poured on the master and allowed to cure. After curing, the master and the fully cured CAFÉ material are inverted and the parting board removed to yield the two halves of the mold. Subsequently the halves are subjected to secondary machining operations needed to establish ejector holes, etc. If the final part exhibits undercuts, the incorporation of suitable sliders may be needed in earlier steps. This last feature prevents the use of CAFÉ tooling for parts with very complex shapes.

One inexpensive way of improving the durability of epoxy-based composite molds produced by RTV-SRM is to deposit a thin layer of metal on the mold surface by means of arc spraying. Usually the metal is a zinc-based alloy (called kirksite) with excellent wear resistance. It is also possible to spray other metals. When metal spraying is adopted, an epoxy or low-melt alloy is sometimes used to create a backfill, so the cooling rate is improved. The advantages of metal spraying include improved mold durability and shortened injection cycle times. The limitation is that it is mainly applicable to large-sized molds of low-to-medium complexity.

Armstrong Mold Corp., NY, has developed a different approach so the production of kirksite shells of more complex shapes. The approach is similar to CAFÉ except that a rubber or urethane material is cast against a shrinkage-compensated master pattern. The coated master is then used to create the cavity and core patterns for casting with kirksite following an intermediate step of casting with plaster. Finally, the plaster is broken away to yield the kirksite core and cavity set that can be fit into a mold base. The resulting tooling is generally capable of producing up to 1000 thermoplastic parts. However, a disadvantage is that the mold is not as accurate as an epoxy or spray metal mold. Hence, it might be necessary to resort to some machining.

Another approach involving 'plaster casting' is Rubber Plaster Molding (RPM) adopted by Armstrong Mold Corp., NY, and American Precision Cast, IL (Wohlers 2002). In this approach, a stereolithography or other type of R P m odel i s u sed as t he p attern t o c reate a silicone r ubber m old f rom which thousands of plaster molds can be produced. If necessary, cores are incorporated. Molten metal is then poured into the plaster molds. The plaster molds are subsequently destroyed to facilitate the removal of the castings. An advantage of this approach is that thin-walled parts with complex shapes including undercuts can be produced. The approach is therefore preferred for low-volume production runs requiring good aesthetics.

A promising alternative to arc spraying of metals is the Rapid Tool Solidification (RSP) process developed by Idaho National Engineering and Environmental Laboratory (INEEL). This process involves the creation of a ceramic reversal of the pattern followed by spraying of tooling steel such as D2, H13, or P20. Molten tool steel is fed into a spraying nozzle under pressure, while the pattern is held in position by means of a 6-axis manual gripper. A high-velocity gas jet atomizes the spray. Metal deposition rates of over 200kg/h are possible. Spraying is continued until the required metal layer thickness is achieved. If required, the exterior walls are machined.

Another noteworthy approach to spraying of steel onto a ceramic pattern is the Sprayform process developed by Sprayform Holdings Ltd. of UK. This p rocess u ses t win w ire m etal a rc s pray g uns t o d eposit c arbon s teel. Investigations to find approaches to eliminate the requirement for an intermediate ceramic pattern have been in progress since the company's takeover by Ford Motor in 1999. Another goal is to be able to interrupt the process with a view to embedding CCC.

Another indirect approach applicable to rapid bridge tooling for injection molding etc. is the use of ceramic composite (NCC) tooling (Wise 1996, Hilton and Jacobs 2002). In such tooling, the hard active surfaces of the mold are made of electroformed nickel backed by a high-strength ceramic while the composite is held in a containment unit. The advantages of using nickel include good thermal conductivity, corrosion resistance and polishing ability while ensuring reasonable abrasion resistance. The nickel shell, ceramic backing, and containment unit move together during injection molding, so delamination is prevented. Amongst the various NCC implementations, the 'two-sided single model' approach is particularly noteworthy (see Figure 9-1) (Hilton and Jacobs 2000).

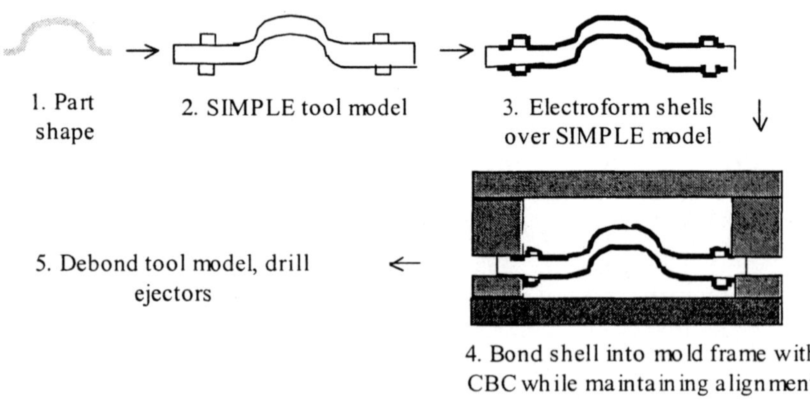

Figure 9-1. NCC tooling, adapted from (Wise 2000).

The process starts with the definition of the parting surface from the CAD model of the part. The parting surface also includes the necessary mold alignment features. The parting surface is then expanded in a linear fashion in the z-direction to determine the geometry of a 'single integrated matching plate electroforming (SIMPLE)' tool model. The amount of z-expansion is set to ensure good stability during subsequent electroforming. Next, the nickel shells are electroformed on the SIMPLE mandrel using conditions favoring good dimensional stability. The term 'mandrel' essentially denotes a pattern used in a process involving certain processes such as electroforming. In view of the good support provided by the ceramic backing, the plating thickness is set somewhat smaller than that usually applied in conventional nickel electroforming. The mandrel with electroformed nickel on both sides is next attached to the mold frame using a set of clamp-ring buss bars. Next, the ceramic is vacuum cast through a small opening in the back of the mold frame. This step is performed in four stages: casting on one side, waiting overnight, flipping over, and casting on the other side. Next the two sides are separated, the nickel surfaces polished if needed, and the ceramic hydrothermally cured. Some post-processing may be necessary for creating holes for ejector pins, etc. Since nickel is corrosion resistant, NCC tooling can be used for molding PVC and glass-reinforced materials. The process is also capable of handling large parts (up to 400mm). On the other hand, nickel is softer than most tool steels. This is the reason for including the process in this section on bridge tooling. However, cases of NCC tooling producing more than 46,000 shots have been reported (Hilton and Jacobs 2000).

Another nickel-based indirect approach to bridge tooling is the Nickel Vapor Deposition (NVD) process originally developed in the late 19th century by Ludwig Mond in Wales, UK. The key to NVD is nickel carbonyl gas, $Ni(CO)_4$ that has the useful property of rapidly breaking down into solid nickel (Ni) and carbon monoxide (CO) gas when heated to 110-190°C. Thus by heating a mandrel to the required temperature in a sealed chamber and passing carbonyl gas (produced previously by mixing nickel powder and carbon monoxide in a heated container) over it, an exact negative of the mandrel can be obtained. The deposition rate of nickel is in the range 0.05 to 0.75mm/h. Provided that the mandrel temperature is uniformly maintained over the entire mandrel surface, the nickel layer obtained via NVD is remarkably more uniform than that obtained via electroforming. Further NVD nickel is much more dens and ductile than NCC nickel. Also it polishes much better. However, a major problem is that few RP models (mandrels) are able to withstand the NVD temperatures. No definitive solution to this problem has so far emerged. Current solutions mainly use a mandrel consisting of a machined metal or epoxy mandrel backed by a composite with matched coefficient of thermal expansion. Another disadvantage of the process is that the gases involved are highly toxic, so they need to be handled carefully. However, overall, the process appears to have great promise in view of its accuracy.

9.3.2 Indirect Methods for Production Tooling

Tooling enabling normal production quantities is called production tooling. This means that production tooling must be much more durable than soft and bridge tooling. A widely used indirect RT process for production tooling is a powder metal process called Keltool™. The process was originally developed as early as in 1976 by 3M. The technology was sold to Keltool Inc., which, in turn was purchased by 3D Systems Inc. in 1996. Since that time, the process has been refined in many ways. As a result, the process is now called 3D Keltool. Today 3D Keltool is recognized as an extremely fast method of producing hard tool inserts for injection molding or die casting. Many companies have used the method to produce hundreds of thousands of parts (Wohlers 2002). However, only broad details of the process are generally known because of the proprietary nature of the process.

. The process starts with a master pattern (the core and cavity) produced by means of SL. Next, the master is finished, polished, and subjected to room temperature vulcanizing (RTV) in a platinum-base, silicone rubber, soft mold. When the SL-master is taken out of the rubber mold, it leaves behind the negative pattern corresponding to the master. Next, a mixture of A-6 powder steel slurry and a binder is poured into the soft mold. Once the slurry has settled into the rubber and hardened, the insert is taken out and placed in a furnace so as to melt the binder out and bring the insert to a green state. Once all the binder is out, the insert is copper impregnated. This brings the tool to the consistency of 70% A-6 steel and 30% copper. Keltool is capable of producing tool inserts with tolerances of the order of 50μm that can be delivered within 8 to 10 days. Through appropriate heat treatment, the inserts can be hardened to 50 Rockwell C. However, the inserts still need to have the parting lines matched, core pins and ejector pins added, and be ground. However, insert dimensions cannot exceed 150mm because of the use of room temperature vulcanizing (RTV) silicone rubber (Hopkinson and Dickens 1997). The Keltool process has also found applications in the production of copper-tungsten EDM electrodes of complex geometry.

A problem with the Keltool process is that the extraction of the green compact from the RTV mold can be quite difficult (Zhou and He 1999). One way is to cut the rubber along the parting line. Often, this is not a trivial task. Another way is to place the whole mold in a furnace and burn out the silicone rubber. This causes pollution.

Zhou and He have explored a technique called rapid prototyping based powder sintering (RPBPS) to avoid the rubber molding step. RPBPS starts with a pattern made on a 3D Printing RP machine of Sanders Prototype, Inc. (now Solidscape, Inc.). This step is essentially similar to the Keltool process, except that the pattern material has a melting point of 105°C. Further, instead of casting a rubber mold around it as in Keltool, the pattern is placed in a flask or a box and a thermosetting powder binder mixture is cast around it. The pattern with the low melting point is then easily removed by heating the green compact (the cast). After the green part is thus free of the pattern, it is sintered and/or infiltrated as with the Keltool process.

Another powder-based production tooling process is Polysteel I™ developed originally by Dynamic Tooling. In this process, mold inserts are produced from a semi-solid polymer/steel compound that is formed directly over a stereolithography or other rapid prototype master. The pouring process is done separately on the two sides of the parting line block. Since a mirror image of the pattern is produced, there is no need for secondary machining or polishing. Since it is a one step process, inaccuracies generally associated with other multi-step processes are eliminated. PolySteel has a very low shrinkage and is 300-400% stronger and more thermally conductive than conventional aluminum-filled epoxy and more abrasion resistant than 6061T6 aluminum. The process is mainly applicable to high volume production. It is possible to include conformal cooling channels in PolySteel molds.

A process similar to PolySteel, called EcoTool, has been developed by the Danish Technological Institute (DTI). This process involves the pouring of EcoTool material on either side of the parting line block produced by SL, LOM, or a RP wax pattern-making process. The poured material hardens at room temperature, but has a very high melting point. Porosity of the material can range up to 50%, so it is possible to infiltrate metal alloys directly into the inserts without requiring subsequent sintering. The process has been used to create tooling for glass-filled plastic injection molding, blow molding, vacuum forming, and aluminum pressure die-casting (Wohlers 2002).

Another major technology for direct fabrication of production tooling (100,000 to 270,000 shots) is the Expert Tool process developed by Hasbro Corp. (PI) and Laser Fare, inc., PI, following a collaboration initiated in 1992 (Hilton and Jacobs 2000). Since Hasbro is a toy making company, the process has been developed mainly for the fabrication of plastic injection molds. The process starts with a CNC-machined mandrel. Usually an easily machined material such as Ciba REN Shape 540 tooling board, or Ciba Express aluminum-reinforced polymer board is selected as the mandrel material. It is possible to produce such a mandrel in the same overall time as required to build a similar one using an existing RP process (Filipak 1998). Next the mandrel is coated with an electrically conductive material. The favored coating processes include spraying a silver paint or a mixture of silver nitrate and a reducing solution. The mandrel is next subjected to nickel electroforming using a set of empirically optimized conditions. The typical thickness of nickel layer is about 2mm. The nickel shell along with the mandrel is inverted and any CCC needed are bent and positioned. With a view to reducing part distortion and injection molding cycle time, the layout of the CCC is determined by a finite-element program that minimizes the active mold surface temperature difference. Next the entire assembly is placed in a secondary electroforming bath so as to form a layer of copper on the nickel shell. The superior thermal conductivity helps in thermal management of the mold. Moreover, electroforming of copper proceeds much faster than that of nickel. Next, the mandrel-nickel shell-copper layer-CCC ensemble is backed by a material such as an epoxy with good compressive strength and rapid curing properties. The final operations include positioning the core and cavity inserts in the mold frame and drilling the ejector holes.

9.3.3 Direct RT Methods for Soft and Bridge Tooling

When the glass transition temperature of the part to be built is low enough, it is possible to use SL for building soft tooling. For instance Colton *et al.* successfully produced parts made of Chevron MC3600 polystyrene from SLA-produced injection molds made of Ciba Tool 7510 built on 3D Systems SLA 3500 (Colton *et al.* 2001). However, they found that the mechanical properties of the mold deteriorated at higher molding temperatures. In short, unless the temperatures to be withstood are not sufficiently low, SL is not suitable for direct fabrication of soft tooling.

Many industries are using SL tooling for low volume production. For instance, Align Technologies uses one-off SL moulds to produce thousands of orthodontic aligners (Hopkinson and Dickens 2001). A popular direct approach to bridge tooling is Direct ACES Injection Molding (Direct AIM) demonstrated first in 1995. As already noted in Chapter 6, ACES is a stereolithography part-building style that aims to produce accurate, clear, epoxy, solid parts with highly improved dimensional stability. The method is said to be 'direct' because the core and cavity inserts of a plastic injection mold are directly built on an SLA machine. Although the glass transition temperature of the SL material (e.g., SL 5170 epoxy resin) used for making the insert is just 85°C, it is possible to inject thermoplastics at up to 300°C without damaging the parts (Hilton and Jacobs 2000). In some cases, injection at even higher temperatures (400°C) has been successfully performed. Thus, interestingly, ACES-built tooling can sometimes outperform even metal molds (Wohlers 2002), so it is possible to make parts of LDPE, HDPE, PS, PP, and ABS from ACES-built cavity and core sets. The main advantage of Direct AIM is that no secondary processes are required. The disadvantages include possible damage to inserts during ejection, the need for long cycle times owing to the poor thermal conductivity of SL resins, and poor strength and abrasion resistance of the inserts. As a result, the technology is applicable to production runs of between 20 and 50 parts only.

A recent review of the effects of various RT routes on part properties has indicated conflicting observations concerning the performance of SL tooling relative to conventional steel tooling for the same part material (Segal and Campbell 2001). For instance, whereas Dusel *et al.* found that parts molded in SL resin tooling had a higher UTS than those molded in steel tools (Dusel *et al.* 1998), Dawson and Muzzy found exactly the reverse just an year later (Dawson and Muzy 1999).

9.3.4 Direct RT Methods for Production Tooling

Selective laser sintering (SLS) has been used to produce durable tooling components directly from CAD data. The University of Texas at Austin, TX, had introduced the first metal-based rapid tooling process, called RapidToolT^{M} (RT) in 1995 (Hejmadi and McAlea 1996)). Since we have already described the SLS process in Chapter 7, there is no need to describe it here.

The original materials for this process were thermoplastics such as PVC, PC, nylon, and wax (for investment casting). Subsequently DTM Corp, Austin, TX, extended SLS to include an acrylic-based powder called TrueForm PM, and RapidSteel$^{®}$ consisting of 1080 carbon steel powder of mean particle size 55mm that is coated with 0.8wt% polymer. When the latter material is used, the polymer melts or softens under laser radiation and produces green parts consisting of metal particles held together by polymer-polymer bonds. Both the densities of the powder bed and green parts are equal (4.3g/cm^{3}), i.e., there is little shrinkage during green part formation. The green part strength is the order of 2.8MPa, so the part is easily handled and features as small as 0.5mm can be built and cleaned. The powder is easily brushed off from the green part and reused while making further parts. The green parts are usually infiltrated with bronze in a subsequent operation. Although RapidSteel is expensive, it exhibits high thermal conductivity and has sufficient wear resistance to produce over 100,000 plastic parts and prototyping quantities of pressure die casting parts (King and Tansay 2002). The powder consists of polymer-coated stainless steel that is subsequently infiltrated with bronze.

Tooling for short run production of plastic parts can also be fabricated from copper polyamide (PA). Mold inserts made of this commercially available material have been able to produce 100-400 parts in polyethylene, straight or glass-filled polypropylene, polystyrene, and ABS. From these materials, SLS is capable of creating tooling features such as runners, gates, conformal cooling lines, ejector pin guides, sprues, and sprue pullers provided they have been included in the .STL file. It is also possible to incorporate CCC. Compared to many other rapid tooling processes, SLS has been found to be associated with the highest tool life, the least cost, and the lowest time (Chua *et al.* 1999). A costing model for injection mold tooling produced by indirect SLS is available in (Dalgarno and Stewart 2001).

DTM is offering two further SLS-RT materials Sandform^{TM} Si and Zr (Agarwala *et al.*1999). The particles in these materials are polymer coated and require post processing at temperatures between 50 and 150°C. Another popular material is LaserForm ST-100 consisting of particles of 420 series stainless steel coated with thin layers of a polymer binder. Green parts built from the steel powder are sufficiently porous to permit 40wt% bronze infiltration.

Another noteworthy SLS-RT material is the nickel-bronze mixture used on EOS machines for direct metal laser sintering (DMLS). The main advantage is its very low shrinkage, so high dimensional accuracy of the order of ±0.05mm is easily achieved. However, the fact that the green parts are infiltrated with epoxy causes some limitations with respect to tool life and the application of secondary processes such as electric discharge machining (EDM) (Radstok 1999).

A problem with SLS-RT is that there can be feature distortion and variable rates of settlement in different part regions owing to gravity-induced creep of the lubricated part structure during a furnace cycle. To avoid this, the green parts may be soaked in a dilute aqueous acrylic emulsion containing a cross-linker. Sometimes, the cross-linker is directly added to the polymer. The green parts are then placed in a high temperature furnace to remove the polymer and sinter the metal particles directly through metal-metal bonds. The furnace atmosphere consists of 30% hydrogen that is helpful in reducing oxides, and 70% nitrogen that is helpful in maintaining carbon levels and preventing copper-induced swelling, if any. The polymer is removed during two temperature holds, the first for five hours at 350°C and the second for a further five hours at 450°C. The degraded polymer products are removed by means of a circulating gas mixture. In this state, the particles are held together by friction. Next the part is held at 1000°C for 8 hours so as to sinter at a significant rate. Alternatively, other processes such as HIP can be applied. In either case, it is possible to achieve full densification, but with the undesirable consequence of high shrinkage and part inaccuracy. A better method is to aim at partial sintering and fill the pores subsequently by means of infiltration of a low melting metal.

9.4 OTHER RT APPROACHES

Several other commercial or patented approaches RT methods are available in addition to those described above (Wohlers 2002). These include indirect RT processes such as Swiftool of Swift Technologies Ltd, UK; PHAST (Prototype Hard and Soft Tooling) of Proctor & Gamble; and a form of sand casting called the V-process. Likewise, there are direct RT processes such as Direct Metal Deposition (DMD) of POM Group Inc., MI, and an ultrasonic welding technology developed by Solidica Inc., MI. Outlines of these processes are available elsewhere (Hilton and Jacobs 2000, Wohlers 2002).

An interesting flexible solution to rapid bridge tooling is Space Puzzle Molding (SPM) of Protoform North America, NY. A SPM mold is designed with many parting lines and built from a set of premachined aluminum or steel tooling components. Some components are produced by Direct AIM or epoxy casting. The components are held together by a patented mold-holding device. Depending on mold size, component loading can be manual or mechanically assisted. The molds need no ejector pins and are more compact than conventional molds. However, production volume per SPM is usually in the range 500 to 1000 shots.

Chapter 10

APPLICATIONS OF RP

RP is applicable in any industry engaged in the production of consumer or engineering products, or tooling. Thus, RP and associated processes are used extensively for automotive, aerospace, aeronautics, railway and electronics applications. Several RP and RT processes and process routes and some areas of application have been described in previous chapters.

RP is of strategic importance wherever product innovation or time-to-market are important. An example of such an industry is the automotive industry, so RP/RT plays a key role in maintaining the competitive advantage of every automobile manufacturing company. For instance, as early as the start of 1990, Daimler-Benz AG had started replacing its traditional model-making approaches by RP techniques such as streolithography and laser sintering processes. By 1996, the RP activity in the company had grown to such a level and complexity as to prompt it to form a rapid prototyping consortium aimed at coordinating internal RP activities, monitoring international projects, and exchanging information among members (Widemann and Jantzen 1999). As a result, the company's RP portfolio expanded to include reverse engineering, LOM (e.g., for mock-ups of exhaust systems and engines), rapid tooling, precision casting from SLS models (e.g., for aluminum engine bearers), and sand sintering (e.g., for water jackets).

This chapter summarizes emerging applications of the various RP processes described in previous chapters in areas beyond rapid fabrication of engineering parts and tooling. In fact, the term 'part' can be misleading since it denotes a component satisfying a specified function in an assembly but not necessarily to, say, a biological model of a virus, a human 3D portrait, or a 3D 'map' of the surface of the moon. This chapter will also refer to such objects. However, the important area of virtual prototyping through collaborative design via the Internet, etc. will not be covered.

10.1   HETEROGENEOUS OBJECTS

Many objects contain regions with different materials. As early as in 1995, it was anticipated that RP would quickly progress from single material parts to multi-material parts (Chandru *et al.* 1995). We have already noted that some RP processes such as SDM (see Chapter 8) allow heterogeneous external objects such as sensors to be embedded while the part is being fabricated in a layered manner. Heterogeneous objects are those with clear material domains or with continuous material variations as in functionally graded materials (FGM). The latter are usually composites characterized by a continuously varying property such as composition, morphology, and crystal structure so as to attain a desirable response to mechanical, thermal, electromagnetic, or biochemical loading (Shishkovsky 2001). For instance, for FG filter elements, a structure and a porous connectivity can be defined to set up a preferred direction for fluid flow.

Traditional FGM fabrication processes include thermally sprayed coating, spark plasma sintering, self-propagated high temperature synthesis (SHS), c hemical v apor d eposition, p hoto-lithography, g alvanoplastics, s ol-gel processes, epitaxial growth, and etching. Owing to their analog nature, such processes are not easily automated. In contrast, the digital nature of most layered or voxel-based RP processes makes them highly suited for the production of FGM. Several of the RP processes described in previous chapters have a lready been applied f or the f abrication of FGM, e.g., SLS (Gureev *et al.* 1999 and Shishkovsky *et al.* 2001), LENS (Griffith *et al.* 1996, Griffith *et al.* 1997 and Hofmeister *et al.* 1999), DLF (Lewis *et al.* 1997 and Lewis and Schlienger 2000), DMD (Mazumdar *et al.* 1999), FDC (Safari 1999 and Haffiangadi and Bandyopadhyay 2000), and SDM (Fessler *et al.* 1997). Non-powder methods such as LOM, laser-stereolithography and Z-printer can also be used for FGM fabrication. For instance, parts with gradient color tones are very useful in medicine, architecture and art. Amongst the currently available RP processes, 3D Printing has been found to be particularly flexible owing to features such as drop-on-demand.

The majority of current RP systems utilize CSG or B-Rep part representations that assume uniform material distribution. Further, when the CAD model is converted into the .STL format, the file contains only the facet information without providing any clue regarding the nature of part interior. In view of these considerations, considerable effort has been put in recent years to extend current part representations to include FGM requirements. The first noteworthy step in this direction is the modeling of multi-material objects as generalized B-rep models through *r*-sets extended to include composition r_m-sets with accompanying Boolean operators (Kumar and Dutta 1997). The r_m object model is a set-based approach to the exact representation of part geometry and associating a material distribution function with this geometry. Within the Euclidean space $\mathbf{E}^3$, the product space $\mathbf{T}=\mathbf{E}^3\times\mathbf{R}^n$ forms the mathematical space to model object $\mathbf{S}$. At the same time, a material space $\mathbf{V}\subset\mathbf{R}^n$ restricting material points is defined. Each point p in $\mathbf{S}$ is a combination of n primary materials and is specified by the volume fractions of these primary materials. Thus, each point $p\in S$ can be modeled as a point $(x\in\mathbf{E}^3, v\in\mathbf{V})$ in $\mathbf{T}$, where x and v represent the geometric and material points. A method of combining r_m-sets with ISO 10303 (informally known as STEP—Standard for Exchange of Product Data Model) has recently been described in (Patil *et al.* 2002). However, such decomposition models do not maintain topological information about the model and are not as memory efficient as the more commonly used B-rep methods incorporating NURBS.

Material distributions can also be modeled via irregular tetrahedral decompositions of the finite-element (FE) type (Jackson *et al.* 1998). Many state-of-the-art CAD systems have features enabling the creation of tetrahedral meshes while maintaining topology through a 'graph of cells' describing the cell-tuple structure (Sachs *et al.* 1998). Every cell can then be associated with information about part composition and the geometry. Material space spanning the d_m materials available can be defined as M, and the material composition as a vector valued function $m(x)$ defined over the interior of the model, thus permitting specification of the overall variation in terms of the distance from a particular feature. This expression can then be used to obtain the volume fractions at the vertices of each tetrahedral element. The composition in the interior of the cell is then obtainable in terms of a set of control points and control compositions blended with barycentric Bernstein polynomials. Further, to enable simultaneous modeling of part geometry, topology, and composition, the solid model can be divided into sub-regions with each region being associated with analytic composition blending functions and distance functions (Jackson *et al.* 1999). In this local composition control (LCC) strategy, blending functions define the composition throughout the model as mixtures of the primary materials available to the RP process. Distance functions enable the specification of composition as a function of the distance from the surface of a part. It is also useful to apply a set of design rules restricting maximum and minimum concentrations.

Another method of accommodating both material and structure data has recently been developed at University of Hong Kong (Siu and Tan 2002). This approach uses a contour subdivision algorithm to decompose each layer into sub-contours according to different gradient variations. A 'gradient step-width' parameter is then defined to control the number of contours as well as the resolution of the grading. The approach has so far been applied to capture 1D variations. Issues concerning 3D variations are waiting to be clarified.

10.2 ASSEMBLIES

An assembly is an end product consisting of solids in point, line, or surface contact. The solids may be made of different materials. The contacts can be stationary, rolling, or sliding. Assemblies can be articulated by having moving (rolling or sliding) contacts.

The traditional approach to assembly fabrication involves producing the parts first and then assembling them one by one. The process of assembly itself can be manual or automated through mechanized or robotic automation. Much of the progress in assembly modeling has occurred from this traditional viewpoint. Many representational and reasoning (algorithmic, knowledge-based, expert systems-based, ANN-based, etc) schemes have been proposed to address issues such as tolerancing and mating (Giacometti and Chang 1990), connectivity and precedence (Lee and Gossard 1985) and assembly directions (Venuvinod 1993). Almost all these issues are closely related to the mating features on the parts. Hence, progress in assembly modeling has depended much on progress in geometric and technological feature recognition (Venuvinod and Yuen 1994, Venuvinod and Wong 1995, Venuvinod 1999, Yuen and Venuvinod 1999 and Yuen *et al.* 2002) in accordance with the precedence and assembly direction (Venuvinod 1993) requirements.

Following the remarkable strides made by rapid prototyping in recent years, the world of assembly is now poised to change or expand dramatically. Almost any RP process that can deposit more than one material in each layer and with supports easily removed can be modified to build MM assemblies. When a multi-material (MM) assembly is made by means of a layer-based RP process, all component regions within each layer are built within the corresponding layer-build cycle. Hence many traditional assembly issues such as precedence and assembly direction become irrelevant. However, information regarding location dimensions, tolerancing, mating relationships, and mating types continue to be highly important. Moreover, additional parameters effecting tool path generation (control of liquefier heads, etc.) need to be assimilated into the representational scheme. It is only recently that the problem of integrating such information into the representational schemes for MM objects has started receiving attention.

Zhu and Yu recently studied the problem of fabricating articulated assemblies using layer-based fabrication methods (Zhu and Yu 1998, Zhu and Yu 2000 and Zhu 2002). In their approach, a graph model organizes components in the given assembly by their boundary mating information to create an *n*-manifold (NM) representation An NM is a connected point set where every point has a neighborhood topologically equivalent to an open ball of E^n (Flegg 1974). Local mating information is also explicitly represented, so slicing operations can be performed directly on the model. In particular, slicing is performed using a dexel encoding approach that eliminates the need for intermediate tessellation of curved bounding surfaces.

The dexel approach works as follows. Firstly, the object is intersected by a set of parallel and equidistant rays (i.e., infinite straight lines). Next the points of intersection of each ray with the object are noted. Each pair of points defining a straight-line segment that is totally inside a given object material makes up a dexel. Dexels corresponding to different material regions within a ray via are identified through an algorithmic approach. The dexels so obtained are next fleshed out into boxes, so the object can be approximated by the resulting collections of boxes corresponding to different material regions. Thus, after slicing the MM assembly, there will be different 2D regions of heterogeneous materials. The dexel model of the object is then combined with the 3D assembly model to assimilate mating information. Tolerances are appended to the end points of each dexel such that positive, negative, and zero values indicate interference, clearance, and neutral fits respectively. Collision-free tool paths (paths of liquefier heads in FDM) are calculated on the basis based of a longest first scheduling approach. This novel approach facilitates filling of MM regions simultaneously and efficiently.

Zhu and Yu attempted to demonstrate the above approach by fabricating an articulated universal joint with nine components on a FDM 2000 machine. This machine was selected because it had the potential to accommodate more than one deposition head. However, all components were made from a single material. The finished assembly was generally satisfactory except for the important fact that the clearances at the various mating surfaces could not be properly realized owing to the low resolution of the FDM process. This result indicates that current RP technologies need to evolve towards better precision if they are to be able to produce articulated assemblies.

10.3 MEMS AND OTHER SMALL OBJECTS

Researchers round the world were enthralled by the prospect of miniaturization when Kurt Peterson of IBM published a paper in 1982 entitled "Silicon as a Mechanical Material'. Armed with the insight that silicon has remarkable mechanical properties including a hardness and tensile yield strength similar to stainless steel, researchers started developing an entirely new field called MicroElectroMechanical Systems (MEMS) so as to be able to study and manipulate the microscopic world. MEMS combine mechanical and electrical components in packages ranging in size from sub-microns to centimeters. Hence, they are also known as micro-machines. MEMS devices are being designed for a wide range of applications such as sensors, controllers and actuators. They have also started playing prominent roles in fields such as biotechnology and machine intelligence. For instance researchers at the University of Connecticut have produced nanometer scale structures by directly cross-linking proteins through multi-photon photo-polymerization.

Traditional bulk-manufacture of MEMS has used machining (etching) techniques as well as IC (integrated circuit) fabrication methods from a multitude of materials including silicon, nickel, quartz, GaAs, ZnO, and TiNi. However, such methods need very expensive plants and are limited to materials that are not particularly desirable from a mechanical viewpoint. There also exist restrictions on device complexity. Many of these constraints disappear when additive processes are used. RP methods are particularly suitable for customized miniature parts exhibiting high geometric complexity. Some examples of devices produced via 'micro-RP' are micro-fluidic devices, optical and chemical sensors, actuators and photonic band-gap structures (Kathuria 2001). We have already described the micro-fabrication capabilities of some RP processes in previous chapters. We will now outline some further RP technologies specifically directed towards MEMS and other very small objects.

Stereolithography for MEMS was first reported in 1993 (Ikuta and Hirowatari 1993). Micro-stereolithography (μSL) is a relatively new process capable of producing 3D complex shapes in small-size polymer structures. The process has also been used for fabricating articulated micro-assemblies (Ikuta 1998). As in conventional SL, the process produces objects layer-by-layer through light-induced polymerization of a liquid resin. A two-photon photo-polymerization process has also been developed to achieve sub-micron resolution. By using a modified μSL process, even articulated micro-assemblies can be produced (Ikuta *et al.* 1998).

Zhang *et al.* have investigated a liquid monomer-based μSL system comprising an Ar^+ laser, a beam delivery system, computer-controlled precision x–y–z stages, a CAD design tool, and an in situ process monitoring systems (Zhang *et al.* 1998). Green parts produced by the system exhibit low flexural moduli. However, part stiffness can be increased to the bulk modulus t hrough p ost-fabrication e xposure to U V r adiation (Manias *e t a l.* 2001). With a resolution of around 1.2 μm, the system is capable of producing micro-tubes with high aspect ratio of 16, and real 3D micro-channels and micro-cones on silicon substrate. In addition, the system has successfully produced micro-gears. Ceramic components can be fabricated by mixing the UV-curable resin with concentrated ceramic powder. The resolution achievable w ith s uch m ixtures i s s trongly i nfluenced by o ptical scattering during the interaction of the laser beam with the mixture. However, it is possible to dope the mixture, so light scattering is minimized and the process precision and control are improved (Sun and Zhang 2002).

μSL machines are now available in the tabletop form. For instance, the apparatus developed at École Polytechnique Fèdèrale De Lausanne (EPFL), Switzerland is a tabletop machine using .STL files. A pattern generator is used to shape the light beam so that it contains the image of the layer to be built. This image is focused and projected on the surface of a photo-polymerizable resin. A shutter occults the light when the layer has been cured. The object is lowered and a new layer of fresh resin is spread on the surface of the partially built object. The process is repeated after the liquid surface has stabilized. The fabrication speed is about 3 to 6 layers per minute, which corresponds to about 1.5mm per hour.

A technology somewhat similar to μSL but with different capabilities is RMPD (Rapid Micro Product Development) commercialized by microTEC, Duisberg, Germany. This technology comes in two types. The first splits a laser into a number of parallel beams that are moved in a vectorial manner over the surface of a photopolymer. In the second type, an expanded light beam cum mask system simultaneously exposes multiple cross-sections over the entire surface of the photopolymer. It is possible to produce as many as 150,000 parts per hour on a single machine provided that the parts have relatively uniform cross-sections along their length. Both technologies use acrylic and epoxy photopolymers and are applicable to batch production of several parts. Multiple materials can be used in a single part. It is also possible to interrupt the process to embed tiny components of various kinds. Part size can be from a few microns to 50mm. At the same time, layers can be as thin as 1μm. Further, by introducing certain post-processing operations, surfaces with nano-scale levels of roughness and near-optical quality can be attained. These features are particularly useful in medical applications as well as for miniature sensors, optics and telecommunication. Since RMPD can create very fine channels in all three spatial directions, it can be used to produce medical products, biochemical sensors or fluidic structures. By itself, RMPD can only process plastics. However, in principle, the process can be combined with other technologies such as microinjection molding (MIM) to produce metal and ceramic parts.

A promising micro-fabrication process called Electrochemical Fabrication (EFAB^{TM}) was patented in 2000 by Adam Cohen of University of Southern California, Los Angeles, CA (Cohen 2000). In 2002, MEMGen, CA, started commercializing the technology. In this process, micro-devices are built tacking tens of thousands of layers of a few microns thickness. Each layer consists of a structural material and a sacrificial support material. Both are built by electroplating. The device is composed of the structural material and imbedded within the sacrificial material. The supporting sacrificial material is removed after the device has been built. The device is first selectively deposited using Instant Masking™ electroplating, so material is deposited on the substrate precisely where it is required. Next, the support material is blanket-plated and the layer is planarized. Finally, the second material is removed leaving behind a 3D multi-layer structure. The currently preferred structural material is nickel. The production volume can range from a few parts to millions.

Till recently, a neglected middle ground was the fabrication of mesoscale devices such as resistors, capacitors and inductors. To fill this gap, the Department of Defense, USA, is funding an R&D program called MICE (Mesoscopic Integrated Conformal Electronics) to develop a technology capable of directly writing miniaturized mesoscale electronics on any surface (Wohlers and Grimm 2002). The goal is to be able to use almost any material as a substrate, while eliminating conventional printed wiring boards, high-temperature processing, and noxious chemicals.

A recently developed technology filling this gap is Maskless Mesoscale Material Deposition (M^3D) of Optomec, Albuquerque, NM. The technology is similar to LENS technology offered by the same company, but the laser is not used to bind the powder. Powder feeding is achieved by two alternative means. Laser particle guidance (LPG) uses laser radiation pressure to move particles to the substrate via a hollow fiber. The technique is capable of depositing about 10,000 particles per second with an accuracy that can reach sub-micron levels. In contrast, flow guidance uses co-flow gas assist to deposit about a billion particles per second. With an accuracy of the order of 25µm, the latter deposition technique is much less accurate than LPG. Once the precursor materials are deposited by either method, they are densified and annealed to produce the best possible electrical and mechanical properties. If a high-temperature substrate is being used, the materials can be fired in an oven. However, M^3D can also deposit the materials and laser-sinter them on low-temperature substrates. The technology is capable of writing lines of electronic materials less than 25µm wide onto polymer, glass or ceramic substrates without using any masks or resists while manufacturing printed wiring boards (PWB) and other devices.

Another promising mesoscale technology is the system developed by Sciperio, Inc., OK. This technology utilizes direct-write deposition by a nozzle and quill-pen dispensing system. Line widths can range fro the size of a hair to that of a wire. Likewise, plastic cellular phone covers to living organisms can be written on any surface.

10.4   MEDICINE

In recent years, several RP technologies including stereolithography (SL) have started to be applied in pre-surgical planning through the fabrication of models of human organs, customized prostheses, tracheobronchial stents, etc. In the application of any RP process, one starts with the CAD model of the part. The CAD model itself can have many uses for doctors. They can manipulate the CAD images to potential scenarios. For instance, mirroring procedures can be applied to utilize body symmetry for diagnostics and for defining prostheses for correcting cranial defects due to trauma, etc. The prostheses can be fabricated just prior to surgery so as to simplify doctor-patient, doctor-doctor, doctor-insurance agent communications. If the models are accurate enough and the materials are right, they can be used as prostheses or surgical aids. Medical modeling and RP processes such as SL are proving to be very useful for surgeons, medical companies and medical researchers alike.

As early as in 1996, Swann presented a case study on the preparation of SL models of a human brain from magnetic resonance image (MRI) data (Swann 1996). There exist many similarities between the slicing procedures used in SL and the scan data represented captured in computer tomography (CT) and MR images. Some procedures for exploiting these similarities are described in (Lightman 1998). In CT scans, the pixel density is usually scaled with the average x-ray density of the medium being measured in a volume (voxel) of size equal to the distance between the scan planes. This image is filtered using a variable threshold to differentiate regions of interest, e.g., bone from tissue. Next, the resulting 'edgels' are spliced together using a smoothing interpolation algorithm so as to produce a set of continuous contours. The contours are then knitted together into a smooth 3D model with the aid of a surface-smoothing algorithm. From the 3D model, it is an easy step to produce an .STL file and input it for the production of the master pattern. The master pattern can be used to produce a plaster mold from which the required model or silicone rubber prostheses can be cast. A description of such a procedure for the fabrication of facial prosthetics is available in (Chua *et al.* 2000).

Tracheobronchial stents are implants providing scaffolding for stenosed conduit so as to help patients suffering from cancer growths narrowing the tracheobronchial system to breath easily. Lim *et al.* compared the fabrication of master patterns for covered tubular silicone stents using SL, SGC, and

SLS and found that only SL was able to produce wall thickness values as low as 0.5mm (Lim *et al.* 2002). The SL-produced stents were also 25-50% thinner than commercially available stents. SL also yielded the best surface finishes and the shortest fabrication times.

SL may also be applied for replicating human organs to facilitate structural testing. For instance, the bone masses of many females decrease substantially after menopause. Owing to the resulting architectural deterioration of bone tissue, there is increased bone fragility leading to proneness to fracture. This condition, osteoporosis, occurs mainly at cancellous bone sites, i.e., at sites exhibiting a porous structure made up of an array of bone fibers interspersed with bone marrow. The traditional way of s tudying t his d isease i nvolves d estructive t esting of r eal b one s amples. With the development of RP, it has now become possible to make physical models with porous structures similar to the real bones. In particular, SL-produced models have been found to be useful for the validation of ultrasound assessment (Langton *et al.* 1997) and the development of phantom images for MRI (Issa *et al.* 1998). The voxel data required for model building may be obtained by means of micro-computed tomography (μCT). However, when such data are converted to the .STL format needed for SL implementation, the resulting files turn out to be of sizes larger than the memory capacities of current RP systems. Algorithmic solutions to this problem have been discussed in (Sisias *et al.* 2002).

Ma *et al.* investigated the application of selective laser sintering (SLS) for fabricating drug delivery devices and models of human organs with complex geometry, such as femur and brain, from magnetic resonance imaging (MRI) data (Ma *et al.* 2001a and Ma *et al.* 2001b). In order to meet the requirements of modeling the complex surfaces of natural objects a new NURBS-based approach to complement the conventional voxel-based approach has been developed. The new approach is capable of modeling natural, artificial, as well as hybrid objects. A 3D medical object is divided into several components, so each component could be modeled independently from the corresponding MRI data through appropriate contour generation, surface fitting, and NURBS-based volume generation. The components are then combined through a voxelization process and Boolean operations. Empirical data are also combined at this stage. Different materials inside the object are colored differently. It is also possible for the surgeons and doctors to cut a part and add new geometric or artificial objects to the image. In order to meet the requirements of modeling, the voxelized data of all the NURBS-volumes are stored in the same dataset without any distinction. An isosurface extraction procedure based on the notion of 'marching cubes' (Lorence and Cline 1987) is applied with a view to generating an .STL file of the normal format, so it can be transferred directly to a SLS machine such as Sinterstation 2500 of DTM Corporation. The resulting models of femur and brain have been found to be satisfactory except for some coarseness and unwanted protrusions of about 1mm length owing to noise in MRI data and limitations of the SLS process. The system has also been used to fabricate drug delivery devices that enable body fluids to flow into the device through interconnected pores, hence dissolving the drug. The porosity in the prototype device so fabricated consisted of 1,024 hollow spheres of different voxel values and radii ranging from 1 to 2.5mm.

More recently, Liew *et al.* applied a different SLS method to fabricate drug delivery devices and other kinds of dual material devices (Liew *et al.* 2002a). In principle, the method is extendible to multi-material devices. The method depends on the principle that sintering leads to densification and, hence, to decrease in volume. Consequently, if a small region in the powder is sintered to a greater degree through multiple laser passes and/or increased laser power, that region will be 'depressed' in comparison to the surrounding sintered part. The depressed regions can be designed to for 'channels' within the part that can be filled subsequently with a second material (the drug). The principle can be implemented on a normal SLS machine by creating separate .STL files for the overall device and the channels and then using a part-sorting function to overlap the channel files on to the main .STL file. In a follow-up paper, Liew *et al.* investigated several techniques including electrography for depositing the secondary material into the channels (Liew *et al.* 2002b). However, the techniques were not integrated into an SLS machine.

There have also been medical applications utilizing RP methods other than SL and SLS. For instance, Nelson and Bailey developed methods for converting 3D Ultrasound (3DUS) data to .STL format so as to create anatomical visualization models of human embryos within the womb (Nelson and Bailey 2002). While preserving scaling information, the data were segmented using threshold and local operators to extract features of interest. Next the voxel raster coordinate data were converted to a set of polygons representing an isosurface. The .STL files were then transferred to an LOM machine and a drop-deposition (DD) machine used to selectively bond powder material with the aid of a liquid adhesive. Such fabrication requires 30 to 60 minutes depending on object size and complexity.

Most applications of medical models allow flexibility in material selection. However, models for intra-operative reference and implant design and testing during actual surgery require 'medical-grade' materials that allow sterilization and limited *in vivo* exposure to human tissue (Wohlers 2002). Several special materials facilitating such applications are commercially available: colorable Stereocol for SL from Vantico, Duraform

Polyamide for SLS from 3D Systems, and medical grade ABS for FDM from Stratasys.

More recently, commercial software such as SurgiCase of Materialise are becoming available for bringing surgeons closer to RP (Wohlers 2002). The system is being used by several periodontists, oral surgeons, and prosthodontists for simulations of implant placements, etc. Other medical RP software systems from Materialise include CT-Modeller, MedCAD, and Mimics/CTM. Such software system enables the user to import various medical imaging modalities, process the images, and export to .STL and other RP formats. Similar software systems from other sources include Velocity from Image3, and Analyze from Analyze Direct.

10.5   MISCELLANEOUS AREAS INVOLVING ART

Although it is misleading to divide human actions into 'science', 'art', and 'technology', one may say that 'art' is mainly the pursuit of pure aesthetic experience through creative effort. There is art in engineering as there is engineering in art. Both use technology to some degree to reach their respective goals. Just as an engineer strives to deliver some utilitarian value to some customers, a professional artist agonizes to deliver some aesthetic values to some customers. For some artists though, the immediate customer is the self.

A niche application of RP is in the creation of geographic information system (GIS) objects, i.e., 3D maps, using a color 3D Printer available from ZCorp. This machine is capable of applying three different adhesives containing pigments in the subtractive base colors cyan, magenta and yellow. When the three adhesives fuse, the result is a multicolored part. One problem though is that the .STL file format is unable to transmit color information. However, it is possible to sidestep the problem by using Virtual Reality Markup Language (VRML) that enables one to assign a color to each triangle (tessellation), parts of the model, or the entire model. Thus one can produce 4D (three spatial dimensions plus the color dimension) choropleth maps. Such maps are more effective in communicating geographic information than tables, contour maps or perspective drawings. Through lifelong experience of evaluating 3D scenes, humans have learnt to estimate distances in a 3D scene by stereopsis and minor changes of the eye point. A disadvantage of 3D/4D maps, however, is that other types of information need to be conveyed through the addition of legends.

Another niche application is in the reconstruction of fossils from fragmentary materials, an activity requiring morphological as well as artistic talents. Paleontologists traditionally create such objects by forming latex molds from actual fossils followed by epoxy casting. RP now provides a more elegant alternative as demonstrated by the report from Zhang *et al.* on the reconstruction of a homunculous skull (Zhang *et al.* 2000). They reverse-engineered three pieces of sharp epoxy casts from the left facial skull and the mandible by means of 3D laser scanning (Surveyor 2000 from Laser Design Inc.). A commercial software system (DataSculpt) was then used to develop models of the fully reconstructed face and lower jaw. This software is capable of converting the 3D coordinate data into image data, and transforming the shape information into a system of cross-sectional contours (polylines) that can be edited. As a result, one can remove unwanted artifacts such as drop-offs that record material outside the actual boundary of the object, or error spikes that are a reflection of dramatic geometrical discontinuities. It is also possible to replace distorted portions of an object. Finally, an SLA 250 machine was used to produce the physical models. A similar approach starting from laser scanning and ending up in an RP process is being used at I.I.T. Kanpur, India to record and archive morphological features of heritage objects. Such technology should also be useful to museums and archeologists.

Artistic creativity is of particular importance in the jewelry industry. Several RP manufacturers including Meiko in Japan and Solidscape the U.S. are concentrating on developing RP techniques to support jewelry industry. Digital design software such as Digital Jeweler™ and RP equipment such as ModelMaker II are being used by several companies. There are also service bureaus and university programs emphasizing the design and manufacture of jewelry using RP technology. For instance, Soo and Yu of the Hong Kong Polytechnic University note that contemporary CAD/CAM/RP systems are developed for Euclidean geometry and are incapable of handling aesthetic patterns exhibiting fractal (self-similar) geometry (Soo and Yu 2001). To address this problem, they developed the Radial-Annular Tree (RAT) data structure that is capable of representing iterated function systems (fractal curves). RP tool-paths are then generated from the RAT data structure. Geometry modeled in the RAT data structure can also be combined with Euclidean geometry derived from traditional CAD systems.

Most often, the output from a rapid prototyping system is a pattern for lost-wax, or other types of casting methods in jewelry manufacture. Direct manufacture of jewelry is not possible at present since precious materials cannot yet be effectively processed by the current RP methods. However, a few artists are exploring the use of the existing materials for processes such as selective laser sintering and stereolithography as final media.

While much has been written about the importance of drawing in architecture (Robbins 1994), the role of models and RP have started to receive attention only recently (Gibson *et al.* 2002). In the early stages of an architectural design cycle, quickly fabricated models are used to evaluate different design ideas. In later stages, one uses more elaborate models to communicate design ideas and details to colleagues, clients, and decision-making bodies. In both stages, models are used to supplement drawings since mass-void relationships and spatial sequences are more legibly communicated through the former (Koepke 1988). Models are also used to test constructibility and structural integrity. Models come in various sizes, large and small. Model material must be selected to serve the model's purpose. Large landscaped models may need to be built in parts and 'jigsawed' together. Architectural models are often deliberately made in different materials and colors to emphasize different aspects. Hence, RP processes that enable multiple materials and coloring are preferable. FDM is particularly attractive from the latter viewpoint because it can use a range of colored ABS plastic materials, or actually make bi-color models. It is also possible to place a color printer mechanism inside an SLS (Ling and Gibson 1998). However, of all existing RP processes, color 3DP is the most attractive in view of its color discrimination ability.

In addition to speed, the attraction of RP as a model-making tool lies in its ability to meld the digital with the physical. Currently SL, SLS and FDM are the technologies used by architects. These processes are fairly expensive and they may need to be obtained through a service vendor or bureau. For these reasons RP has not yet been applied to the fullest possible extent by architects in general. However, this is likely to change with the advent of less expensive and office-friendly desktop RP technologies such as 3DP.

Among all forms of pure art, RP is most immediately relevant to sculpture. Sculpture is the creation of 3D objects for artistic rather than utilitarian purposes. Michael Rees, a respected sculptor, thinks that RP "is the most powerful and flexible tool for sculptors that has ever been invented. Aside from what is known about rapid prototyping—that it is WYSIWYG [what you see is what you get] and an additive process—research into what can actually be accomplished in sculpture is in its infancy " (Rees 1999). Another sculptor has seen RP as a means for concretizing abstract specification of form.

RP bridges the gap between the digital world and the physical world. The digital world is particularly prone to mathematical manipulation. Thus, RP has the potential to link two fundamentally different fields, mathematics and art. There have been collaborative efforts between computer scientists and sculptors to produce a class of sculptures inspired by the mathematics of minimal surfaces. The shape complexity of some of these sculptures precludes them from being created by a non-RP process. Further, color 3DP can reach nooks not reachable by the paintbrush. The main RP processes used so far by sculptors are stereolithography and wax investment casting directly from .STL files.

Rees says that it was not until his introduction to additive fabrication that the full import of computer modeling became clear to him (Rees 1997). With digital-additive technology, he is able to imagine a convoluted surface, describe it within a computer using a 3D modeling program, and make it corporeal. He is deeply compelled by the aesthetics of additive fabrication. "The Aqualine creature" is a widely acclaimed sculpture created by him. As he has put it, this sculpture is a surrealist organism that combines the anatomical illustration of western science and the metaphysical illustration of eastern mysticism. SL afforded some interesting tools to realize this strange anatomy. In his imagination, he had conceived layers upon layers of objects that were reminiscent of an organic anatomy. Concurrently, he wanted the freedom to play with that anatomy and stretch its implications. He could solve all the problems of realizing these by using SL. The particular qualities of SL that enabled this sculpture were semi-transparency, staircase effect (Rees finds it "beautiful" whereas engineers generally strive to get rid of it!), and shaped displacement shading that can create a infinite variety of textures. Finally, he found that "[c]onvoluted shape, texture, and nesting all contribute to an accuracy of form which is unparalleled in sculpture (Rees 1997)."

Notwithstanding the enthusiasm exhibited by sculptors such as Rees, it appears from reports on the forum titled "RP for Art and Conceptual Design" and conducted as a part of the 2001 Web conference on "Future of RP" that the "use of RP in art has already reached a plateau and that RP is accepted just as a 'print-out" that enables the creation of sculptures that "seem to have sprung directly from the artist's mind" (Wai 2001). As some artists said at the forum, "from bits to atoms" might be attractive to engineers engaged in conceptual design, but it is not the way for artistic exploration and recursion.

However, it is quite conceivable that the emerging interaction between RP and art will actually benefit RP more than art. The world of RP has so far been dominated by engineers thinking within their narrow areas of interest in a goal-oriented and structured manner. In contrast, artists are spontaneous and driven by inspirational flashes. The involvement of artists should therefore come as a breath of fresh air to the field of RP. Perhaps RP engineers will start asking (and answering) such questions as those posed in (Rees 1999): "How does a sculptor build meaning in a work that is relevant to a scientist's language and vice versa?" "What will the availability of high level manufacturing technology at low-level prices mean to the CAD-literate community and to sculptors? Will the underprivileged have access to this technology? Will various disciplines be encouraged to work together to produce unusual applications and new proprietary technologies? Will the artist, the scientist and the engineer come closer together—perhaps practicing their art in a way that is more akin to the renaissance?"

References

3D Systems, Inc. (1989). *StereoLithography Interface Specification*, 3D Systems, Inc., Valencia, CA.

3D Systems, Inc. (1994). *SLC File Format Information*, p/n 50065-S03-00, 3D Systems Inc., Valencia, CA.

Abe, E., K. Osakada, M. Shiomi, K. Uematsu and M. Matsumoto (2001). The manufacturing of hard tools from metallic powders by selective laser melting. *J. Mat. Processing Tech.*. Vol. 1, pp. 210-213.

Ajh, S.J., W. Rauh, H.J. Warnecke (2001). Least squares orthogonal distances fitting of circle, sphere, ellipse, hyperbola, and parabola. *Pattern Recognition*, Vol. 34, pp. 2283-2303.

Agarwala, M., D. Bourell, J. Beaman, H. Marcus and J. Barlow (1995a). Direct selective laser sintering of metals. *Rapid Prototyping J.*, Vol. 1, No. 1, pp. 26-36.

Agarwala, M., D. Bourell, J. Beaman, H. Marcus and J. Barlow (1995b). Post-processing selective laser sintered metal parts. *Rapid Prototyping J.*, Vol. 1, No. 2, pp. 36-44.

Agarwala, M.K., R. van Weeren, R. Vaidyanathan, A. Bandyopadhyay, G. Carrasquillo, V. Jamalabad, N. Langrana, A. Safari, S.H. Garofalini and S.C. Danforth (1995c). Structural ceramics by fused deposition of ceramics. *Proc. 6th Solid Freeform Fabr. Symp.*, The University of Texas at Austin, Austin, TX, Vol. 6, pp. 1-8.

Agarwala, M.K., V.R. Jamalabad, N.A. Langrana, A. Safari, P.J. Whalen, P.J. and S.C. Danforth (1996a). Structural quality of parts processed by fused deposition. *Rapid Prototyping J.*, Vol. 2, No. 4, pp. 4-19.

Agarwala, M.K., R. van Weeren, A. Bandyopadhyaya, A. Safari and S.C. Danforth (1996b). Fused deposition of ceramics and metals: An overview. *Proc. 7th Solid Freeform Fabr. Symp.*, The University of Texas at Austin, Austin, TX, Vol. 7, pp. 385-392.

Agarwala, M., D. Klosterman, N. Osborne, A. Lightman, R. Dzugan, G. Rhodes and C. Nelson (1999). Hard metal tooling via SFF of ceramics and powder metallurgy. *Proc. 10th Solid Freeform Fabr. Symp.*, The University of Texas at Austin, Austin, TX, 1999.

Ahn, S.-H., M. Montero, D. Odell, S. Roundy, S. and P.K. Wright (2002). Aniosotropic material properties of fused deposition modeling of ABS. *Rapid Prototyping J.*, Vol. 8, No. 4, pp. 248-257.

Alexander, P., S. Allen and D. Dutta (1998). Part orientation and build cost determination in layered manufacturing, *Computer-Aided Design*, Vol. 30, No. 5, pp. 343-356.

Allcock, H.R. and F.W. Lampe (1990*). Contemporary Polymer Chemistry.* 2nd Ed., Prentice Hall, Englewood Cliffs, NJ.

Allen, S and D. Dutta (1994). On the computation of part orientation using support structures in layered manufacturing. *Proc. 5th Solid Freeform Fabr. Symp.*, The University of Texas at Austin, Austin, TX, pp. 259-69.

Allen, S and D. Dutta (1995). Determination and evaluation of support structures in layered manufacturing. *J. Design and Manufacturing*, Vol. 5, pp. 153-162.

Amon, C., S. Finger, R. Merz, F.B. Prinz, K. Schmalz and L. Weiss (1997). Shape deposition manufacturing with microcasting. *J. Japan Welding Soc..* Vol. 66, No. 4, pp. 64-69.

Anitha, R., S. Arunachalam, S. and P. Radhakrishnan (2001). Critical parameters influencing the quality of prototypes in fused deposition modeling. *J. Mat. Processing Tech..* Vol. 118, pp. 385-388.

Anon. (1976). *Instructions: DSC-2 Differential Scanning Calorimeter.* Perkin-Elmer Corp.

Anon. (1993). Allied Signal, *Preliminary Test Results for Exactomer 2202 SF.* Allied Signal Corp.

Anon. (1995). Rapid prototypes. *Mech. Eng.*, Vol. 117, No. 5, pp. 20.

Anon. (1995a). *SCR-300 Resin Catalogue: UV Curable Resin.* Nippon Synthetic Rubber Ltd., Japan, 1995.

Anon. (1995b). *Surface Finishing of SLS Parts.* DCN:8275-00541, DTM Corp.

Anon. (1996a). Rapid concept modelers. *Mech. Eng.*, Vol. 118, No. 1, pp. 64-66.

Anon. (1996b). A new dimension for office printers. *Mech. Eng..* Vol. 118, No. 3, pp. 112-114.

Anon., Ennex Fabrication Technologies (1996c). Ennex Fabrication Technologies, Conveyed-adherent autofab. *Rapid Prototyping Report*, Vol. 6, No. 8, pp. 2–4.

Anon. (1997). *Rapid Prototyping in Europe and Japan.* WTEC Inc., World Technology Evaluation Center, Baltimore, MD.

Anon. (1999). Sanders prototype rapid PatternMaker. *Rapid Prototyping Report.* Vol. 9, No. 5, pp. 6-7.

Anon. (2001). *Hardware Reference Manual for SLA Systems.* P/N 22139-M10-01, Rev. A, 3D Systems Inc.

Apelskog-Killander, L. (1996). Rapid mould: Epoxy-infiltrated laser sintered inserts. *Rapid Prototyping J.,* Vol. 2, pp. 34-40.

Ashley, S. (1991). Rapid prototyping systems. *Mech. Eng..* Vol. 113, No. 4, pp. 34-43.

Au, C. K. and M. M. F. Yuen (1995). Unified approach to NURBS curve shape modification. *Computer-Aided Design*, Vol. 27, No. 2, pp. 85-93.

Azariadis, P.N. (2003). Parameterization of clouds of unorganized points using dynamic base surfaces. To appear in *Computer-Aided Design.* (http://www.sciencedirect.com/).

Bablani, M. and A. Bagchi (1995). Quantification of errors in rapid prototyping processes, and determination of preferred orientation of parts. *Trans. of the NAMRI of SME.* Vol. 23, pp. 319-324.

Badrinarayanan, B. (1995). *Study of the Selective Laser Sintering of Metal-Polymer Powders.* Ph.D. Dissertation, The University of Texas at Austin, Austin, TX.

Baese, C. (1904*). Photographic Process for the Reproduction of Plastic Objects.* US Patent 774,549.

Balasubramanian, B. and J.W. Barlow (1990). Prediction of thermal conductivity of beds which contain polymer coated metal particles. *Proc 1st Solid Freeform Fabr. Symp.*, The University of Texas at Austin, Austin, TX, Vol. 1, pp. 91-99.

Balasubramanian, B. and J.W. Barlow (1991). Selective laser sintering of a copper-PMMA System. *Proc. 2nd Solid Freeform Fabr. Symp.*, The University of Texas at Austin, Austin, TX, Vol. 2, pp. 245-250.

Balasubramanian, B. and J.W. Barlow (1992). Metal parts from selective laser sintering of metal-polymer powders, *Proc. 3rd Solid Freeform Fabr. Symp.*, The University of Texas at Austin, Austin, TX, Vol. 3, pp. 141-150.

Balasubramanian, B. and J.W. Barlow (1995). Effect of processing parameters in SLS in metal/polymer powders. *Proc. 6th Solid Freeform Fabr. Symp.*, The University of Texas at Austin, Austin, TX, Vol. 6, pp. 55-63.

Bandyopadhyay, A., R. Panda, R., Janas, M.K. Agarwala, S.C. Danforth and A. Safari (1997). Processing of piezocomposites by fused deposition technique. *J. Amer. Cer. Soc..* Vol. 80, pp. 1366-1372.

Barequet, G. and Y. Kaplan (1998). A data front-end for layered manufacturing. *Computer-Aided Design*, Vol. 30, No. 4, pp. 231-243.

Barlage, W.B., C.C. Jara-Almonte, A. Bagchi, A.A. Ogale and R.L. Dooley (1992). Fibre/resin composite manufacturing using solid freeform fabrication. *Proc. 3rd Int. Conf. Rapid Prototyping*, Dayton, OH, pp. 15-24.

Barlow, J.W. (1992). Metallic and ceramic structures from selective laser sintering of metal-ceramic powders. *Proc. 3rd Solid Freeform Fabr. Symp.*, The University of Texas at Austin, Austin, TX, Vol. 3, pp. 73-76.

Barlow, J.W. and N.K. Vail (1994). *A Method for Producing High-Temperature Parts by Way of Low-Temperature Sintering.* US Patent 5,284,695.

Barlow, J.W., J.J. Beaman and B. Balasubramanian (1996). A rapid mold making system: Material properties and design considerations. *Rapid Prototyping J.*, Vol. 2, No. 3, pp. 4-15.

Beaman, J.L., J.W. Barlow, D.L. Bourell, R.H. Crawford, H.L. Marcus and K.P. McAlea (1997). *Solid Freeform Fabrication: A New Direction in Manufacturing.* Kluwer Academic Publishers, London, UK.

Beardsley, H.E. and R. Kovacevic (1998). Controlling heat input and metal transfer for 3D welding-based rapid prototyping. *Proc. 5th Int. Conf. Trends in Welding Res.*, Pine Mountain, GA.

Benkó, P. and T. Várady (2001). Algorithms for reverse engineering boundary representation models. *Computer-Aided Design*, Vol. 33, No. 11, pp. 839-851.

Benkó, P. and T. Várady (2003). Segmentation methods for smooth point regions of conventional engineering objects. To appear in *Computer-Aided Design*. (http://www.sciencedirect.com/).

Bennett G. (1997). *Rapid Prototyping Casebook*, Mechanical Engineering Publications, London.

Bernard, A. and G. Taillandier (1998). *Le Prototypage Rapide.* Hermes Science Publications, France.

Bertoldi, M., M.A. Yardimci, C.M. Pistor, C.M., S.I. Guceri and G., Sala (1998). Mechanical characterization of parts processed via fused deposition. *Proc. 9th Solid Freeform Fabr. Symp.*, The University of Texas at Austin, Austin, TX, Vol. 9, pp. 557-565.

Binnard, M. (2001). *Design by Composition for Rapid Prototyping,* Kluwer Academic Publishers, Boston, MA.

Bird, R.B., W.L. Stewart and E.N. Lightfoot (1960). *Transport Phenomena.* John Wiley & Sons, New York, NY.

Bird, R.B., R.C. Armstrong and O. Hassager (1977). *Dynamics of Polymer Fluids,* Vol. 1, Wiley, New York, NY.

Billimeyer, F.W. Jr. (1962). *Textbook of Polymer Science*. Interscience Publishers, New York, NY.

Bjørke, Ø. (1991). How to make stereolithography into a practical tool for tool production. *Annals of CIRP*. Vol. 40, No. 1, pp. 175-177.

Bjørke, Ø. (1992). *Layer Manufacturing*. Tapir Publisher.

Blair, B.M. (1998). *Post-build Processing of Stereolithography Molds*. MS Thesis, Dept of Mech. Eng., Georgia Institute of Technology, Atlanta, GA.

Blanther, J.E. (1892). *Manufacture of Contour Relief Maps*. US Patent 473,901.

Bloomenthal, J., C. Bajaj, J. Blinn, M.-P. Cani-Gascuel, B. Wyvill and G. Wyvill (1997). *Introduction to Implicit Surfaces*, Morgan Kaufmann Publishers, Inc., San Francisco, California.

Bohn, J.H. (1997). File format requirements for the rapid prototyping technologies of tomorrow. *Proc. Int. Conf. Manufacturing Automation*, University of Hong Kong, Vol. 2, pp. 878-83.

Boillat and E.R. Glardon (2000), Numerical simulation of the selective laser sintering process. *Proc. 16th IMACS World Congress*, pp. 1-7.

Bourell, D.L., H.L. Marcus, J.W. Barlow and J.L. Beaman (1992). Selective laser sintering of metals and ceramics. *Int. J. Powder Metallurgy*. Vol. 4, pp. 369-381.

Bovey, F.A., I.M. Kolthoff, A.I. Medalin and B.J. Meehan (1955). *Emulsion Polymerization*. Interscience Publishers, New York, NY.

Boyd, D..P. and M. J. Lipton (1983). *Proc. IEEE,* Vol. 71, pp.298-307.

Brandrup, J. and E.H. Immergut (1989). *Polymer Handbook*. 3rd Ed., John Wiley & Sons, New York, NY.

Bugeda, G.B., M. Cervera and G. Lombera (1999). Numerical prediction of temperature and density distribution in selective laser sintering processes. *Rapid Prototyping J.,* Vol. 5, No. 1, pp. 21-26.

Bunnell, D.E., S. Das, D.L. Bourell, J.J. Beaman and H.L. Marcus (1995). Fundamentals of liquid phase sintering during selective laser sintering. *Proc. 6th Solid Freeform Fabr. Symp.*, The University of Texas at Austin, Austin, TX, Vol. 6, pp. 440-447.

Burns, M. (1993). *Automated Fabrication: Improving Productivity in Manufacturing*. Prentice Hall, Englewood Cliffs, New Jersey, NJ.

Burns, M. and J. Howison (2001). Internet delivery of physical products. *Rapid prototyping J.*, Vol. 7, No. 4, pp. 194-196.

Byun, H.-S. H.-J. Shin and K.H. Lee (2002). Design of benchmarking part and selection of optimal rapid prototyping processes. *Proc. 2nd Int. Conf. Rapid Prototyping and Manuf.*, 2002, Beijing, China, pp. 469-477.

Carrión, A. (1997). Technology forecast on ink-jet head technology: Applications in rapid prototyping. *Rapid Prototyping J.,* Vol. 3, No. 3, pp. 99-115.

Carslaw, H.S. and J.C. Jaeger (1959). *Conduction of Heat in Solids*. 2nd Ed., Oxford at Clarendon Press, London, UK.

Carter, W.T. and M.G. Jones (1993). Direct laser sintering of metals. *Proc. 4th Solid Freeform Fabr. Symp.*, The University of Texas at Austin, Austin, TX, Vol. 4, pp. 51-59.

Catmull, E. and J. Clark (1978). Recursively generated B-spline surfaces on arbitrary topological meshes. *Computer-Aided Design*, Vol. 10, No. 6, pp. 350-355.

Chalasani, K., L. Jones and L. Roscoe (1995). Support generation for fused deposition modeling. *Proc. 6th Solid Freeform Fabr. Symp.*, The University of Texas at Austin, Austin, TX, Vol. 6, pp. 229-241.

Chandru, V., S. Manohar and C.E. Prakash. (1995). Voxel-based modeling for layered manufacturing. *IEEE Computer Graphics and Applications,* pp. 42-47.

Charan, R., T. Renault, A.A. Ogale and A. Bagchi (1994). Automated fiber-reinforced composite prototypes. *Proc. 5th Int. Conf. Rapid Prototyping*, Dayton, OH, pp. 91-97.

Chartier, T., C. Chaput, F. Doreau and M. Loiseau (2002). Stereolithography of structurally complex ceramic parts. *J. Material Sci.*. Vol. 37, pp. 3141-3147.

Chartoff, R.P., B. Priore, D.A. Klosterman and S.S. Pak (1996). Composite tooling via laminated object manufacturing, A rapid and affordable method. *Proc 28th Int. SAMPE Tech. Conf.*, Seattle, WA.

Chartoff, R.P. and J. Du (1999). Photopolymerization reaction rates by reflectance real time infrared spectroscopy: Application to stereolithography resins. *Proc. 10th Solid Freeform Fabr. Symp.*, The University of Texas at Austin, Austin, TX, Vol. 10, pp. 298-304.

Chartoff, R.P., J.W. Schultz and J.S. Ullett (1998). Processing and properties of novel liquid crystal resins for stereolithography. *RadTech Report*, RadTech Int., North America, Vol. 12, No. 5, pp. 20-25.

Cheah, C.M., J.Y.H. Fuh, A.Y.C. Nee and L. Lu, (1999). Mechanical characteristics of fiber-filled photo-polymers used in stereolithography. *Rapid Prototyping J.*, Vol. 5, No. 3, pp. 112-119.

Cheah, C.M. (2001).*Mechanical Characterization of Photosensitive Material Used in Rapid Prototyping*, Ph.D. Dissertation, National University of Singapore, Singapore.

Chen, C.C. and P.A. Sullivan (1996). Predicting total build-time and the resultant cure depth of the 3-D stereolithography process. *Rapid Prototyping J.*, Vol. 2, No. 4, pp. 27-40.

Chen, Y. H. and C. Y. Liu (1997). Robust segmentation of CMM data based on NURBS. *The Int. J. Adv. Manuf. Tech.*, Vol. 13, pp. 530-534.

Cheng, W., J.Y.H. Fuh, A.Y.C. Nee, Y.S. Wong, H.T. Lo and T. Miyazawa (1995). Multi-objective optimization of part-orientation in stereolithography. *Rapid Prototyping J.*, Vol. 1, No. 4, pp. 12-23.

Childs, T.H.C. and N.P. Juster (1994). Linear and geometric accuracies from layer manufacturing. *Annals of CIRP*, Vol. 43, No. 1, pp. 163-166.

Childs, T.H.C., M. Berzine, G.R. Ryder and A.E. Tontowi (1999). Selective laser sintering of an amorphous polymer—Simulation and experiments. *J. Eng. Manufacture*, Proc. I. Mech. E., Part B, Vol. 213, No. B4, pp. :333-349.

Childs, T.H.C., C. Hauser, C.M. Taylor and A.E. Tontowi (2000). Simulation and experimental verification of crystalline polymer and direct metal selective laser sintering. *Proc. 11th Solid Freeform Fabr. Symp.*, The University of Texas at Austin, Austin, TX.

Childs, T.H.C. and A.E. Tontowi (2001). Selective laser sintering of a crystalline and glass-filled crystalline polymer: Experiments and simulations. *Proc. Inst. Mech. Engrs*, Vol. 215, No. B, pp. 1481-1495.

Chivate, P. N. and A.G. Jablokow (1993). Solid-model generation from measured point data. *Computer-Aided Design*, Vol. 25, No. 9, pp. 587-600.

Chivate, P. N. and A. G. Jablokow (1995). Review of surface representations and fitting for reverse engineering. *Computer Integrated Manuf. Systems*, Vol. 8, No. 3, pp. 193-204.

Cho, I., K. Lee, W. Choi and Y.-A. Song (2000). Development of a new sheet deposition type rapid prototyping system. *Int. J. Machine Tools & Manufacture*, Vol. 40, pp. 1813-1829.

Chua C.K. and L.K. Fai (1997). *Rapid Prototyping: Principles & Applications in Manufacturing*. John Wiley & Sons, New York, NY.

Chua, C.K., K.H. Hong and S.L. Ho (1999). Rapid tooling technology, Part I: A comparative study. *Int. J. Adv. Manuf. Tech.* Vol. 15, pp. 604-608.

Chua, C.K., C.S. Meng, L.S. Ching, L.S. Teik and S.C. Aung (2000). Facial prosthetic model fabrication using rapid prototyping tools. *Integrated Manuf. Sys.*. Vol. 11, No. 1, pp. 42-53.

Chua, C.K., K.F. Leong, C.M. Cheah and S.W. Chua (2003). Development of a Tissue Engineering Scaffold Structure Library for Rapid Prototyping. Part 2: Parametric Library and Assembly Program. *Int. J. Advanced Manuf. Tech.*, Vol. 21, pp. 302–312.

Cielo, P. (1988). *Optical techniques for Industrial Inspection*, Academic Press, Boston.

Cima, M., J. Yoo, S. Khanuja, M. Rynerson, D. Nammour, B. Giritlioglu, B., J. Grau and E. Sachs (1995). Structural ceramic components by 3D printing *Proc. 6th Solid Freeform Fabr. Symp.*, The University of Texas at Austin, Austin, TX, Vol. 6, pp. 479-488.

Ciraud, P.A. (1972). *Process and device for the Manufacture of Any Objects Desired from Any Meltable Material.* FRG Disclosure Publication 2,263,777.

Clarke, T., S. Robson and J. Chen (1993). A comparison of the three measurements for the 3-D measurement of turbine blades. *Proc. SPIE - Measurement Technology and Intelligent Instruments*, Vol. 2101, pp.1-12.

Cohen, A. L. (1991). Technology focus: Laminated object manufacturing. *Rapid Prototyping Report* Vol. 1, pp. 6- 8.

Cohen, A.L. (2000). *Method for electrochemical fabrication.* US Patent 6027630.

Colton, J. and B. Blair (1999). Experimental study of post-build cure of stereolithography polymers for injection molds. *Rapid Prototyping J.*, Vol. 5, No. 2, pp. 72-81.

Colton, J.S., J. Crawford, G. Pham and V. Rodet (2001). Failure of rapid prototype molds during injection molding. *Annals of CIRP.* Vol. 50, No. 1, pp. 129-132.

Comb, J.W. and W.R. Priedeman (1993). Control parameters and material selection criteria for rapid prototyping systems. *Proc. 4th Solid Freeform Fabr. Symp.*, The University of Texas at Austin, Austin, TX, Vol. 4, pp. 86-91.

Comb, J.W., W.R. Priedeman and P.W. Turley (1994). FDM technology process improvements. *Proc. 5th Solid Freeform Fabr. Symp.*, The University of Texas at Austin, Austin, TX, Vol. 5, pp. 42-49.

Cooper, A.G., S. Kang, J.W. Kietzman, F.B. Prinz, J.L. Lombardi and L. Weiss (1998). Automated fabrication of complex molded parts using Mold SDM. *Proc. 9th Solid Freeform Fabr. Symp.*, The University of Texas at Austin, Austin, TX.

Cooper, K.G. (2001). *Rapid Prototyping Technology: Selection and Application.* Marcel Dekker, New York, N.Y.

Copley, D.C., J.W. Eberhard and G.A. Mohr (1994). Computed tomography, Part 1: introduction and industrial applications. *JOM*, Vol. 46, No. 1, pp.14-26.

Cormier, D., K. Unnanon and E. Sanii (2000). Specifying non-uniform cusp heights as a potential aid for adaptive slicing. *Rapid Prototyping J.*, Vol. 6, No. 3, pp. 204-212.

Cox, H.L. (1952). The elasticity and strength of paper and other fibrous materials. *British. J. Applied Physics*, Vol. 3, pp. 72-79.

Cox, M.G. (1990). Linear algebra support modules for approximation and other software. In J.C. Mason and M.G. Cox (Eds.), *Scientific Software Systems.* Chapman and Hall, London, pp. 21-29.

Crafer, R.C. and P.J. Oakley (1992). *Laser Processing in Manufacturing.* Kluwer Academic Publishers, Boston, MA.

Crawford, R.H. (1993). Computer Aspects of Solid Freeform Fabrication: Geometry, Process Control, and Design. *Proc. 4thSolid Freeform Fabr. Symp.*, The University of Texas at Austin, Austin, TX, pp. 102-112.

Crivello, J.V. and J.J. Lam (1976a). *Polymer Science.* Vol. 6, pp. 383

Crivello, J.V. (1976b). *Method for Making Certain Halonium Salt Photoinitiators.* US Patent 3,981,897 1976.

Crivello, J.V., J. Lam and C.J. Volante (1977). *Radiation Curing.* Vol. 4, No. 3, pp. 2.

Crivello J.V. and J.H.W. Lam (1978). *J. Polymer Sci.*, Polymer Chemical Edition, Vol. 16, pp. 2441-2451.

Crivello, J.V. and J.J. Lam (1979). *Polymer Sci.*, Polymer Chemical Edition, Vol. 17, No. 4, pp. 2.

Das, S., J.J. Beaman, M. Wohlert, M. and D.L. Bourell (1998). Direct laser freeform fabrication of high performance metal components. *Rapid Prototyping J.*, Vol. 4, No. 3, pp. 112-117.

Dawson, E.K. and J. D. Muzzy (1999). The effect of rapid tooling on mechanical polystyrene properties. *Proc. Time Compression Technologies Conf.*, Nottingham, pp. 205-207.

De Boor, C. (1978). *A Practical Guide to Splines*, Springer-Verlag, Berlin.

Deckard, C.R. and J.J. Beaman (1987). Recent advances in selective laser sintering. *Proc. 14th Conf. Prodn. Res. and Technology*, Michigan, pp. 447-451.

Deckard, C.R. (1988). *Selective Laser Sintering*. Ph.D. Thesis, The University of Texas at Austin, Austin, TX.

Deckard, L. and T.D. Claar (1993). Fabrication of ceramic and metal matrix composites from selective laser sintered ceramic preforms. *Proc. 4th Solid Freeform Fabr. Symp.*, The University of Texas at Austin, Austin, TX, Vol. 4, pp. 215-222.

Delgarno, K.W. and T.D. Stewart (2001). Manufacture of production injection mould tooling incorporating conformal cooling channels via indirect selective laser sintering. *Proc. Inst. Mech. Engrs* Vol. 215, No. B, pp. 1323-1332.

DeWitt, D.P. and G.D. Nutter (1988). *Theory and Practice of Radiation Thermometry.* John Wiley & Sons, New York, NY.

Demkohler, G. (1937). *Fer Chemie-Ingenieur*, Vol. 3, No. 1, pp. 359.

Denton, K.R. (1995). The economics of rapid tooling and rapid prototyping. *Proc. 6th Int. Conf. Rapid Prototyping*, 1995, pp. 179-188.

Dickens, P.M., R. Cobb, I. Gibson, I. and M.S. Pridham (1993). Rapid prototyping using 3D welding. *J. Design and Manuf.*, No. 3.

Diehl, S. (2001). *Distributed Virtual Worlds*. Springer Verlag, New York, NY.

DiMatteo, P.L. (1976). *Method of Generating and Constructing Three-dimensional Bodies.* US Patent 3,932,923.

Dimitrov, D., W. van Wijck, K. Schreve and N. de Beer (20030. An investigation of the capability profile of the three-dimensional printing process with an emphasis on the achievable accuracy. *Annals of CIRP*, Vol. 52, No. 1, pp. 189-192.

Dolenc, A. and I. Mäkelä (1992). LEAF: A data exchange format for LMT processes. *Proc. 3rd Int. Conf. Rapid Prototyping*, Dayton, Ohio, pp. 155-160.

Dolenc, A. and I. Mäkelä (1994). Slicing procedures for layered manufacturing techniques. *Computer-Aided Design*, Vol. 26, No. 2, pp. 119-126.

Doo, D. and M. Sabin (1978). Analysis of the behavior of recursive subdivision surfaces near extraordinary points. *Computer-Aided Design*, Vol. 10, No. 6, pp. 356-360.

Doyle, T.E. (1991). Shape melting technology. *Proc. 3rd Int. Conf. Desktop Manuf.*, 1991.

Duley, W.W. (1983). *Laser Processing and Analysis of Materials*. Plenum Press, New York, NY.

Duley W.W. (1996). UV *Laser: Effects & Applications in Materials Science*. Cambridge University Press, London, England.

Dusel, K.H., J. Eschl, P. Eyerer and T. Lück. (1998), Rapid tooling - Simulation and applications of the injection molding process. *Proc. 7th Eur. Conf. Rapid Prototyping and Manuf.*, Aachen, Germany, July 1998.

Dutta, D., V. Kumar, M.J. Pratt and R.D. Sriram (1998). Towards STEP-based data transfer in layered manufacturing. *Proc. 10th Int. IFIP WG5.2/5.3 Conf., Prolamat '98*, 9-11 September.

Elliott, D.J. (1995). *Ultraviolet Laser Technology and Applications*. Elsevier Science & Technology Books, Amsterdam.

Evans, H. and P.F. Jacobs (1991). The development of photopolymer modulus with actinic exposure, *Proc. 2nd Int. Conf. Rapid Prototyping*, University of Dayton, Dayton, OH, pp. 69-85.

Fadel, G.M and C. Kirschman (1996). Accuracy issues in CAD to RP translations. *Rapid Prototyping J.*, Vol. 2, No. 2, pp. 4-17.

Fan, K.-C. and K.-P. Wen (1993b). Non-contact automatic measurement of free-form surface profiles on CNC machines. *Proc. SPIE - Measurement Technology and Intelligent Instruments*, Vol. 2101, pp.949-958.

Farago, F.T. and M.A. Curtis (1994). *Handbook of Dimensional Measurement*, The 3rd Ed., Industrial Press Inc., New York.

Farin, G. (1992). From conics to NURBS. *IEEE Computer Graphics and Applications*, Vol. 12, pp. 78-86.

Farin, G. (2002). *Curves and surfaces for CAGD: a Practical Guide*, Morgan Kaufmann Publishers, San Francisco.

Farouki, R.T., T. Koenig, K. Tarabanis, J.U. Korein and J.S. Batchelder (1995). Path planning with offset curves for layered fabrication processes. *Journal of Manufacturing Systems*, Vol. 14 No. 5, pp. 355-68.

Ferguson, B.L. and O.D. Smith (1984), Ceracon process. In *Metals Handbook*, 9th Edition, Volume 7: Powder metallurgy, ASM Int.

Ferry, J.D. (1961). *Viscoelastic Properties of Polymers*. John Wiley & Sons, New York, NY.

Fessler, J.R., R. Merz, A.H. Nickel, F.B. Prinz, F.B. and L.E. Weiss (1996). Laser deposition of metals for shape deposition manufacturing. *Proc. 7th Solid Freeform Fabr. Symp.*, The University of Texas at Austin, Austin, TX, Vol.7, pp. 117-124.

Fessler, J., A. Nickel, G. Link, G., F.B. Prinz and P. Fussell (1997). Functional gradient metallic prototypes through shape deposition manufacturing. *Proc. 8th Solid Freeform Fabr. Symp.*, The University of Texas at Austin, Austin, TX, Vol. 8, pp. 521-528.

Festa, R., O. Manca and V.A. Naso (1988). A comparison between models of thermal fields in laser and electron beam surface processing. *Int. J. Heat and Mass Transfer.* Vol. 21, pp. 99-106.

Feygin, M. and B. Hsieh (1991). Laminated object manufacturing (LOM): A simpler process. *Proc. 2nd Solid Freeform Fabr. Symp.*, The University of Texas at Austin, Austin, TX, Vol. 2, pp. 23-130.

Feygin, M. and S. Pak (1997). *Apparatus for Forming an Integral Object from Laminations*. US Patent 5,637,175.

Feygin, M. and S.S. Pak (1999). *Laminated Object Manufacturing Apparatus and Method*. US Patent 5,876,550.

Filipak, K. (1998). Injection molding thermoplastic parts in days in tooling produced from new composite board. *Proc. SME Rapid Prototyping & Manuf. Conf.*, Soc. of Manuf. Engineers, Dearborn, MI, pp. 223-243.

Flach, L. and R.P. Chartoff (1994). Stereolitography process modeling: Shrinkage prediction. *Proc. 5th Int. Conf. Rapid Prototyping*, Dayton, OH.

Fodran, E., M. Koch and U. Menon (1996). Mechanical and dimensional characteristics of fused deposition modeling build styles. *Proc. 7th Solid Freeform Fabr. Symp.*, The University of Texas at Austin, Austin, TX, Vol. 7, pp. 419-442.

Foley, J.D., A. van Dam, S. K. Feiner and J. F. Hughes (1990). *Computer Graphics*, Addison-Wesley Publishing Company, New York.

Forderbase, P., M. Ganninger and K. McAlea (1995). Nylon and nylon composite SLS prototypes. *Proc. 4th Eur. Conf. Rapid Prototyping and Manuf.*, University of Nottingham, Nottingham, UK.

Frank, D. and G. Fadel (1994). Preferred Direction of build for rapid Prototyping Processes. *Proc. 5th Int. Conf. Rapid Prototyping*, Dayton, Ohio, pp. 191-200.

Frank, D. and G. Fadel (1995). Expert system-based selection of the preferred direction of build for rapid prototyping processes, *J. Intelligent Manufacturing*, 1995, No. 6, 339-45.

Freeman, S.K. (1974). *Applications of Raman Spectroscopy*. John Wiley & Sons, New York, NY.

Frenkel, J. (1945). Viscous flow of crystalline bodies under the action of surface tension. *J. Physics* (USSR). Vol. 9, pp. 385-391.

Fuh, J.Y.H., Y.S. Choo, A.Y.C. Nee, L. Lu and K.C. Lee (1995). Improvement of the UV curing process for the laser lithography technique. *Materials & Design.* Vol. 16, No. 1, pp. 23-32.

Fuh, J.Y.H., S.Y. Choo, L. Lu, A.Y.C. Nee, S.Y. Wong, L.W. Wang, T. Miyazawa and S.H. Ho (1997). Post-cure shrinkage of photo-sensitive material used in laser lithography process. *J. Mat. Processing Tech.* 1997, 63: 881-886.

Fuh, J.Y.H., L. Lu, C.C. Tan, Z.X. Shen and S. Chew (1999). Curing characteristics of acrylic photopolymer used in stereolithography process. *Rapid Prototyping J.,* Vol. 5, No. 1, pp. 27-34.

Gargiulo, E.P. (1992). S tereolithography p rocess a ccuracy: User experience. *Proc. 1st Eur. Conf. Rapid Prototyping*, pp. 213-233.

Gasdaska, C., R. Clancy, M. Ortiz, V. Jamalabad, A. Virkar, A. and D. Papovitch (1998). Functionally optimized ceramic structures. *Proc. 9th Solid Freeform Fabr. Symp.*, The University of Texas at Austin, Austin, TX, Vol. 9, pp. 705-712.

Gasvik, K., T. Hovde and T. Vadeth (1989). Moiré technique in 3-D machine vision. *Optics and Laser in Eng.*, Vol. 10, pp.241-249.

German, R.M. (1990). Supersolidus liquid phase sintering. Part II: Densification theory. *Int. J. Powder Metallurgy.* Vol. 26, No. 1, pp. 35.

Giacometti, F. and T.C. Chang (1990). A framework to model parts, assemblies, and tolerances. *Proc. Adv. Integrated Product Design and Manuf.*, ASME, pp. 117-125.

Gibson, I. (2002). *Software Solutions for Rapid Prototyping*, Professional Engineering Publishing, London, UK.

Gibson, I., T. Kvan and L.W. Ming (2002). Rapid prototyping for architectural models. *Rapid Prototyping J.*, Vol. 8, No. 2, pp. 91-99.

Gilman, C.R. and S.J. Rock (1995). The Use of STEP to Integrate Design and Solid Freeform Fabrication. *Proc. 6th Solid Freeform Fabr. Symp.*, University of Texas at Austin, Austin, Texas, pp. 213-220.

Ginnings, D.C. and G.T. Furukawa (1953). Heat capacity standards for the range 14 to 1200°K. *J. Amer. Chem. Soc..* Vol. 75, pp. 522-529.

Godin, G., G. Roth and P. Boulanger (1994). Using laser geometric sensing for rapid product development. *IMS. Proc. Int. Conf. on Rapid Product Development*, Stuttgart, Germany, Jan. 31-Feb. 2, pp.405-416.

Golub, G. H. and C. F. Van Loan (1989). *Matrix Computations*, The Johns Hopkins University Press, Baltimore and London.

Grabowski, R., W. Schweizer, J. Molnar and L. Unger (1989). Three-dimensional pictures of industrial scenes applying an optical radar. *Optics and Laser in Engineering*, Vol. 10, pp.205-226.

Graessley, W.W. (1974). The entanglement concept in polymer rheology. *Advances in Polymer Sci.*. Vol. 16, pp. 1-179.

Greer, C., J. McLaurin and A.A. Ogale (1996). *Proc. 7th Solid Freeform Fabr. Symp.*, The University of Texas at Austin, Austin, TX, pp. 307.

Grffith, M.L. and J.W. Halloran (1996). *J. Amer. Cer. Soc.*. Vol. 79, No. 10, pp. 2601.

Griffith, M., D. Keicher, C. Atwood and J. Romero (1996). Free form fabrication of metallic components using laser engineered net shaping (LENS). *Proc. 7th Solid Freeform Fabr. Symp.*, The University of Texas at Austin, Austin, TX, Vol. 7, pp. 125-131.

Griffith, M., L. Harwell, J. Romero, E. Schlienger, C. Atwood and J. Smugeresly (1997). Multi-material processing by LENS. *Proc. 8th Solid Freeform Fabr. Symp.*, The University of Texas at Austin, Austin, TX, Vol. 8, pp. 387-393.

Gu, P. and X. Yan (1995). Neural network approach to the reconstruction of freeform surfaces for reverse engineering. *Computer-Aided Design*, Vol. 27, No. 1, pp. 59-64.

Guess, T.R. and R.S. Chambers (1995). In situ property measurement on laser-drawn strands of SL 5170 Epoxy and SL 5149 Acrylate. *Proc. 6th Solid Freeform Fabr. Symp.*, The University of Texas at Austin, Austin, TX, Vol. 6, pp. 134-141.

Guisser, L., R. Payrissat and S. Castan (1992). A new 3-D surface measurement system using structured light, Proceedings CVPR'92, Computer Vision and Pattern Recognition, IEEE Computer Society, pp. 784-786.

Gureev, D., A. Petrov and I. Shishkovsky (1999). Biocompatible intermetallics phase fabrication under laser sintering of SHS powder compositions. *Issues of Russian Academy of Science,* Physical part, No. 10, pp. 2077-81.

Haffiangadi, A. and A. Bandyopadhyay (2000). Modeling of multiple pore ceramic materials fabricated via fused deposition process. *Scripta Materialia*, Vol. 42, pp. 581-588.

Hageniers, O.L. (1994). Recent advances in laser triangulation-based measurement of airfoil surface. *Proc. SPIE - Industrial Optical Sensors for Metrology and Inspection*, Vol. 2349, pp. 2-22.

Hague, R., G. D'Costa and P.M. Dickens (2001). Structural design and resin drainage characteristics of QuickCast 2.0. *Rapid Prototyping J.,* Vol. 7, No. 2, pp. 66-73.

Halstead M, M. Kass and T. DeRose (1993). Efficient, Fair interpolation using Catmull-Clark surfaces. *SIGGRAPH Computer Graphics Proc.* , ACM, pp. 35-44.

Hartmann, K., R. Krishnan, R., Merz, G. Neplotnik, F.B. Prinz, F.B., L. Schultz, M. Terk and L.E. Weiss (1994). Robot-assisted shape deposition manufacturing. *Proc. IEEE Int. Conf. Robotics and Automation*, San Diego.

Hausler, G., and W. Heckel (1988). Light sectioning with large depth and high resolution. *Applied Optics*, Vol. 27, No. 24, pp. 5165-5169.

Hease, P (1989). *Selective Laser Sintering of Metal Powders*. MS Thesis, The University of Texas at Austin, Austin, TX.

Hecht, J. (1994). *Understanding Lasers: An Entry-Level Guide*. IEEE Press and John Wiley & Sons Inc., New York, NY.

Hejmadi U. and K. McAlea (1996). Selective laser sintering of metal molds: The Rapid Tool process. *Proc. 7th Solid Freeform Fabr. Symp.*, The University of Texas at Austin, Austin, TX, Vol. 7, pp. 97-104.

Heinzl, J. and C.H. Hinz (1984). Ink-jet printing. *Advances in Electronics and Electron Physics*, Vol. 65, pp. 91.

Heller, T.B. (1991). Three-dimensional lithography: Laser modeling using photopolymers. *SPIE Proc.*, Vol. 1454, pp. 272-282.

Hilton, P.D. and P.F. Jacobs (Eds.) (2000). *Rapid Tooling: Technologies and Applications.* Marcel Dekker, Inc., New York, NY.

Hinkzewski, C., S. Corbel and T. Chartier (1998). Stereolithography for the fabrication of ceramic three-dimensional parts. *Rapid Prototyping J.*, Vol. 4, No. 3, pp. 104-111.

Hitz, C.B., Hecht, J. and J.J. Ewing (2001). *Introduction to Laser Technology.* 3rd Edition, IEEE, New York, NY.

Ho, H.C.H., W.L. Cheung and I. Gibson (2002). Effects of graphite powder on the laser sintering behavior of polycarbonate. *Rapid Prototyping J.*, Vol. 8, No. 4, pp. 233-242.

Hoffmann, C.M. (1989). *Geometric & Solid Modeling: An Introduction.* Morgan Kaufmann, San Mateo, California.

Hofmeister, W., M. Wert, J. Smugereski, J.A. Philliber M. Griffith and M. Eusz (1999). `Investigation solidification with LENS process. *J. of Mat.*, Vol. 51, No. 7.

Hope, R.L., R.N. Roth and P.A. Jacobs (1997). Adaptive slicing with sloping layer surfaces. *Rapid Prototyping J.*, 1997, Vol. 3, No. 3, pp. 89-98.

Hon, K.K.B. and T.J. Gill (2003). Selective laser sintering of SiC/polyamide composites. *Annals of CIRP*, Vol. 52, No. 1, pp. 173-176.

Hopkinson and N., P. Dickens (1997). Thermal effects on accuracy in the 3D Keltool(TM) process. *Proc. 8th Solid Freeform Fabr. Symp.*, The University of Texas at Austin, Austin, TX, Vol. 8, pp. 267-274.

Hopkinson, N. and R. Hague (2001). *3D Printer Benchmarking Project, Final Report.* Neil Hopkinson & Richard Hague Rapid Prototyping Consortium.

Hopkinson, N. and P. Dickens (2001). Rapid prototyping for direct manufacture. *Rapid Prototyping J.*, Vol. 7, No. 4, pp. 197-202.

Hoppe, H., T. DeRose, T. Duchamp, M. Halstead, H. Jin, J. McDonald, J. Cchweitzer and W. Stueltzle (1994). Piecewise smooth surface reconstruction. *SIGGRAPH Computer Graphics Proc.*, ACM, pp. 295-302.

Hoschek, J. and D. Lasser (1993). *Fundamentals of Computer Aided Geometric Design*, A.K. Peters, Wellesley, Massachusetts.

Hoyle, C.E. and J.F. Kristle (1990). *Radiation Curing of Photopolymers.* Washington D.C.: Amer. Chem. Soc..

Hsieh, Jiang (2003). *Computed Tomography, Principles, Design, Artifacts, and Recent Advances.* SPIE Press, Bellingham, Washington, USA.

Hu, Z., K. Lee and J. Hur (2002). Determination of optimal build orientation for hybrid rapid-prototyping. *J. Mat. Processing. Tech.*, Vol. 130-131, pp. 378-383.

Hug, W. (1992). Lasers for rapid prototyping and manufacturing. In P.F. Jacobs (Ed.), *Rapid Prototyping & Manuf.: Fundamentals of Stereolithography.* Soc. of Manuf. Engineers, Dearborn, MI.

Hull, D. (1981). *An Introduction to Composite Materials.* Cambridge University Press, London, England.

Hull, C.W. and C.W. Lewis (1991). Method and apparatus for production of three-dimensional objects by stereolithography. US Patent #4,999,143, March 12.

Hunziker, M. and Leyden, R. (1992). Basic polymer chemistry. In P.F. Jacobs (Ed.), *Rapid Prototyping & Manuf.: Fundamentals of Stereolithography.* Soc. of Manuf. Engineers, Dearborn, MI.

Hur, J. and K. Lee (1996). Efficient algorithm for automatic support structure generation in layered manufacturing. *The 1996 ASME Design Engineering Technical Conferences and Computers in Engineering Conference*, Irvine, CA, August 18-22.

Hur, S.M., K.H. Choi, S.H. Lee and P.K. Chang (2001). Determination of fabricating orientation and packing in SLS process. *J. Mat. Processing Tech.*, Vol. 112, No. 2-3, pp. 236-243.

Hur, J., K. Lee, Zhu-Hu and J. Kim (2002). Hybrid prototyping system using machining and deposition. *Computer-Aided Design*. Vol. 34, pp. 741-754.

Ikuta, K. and K. Hirowatari (1993). Real three dimensional micro fabrication using stereo lithography and metal molding. *Proc. IEEE Micro Electro Mech. Sys.*, Fort Lauderdale, Florida, pp. 42-47.

Ikuta K., S. Maruo and S. Kojima (1998). New micro stereo lithography for freely movable 3D micro structure. *Proc. 11th Int. Workshop on MicroElectroMechanical Sys.*, Heidelberg, Jan 25-29, 1998, pp. 290-295.

Incropera, F.P. and D.P. DeWitt (1985). *Fundamentals of Heat and Mass Transfer.* 2nd Ed., John Wiley & Sons: New York, NY, pp. 740-782.

Inui, E. (2000). PLT rapid prototyping, high accuracy modeling and tooling applications. *8th Int. Conf. Rapid Prototyping*, Session on Machine Trends, Tokyo, Japan.

Ippolito, R. and L. Luliano (1995). Benchmarking of rapid prototyping techniques in terms of dimensional accuracy and surface finish. *Annals of CIRP*. Vol. 44, No. 1, pp. 157-160.

Issa, B., P. Gibbs, R. Hodgskinson, C.M. Langton, C.M. and L.W. Turnbull (1998). Assessment of the pore geometry of stereolithography models by high resolution MRI. *Magnetic Resonance Imaging*, Vol. 16 No. 5–6, pp. 651-3.

Jacobs, P.F. (Ed.) (1992a). *Rapid Prototyping & Manuf.: Fundamentals of Stereolithography*, Soc. of Manuf. Engineers, Dearborn MI.

Jacobs, P.F. (1992b). Fundamental processes. In P.F. Jacobs (Ed.), *Rapid Prototyping & Manufacturing: Fundamentals of Stereolithography*. Soc. of Manuf. Engineers, Dearborn, pp. 79-110.

Jacobs, P.F. (1993). Stereolithograpy 1993: Epoxy resins, improved accuracy in investment casting. *Proc. 4th Eur. Conf. Rapid Prototyping & Manuf.*,. Nottingham, UK, pp. 95-113.

Jacobs, P.F. (1995). QuickCast™ 1.1 and rapid tooling. *Proc. Eur. Conf. Rapid Prototyping & Manuf.*, Nottingham, UK, pp. 95-113.

Jacobs, P.F. (1996). *Stereolithography and Other RP&M Technologies: From Rapid Prototyping to Rapid Tooling*, Soc. of Manuf. Engineers and the Rapid Prototyping Association, New York, N.Y.

Jacobs, P.F. (1999). New frontiers in mold construction. *NASA Tech. Briefs*, Vol. 23, No. 3.

Jackson, T.R., N.M. Patrikalakis, E.M. Sachs and M.J. Cima (1998). Modeling and designing components with locally controlled composition. *Proc. 9th Solid Freeform Fabr. Symp.*, The University of Texas at Austin, Austin, TX, Vol. 9, pp. 259-266.

Jackson, T.R., H. Liu, N.M. Patrikalakis, E.M. Sachs and M.J. Cima (1999). Modeling and designing functionally graded material components for fabrication with local composition control. *Materials & Design*, Vol.20, pp. 63-75.

Jafari, M.A., W. Han, F. Mohammadi, A. Safari, S.C. Danforth and N.A. Langrana (2000). Novel system for fused deposition of advanced multiple ceramics. *Rapid Prototyping J.*, Vol. 6, No. 3, pp. 161-174.

Jakubenas, K.L. and H.L. Marcus (1995). Silicon carbide from laser prolysis of polycarbosilane. *J. Amer. Cer. Soc.*. Vol. 78, No. 8, pp. 2263-2266.

Jamieson, R. and H. Hacker (1995). Direct slicing of CAD models for rapid prototyping. *Rapid Prototyping J.*, Vol. 1, No. 2, pp. 4-12.

Jeng, J.-Y., J.-C. Wang and T.T. Lin (2000a). A New flexible layer method for jet system to accelerate fabrication speed. *Rapid Prototyping J.*, Vol. 6, No. 4, pp. 226-234.

Jeng, J.-Y., J.-C. Wang and T.T. Lin, T.T. (2000b). Fast interior filling of Model Maker models using spraying nozzle to accelerate build speed. *Rapid Prototyping J.,* Vol. 6, No. 4, pp. 235-243.

Jeng, J.-Y. and M.-C. Lin (2001) Mold fabrication and modification using hybrid processes of selective laser cladding and milling. *J. Mat. Processing Tech..* Vol. 110, pp. 98-103.

Johnson, J.L. (1994). *Principles of Computer Automated Fabrication,* Palatino Press Inc., USA, 1994.

Jurrens, K.K. (1999). Standards for the rapid prototyping industry. *Rapid Prototyping J.,* Vol. 5, No. 4, pp. 169 - 178.

Kakunai, S., T. Sakamoto and K. Iwata (1999). Profile measurement using a liquid crystal gratings. *Appl. Opt.,* Vol. 38, No. 13, pp. 2824-2828.

Kalpakjian, S. (1995). *Manufacturing Engineering and Technology.* Addison-Wesley Publishing Company, Reading, MA.

Kalender, W.A. (2000). *Computed Tomography, Fundamentals, System Technology, Image Quality, Application.* Publicis MCD Verlag, Munich, Germany.

Kalender, W., W. Seissler, E. Klotz and P. Vock (1990). Spiral volumetric CT with single breath hold technique, continuous transport and continuous scanner rotation, *Radiology* Vol. 176, pp.181-183.

Kamitani, T. and Y. Marutani (1996). Stereolithography system using multiple spot exposure. *Proc. 7th Solid Freeform Fabr. Symp.,* The University of Texas at Austin, Austin, TX, Vol. 7, pp. 316-326.

Karalekas, D. and D. Rapti (2002). Investigation of the processing dependence of SL solidification residual stresses. *Rapid Prototyping J.,* Vol. 8, No. 4, pp. 243-247.

Karunakaran, K.P., S. Dibbi, P.V. Shamuganathan, D.S. Raju and S. Kakaraparti (2001). Efficient stock cutting for laminated manufacturing. *Computer-Aided Design,* Vol. 34, pp. 281-298.

Kataria, A. and D.W. Rosen (2001). Building around inserts: Methods of fabricating complex devices in Stereolithography. *Rapid Prototyping J.,* Vol. 7, No. 5, pp. 253-261.

Kathuria, Y.P. (2001). An overview of 3D structuring in microdomain. *J. Indian Inst. Sci.,* Vol. 81, pp. 659-664.

Katz, Z. and P.E.S. Smith (2001), On process modeling for selective laser sintering of stainless steel. *Proc. Inst. Mech. Engrs,* UK, Vol. 215, No. B, pp. 1497-1504.

Kelly, A. (1973). *Strong Solids.* 2nd Ed., Oxford University Press, Oxford, UK.

Khaing, M.W., J.Y.H. Fuh and L. Lu (2001). Direct metal laser sintering for rapid tooling: Processing and characteristics of EOS parts. *J. Mat. Processing Tech..* Vol. 113, pp. 269-272.

Khoshnevis, B. (1998), Innovative rapid prototyping process makes large sized, smooth surfaced complex shapes in a wide variety of materials. *Mat. Tech..* Vol. 13, No. 2, pp. 52-63.

Khoshnevis, B., S. Bukkapatnam, H. Kwon, H. and J. Saito (2001). Investigation of contour crafting using ceramic materials. *Rapid Prototyping J.,* Vol. 7, No. 1, pp. 32-41.

Kietzman, J. (1999). *Rapid Prototyping Polymer Parts via Shape Deposition Manufacturing.* Ph.D. Thesis, Stanford University, CA.

Killander, L.A. (1995). Future direct manufacturing of metal parts with freeform fabrication. *Annals of CIRP,* Vol. 44, No. 1.

King, D. and T. Tansey (2002). Alternative metals for rapid tooling. *J. Mat. Processing Tech..* Vol. 121, pp. 313-317.

Kirschman, C.F., C. Namboodri, C.C. Jar-Almonte, A. Bagchi, R.L. Dooley and A.A. Ogale (1992). S tereolithographic s upport s tructure d esign f or r apid p rototyping. *Proc. 3rd Int. Conf. Rapid Prototyping*, Dayton, OH, pp. 259-268.

Kishinami, T., F. Tanaka and N. Narahara (1997). Distortion mechanism of stereolithography parts. *Proc. 8th Int. Conf. Prodn Eng. – Rapid Product development*, Sapporo, Japan.

Klocke, F., T. Celiker and Y.-A. Song (1995). Rapid metal tooling. *Rapid Prototyping J.*, Vol. 1, No. 3, pp. 32-42.

Klcoke, F., H. Wirtz and W. Meiners (1996). Direct manufacturing of metal prototypes and prototype tools. *Proc. 7th Solid Freeform Fabr. Symp.*, The University of Texas at Austin, Austin, TX, Vol. 7, pp. 137-144.

Klocke, F. and C. Wagner (2003). Coalescence behavior of two metallic particles as base mechanism of selective laser sintering. *Annals of CIRP*. Vol. 52, No. 1, pp. 177-180.

Kloosterboer, J.G. (1988). Network formation by crosslinking photopolymerization and its applications in electronics. *Polymers in Electronics*. Vol. 84, pp. 1-61.

Klosterman, D., R. Chartoff, N. Osborne and G. Graves (1997a). Automated fabrication of monolithic and ceramic matrix composites via laminated object manufacturing (LOM). *Proc. 8th Solid Freeform Fabr. Symp.*, The University of Texas at Austin, Austin, TX, Vol. 8, pp. 537- 549.

Klosterman, D., B. Priore, B. and R. Chartoff (1997b). Laminated object manufacturing of polymer matrix composites. *Proc. 7th Int. Conf. Rapid Prototyping*, University of Dayton, 1997, pp. 283- 292.

Ko, H., M. Kim, H. Park and S. Kim (1994). Face sculpturing robot with recognition capability. *Computer-Aided Design*, Vol. 26, No. 11, pp.814-821.

Kochan, D., (Ed.) (1993). *Solid Freeform Manufacturing*, E lsevier Publishers, Amsterdam, The Netherlands.

Kochan, D., C.K. Chua and Z. Du (1999). Rapid prototyping issues in the 21st century. *Comp. in Industry*, Vol. 39, pp. 3-10.

Kodama, H. (1981). Automatic method for fabricating a three-dimensional plastic model with photohardening polymer. *Rev. Sci. Inrum.*, pp. 1770-1773.

Koepke, M .L. (1988). *Model Graphics: Building and Using Study Models*, Van N ostrand Reinhold, New York, NY.

Kovacevic, R. and H.E. Beardsley (1998). Process control of 3D Welding as a droplet-based rapid prototyping technique. *Proc. 9th Solid Freeform Fabr. Symp.*, The University of Texas at Austin, Austin, TX.

Kruth, J.P. (1991). Material increase manufacturing by rapid prototyping technologies. *Annals of CIRP*, Vol. 40/2, pp. 603-614.

Kruth, J. P., and W. Ma (1992). Parameter identification of geometric elements from digitized data of coordinate measuring machines. *Proc. Int. Machine Tool Des. & Res. Conf.*, The Macmillan Press, Ltd, Vol. 29, pp. 277-285.

Kruth, J.P., L. Froyen, M. Rombouts, J. Va Vaerenbergh and P. Mercelis (2002). New ferro powder for selective laser sintering of dense parts. *Annals of CIRP*, Vol. 52, No. 1, pp. 139-142.

Kulkarni, P. and D. Dutta (1996). An accurate slicing procedure for layered manufacturing. *Computer-Aided Design*, 1996, Vol. 28, No. 9, pp. 683-697.

Kulkarni, P., A. M arsan a nd D . Dutta (2000). A r eview o f p rocess p lanning t echniques i n layered manufacturing techniques. *Rapid Prototyping J.*, Vol. 6, No. 1, pp. 18-35.

Kumar, V. and D. Dutta (1997). An approach to modeling multi-material objects. *Proc. 4th Symposium on Solid Modeling and Application*, Atlanta, GA, ACM SIGGRAPH, New York, NY, 336-353.

Kuneids, M. and T. Nakagawa (1984). Development of laminated drawing dies by laser cutting. *Bull. Of JSPE*, pp. 353-354.

Kyser, E.L. (1981). Drive pulse optimization. In J. Gaynor (Ed.), *Advances in Non-Impact Printing*, pp. 1175.

Lacari J. (1965). U.S. Patent, 3,205,157, 1965.

Lakshminarayanan, U. and H.L. Marcus (1991). Microstructural and mechanical properties of Al_2O_3/P_2O_5 and Al_2O_3/B_2O composites fabricated by selective laser sintering. *Proc. 2nd Solid Freeform Fabr. Symp.*, The University of Texas at Austin, Austin, TX, Vol. 2, pp. 205-212.

Lakshminarayanan, U. (1992). *Selective Laser Sintering of Ceramic Materials*. Ph.D. Dissertation, The University of Texas at Austin, Austin, TX.

Lan, P.T., S.Y. Chou, L.L. Chen and D. Gemmill (1997). Determining fabrications for rapid prototyping with stereolithography apparatus. *Computer-Aided Design*, Vol. 29, No. 1, pp. 53-62.

Langbein, F.C., A.D. Marshall and R.R. Martin (2003). Choosing consistent constraints for beautification of reverse engineered geometric models. To appear in *Computer-Aided Design*. (http://www.sciencedirect.com/).

Langton, C.M., M.A. Whitehead, D.K. Langton, D.K. and G. Langley (1997). Development of a cancellous bone structural model by stereolithography for ultrasound characterisation of the calcaneus. *Medical Eng. and Physics*, Vol. 19 No. 7, pp. 599-604.

Lau, P.T., S.Y. Chou, L.L. Chen and D. Gemmill (1997). Determining fabrication orientations for rapid prototyping with stereolithography apparatus. *Computer Aided Design*. Vol. 29, No. 1, 5-61.

Lee, E.T.Y. (1989). Choosing nodes in parametric curve interpolation. *Computer-Aided Design*, Vol. 21, No. 6, pp.363-370.

Lee, G. and J.W. Barlow (1993). Selective laser sintering of biomaterials for implants. *Proc. 4th Solid Freeform Fabr. Symp.*, The University of Texas at Austin, Austin, TX, Vol. 4, pp. 376-380.

Lee, K. and D.C. Gossard (1985). A hierarchical data structure for representing assemblies, Part I. *Computer-Aided Design*, Vol. 17, No. 1, pp. 15-19.

Lee, G. and J.W. Barlow (1994). Selective laser sintering of calcium phosphate powders. *Proc. 5th Solid Freeform Fabr. Symp.*, The University of Texas at Austin, Austin, TX, Vol. 5, pp. 191-197.

Lee, S.-J.J., E. Sachs and M. Cima (1995). Layer position accuracy in powder-based rapid prototyping. *Rapid Prototyping J.*, Vol. 1, No. 4, pp. 24-37.

Lee, K.H. and K. Choi (2000). Generating optimal slice data for layered manufacturing. *Int. J. Advanced Manuf. Tech.*, Vol. 16, No. 4, pp. 277-284.

Lee, S.H., W.S. Park, H.S. Cho, W. Zhang and M.C. Leu (2001). A neural network approach to the modeling and analysis of stereolithography processes. *J. of Eng. Manufacture, Proc. Inst. Mech. Engrs*, Part B, Vol. 215, No. 12, pp. 1719-1733.

Leu, D. (2000). *Handbook of Rapid Prototyping and Layered Manufacturing*, Academic Press, New York, NY.

Leu, M.C., Q. Liu and F.D. Bryant (2003). Study of part geometric features and support materials in rapid freeze prototyping. *Annals of CIRP*. Vol. 52, No. 1, pp. 185-188.

Levi, H. (1991). Accurate rapid prototyping by the solid ground curing technology. *Proc. 2nd Solid Freeform Fabr. Symp.*, The University of Texas at Austin, Austin, TX, Vol. 2, pp. 110-114.

Lewandowski, J., B. Menard and D. Hennequin (1993). Light sectioning with large depth of focus by means of Fresnel diffraction of an edge. *Optical Eng.*, Vol. 32, No. 9, pp. 2181-2184.

Lewandowski, J. and L. Desjardins (1995). Light sectioning with Improved depth resolution. *Optical Eng.*, Vol. 34, No. 8, pp. 2481-2486.

Lewis, G., J. Milewski and D. Thoma (1997). Multi-material Proc. *Proc. 8th Solid Freeform Fabr. Symp.*, The University of Texas at Austin, Austin, TX, Vol. 8, pp. 513-520.

Lewis, G. and E. Schlienger (2000). Practical considerations and capabilities for laser assisted direct metal deposition. *Materials & Design.* Vol. 21, No. 4, pp. 417-423.

Liao, Y.S. and Y.Y. Chiu (2001a). A new slicing procedure for rapid prototyping systems. *Int. J. Advanced Manuf. Tech.*, Vol. 18, No. 8, pp. 579-585.

Liao, Y.S. and Y.Y. Chiu (2001b). Adaptive crosshatch approach for the laminated object manufacturing (LOM) process. *Int. J. Prodn Res.*. Vol. 39, No. 15, pp. 3479-3490.

Liew, C.L., K.F. Keong, C.K. Chua and Z. Du (2002a). Dual material rapid prototyping techniques for the development of biomedical devices, Part 1: Space creation. *Int. J. Adv. Manuf. Tech.* Vol. 18, pp. 717-723.

Liew, C.L., K.F. Keong, C.K. Chua and Z. Du (2002b). Dual material rapid prototyping techniques for the development of biomedical devices, Part 2: Secondary powder deposition. *Int. J. Adv. Manuf. Tech.* Vol. 19, pp. 679-687.

Liu, Q., G. Sui and M.C. Leu (2002). Experimental study on the ice pattern fabrication for the investment casting by rapid freeze prototyping (RFP). *Comp. in Industry.* Vol. 48, pp. 181-197.

Liu, S. (2000). Automatic Segmentation of 3-D Surfaces and Features from CT-Contour Data for CAD Modeling. Ph.D. Thesis, Dept. MEEM, City University of Hong Kong, Hong Kong.

Liu, S. and W. Ma (1999). Seed-growing segmentation of 3D surfaces from CT-contours data. *Computer-Aided Design*, Vol. 31, No. 8, pp. 517-536.

Lightman, A. (1998). Image realization—physical anatomical models from scan data. *Proc. SPIE Symposium on Medical Imaging*, San Diego, CA.

Lim, C.S., P. Eng, S.C. Lin, S.C., C.K. Chua and S.T. Lee (2002). Rapid prototyping and tooling of custom-made tracheobronchial stents. *Int. J. Adv. Manuf. Tech.* Vol. 20, pp. 44-49.

Ling, W.M. and I. Gibson (1999). Possibility of coloring SLS prototype using the Ink-Jet method. *Rapid Prototyping J.*, Vol. 5, No. 4, pp. 152-154.

Litke, N., A. Levin and P. Schröder (2001). Fitting Subdivision Surfaces. *Proceedings of Scientific Visualization*, San Diego, California, pp. 319-324.

Loop, C. (1987). *Smooth Subdivision Surface Based on Triangles.* M.Sc. Thesis, Department of Mathematics, University of Utah.

Loose, K., T. Nakagawa and T., T. Niino (1997). Concept modeling using multiple LED photographic curing. *Proc. 13th Rapid Prototyping Symposium*, Yokohama, pp. 35-40.

Loose, K., T. Niino and T. Nakagawa (1999a). Multiple LED photographic curing of models for design evaluation. *Rapid Prototyping J.*, Vol. 5, No. 1, pp. 6-11.

Loose, K., T. Niino and T. Nakagawa (1999b). Raster-based exposure through multiple parallel beams in stereolithography. *Rapid Prototyping J.*, Vol. 5, No. 3, pp. 103-111.

Lorence, W.E. and H.E. Cline (1987). Matching cubes: A high resolution 3D surface construction algorithm. *Computer Graphics.* Vol. 21, No. 4, pp. 163-169.

Loughlin, C. (1993). *Sensors for Industrial Inspection*, Kluwer Academic Publishers, Boston.

Lu, L., J.Y.H. Fuh and Y.S. Wong (2001). *Laser-Induced Materials and Processes for Rapid Prototyping*, Kluwer Academic Publishers, Boston, MA.

Luo, R.C., Y. Ma (1995). A slicing algorithm for rapid prototyping and manufacturing. *Proc. IEEE Int. Conf. Robotics and Automation*, IEEE, pp. 2841-2846.

Ma, D., F. Lin and C.K. Chua (2001a). Rapid prototyping applications in medicine, Part 1: NURBS-based volume modeling. *Int. J. Adv. Manuf. Tech* Vo. 18, pp. 103-117.

Ma, D., F. Lin and C.K. Chua (2001b). Rapid prototyping applications in medicine, Part 2: STL file generation and case studies. *Int. J. Adv. Manuf. Tech.* Vol. 18, pp. 118-127.

Ma, W., J. P. Kruth and P. Vanherck (1993). CAD modeling of sculptured workpieces from physical models. *Proc. Int. Machine Tool Des. & Res. Conf.*, Vol. 30, pp.511-518, The Macmillan Press, Ltd.

Ma, W. and J. P. Kruth (1994). Mathematical modeling of free-form curves and surfaces from discrete points with NURBS. In P.J. Laurent, A. Le Méhauté, L.L. Schumaker (Eds.), *Curves and Surfaces in Geometric Design*. A.K. Peters Ltd., Wellesley, Massachusetts, pp. 319-326.

Ma, W. and J.P. Kruth (1995a). Parameterization of randomly measured points for least squares fitting of B-spline curves and surfaces. *Computer-Aided Design*, Vol. 27, No. 9, pp. 663-675.

Ma, W. and J.P. Kruth (1995b). NURBS curve and surface fitting and interpolation. In M. Daehlen, T. Lyche and L.L. Schumaker (Eds.), *Mathematical Methods for Curves and Surfaces*. Vanderbilt University Press, Tennessee, pp. 315-322.

Ma W., B. Swaelens and W. Vancraen (1997). CAD Modeling from CT-Images, *Proc. Int. Conf. on Manufacturing Automation,* ICMA'97, The University of Hong Kong, April 28-30, pp. 1021-1026.

Ma, W. and J. P. Kruth (1998). NURBS curve and surface fitting for reverse engineering. *Int J. Adv. Manuf. Tech.*, Vol. 14, No. 12, pp. 918-927.

Ma, W. and P. He (1998). B-spline surface local updating with unorganized points. *Computer-Aided Design*, Vol. 30, No. 11, pp. 853-862.

Ma, W. and P. He (1999). An Adaptive Slicing and Selective Hatching Strategy for Layered Manufacturing. *J. Mat. Processing Tech.*, Vols. 89-90, pp. 191-197.

Ma, W. and N. Zhao (2000). Catmull-Clark surface fitting for reverse engineering applications. *Proc. Geometric Modeling and Processing 2000*, IEEE Computer Society, pp. 174-283.

Ma, W. and N. Zhao (2002). Smooth multiple B-spline surface fitting with Catmull-Clark surfaces for extraordinary corner patches. *The Visual Computer*, Vol. 18, No. 7, pp. 415-436.

Ma, W., X. Ma, S. K. Tso and Z. Pan (2003a). A direct approach for subdivision surface fitting from a dense triangle mesh. To appear in *Computer-Aided Design.* (http://www.sciencedirect.com/).

Ma, W., W.C. But and P. He (2003b). NURBS-based adaptive slicing algorithms with implementation and realization for efficient model prototyping. *Computer-Aided Design*, (submitted), 2003.

Mackenzie, I. and R. Shuttleworth (1949). Phenomenological theory of sintering. *Proc. Physical Soc.*, London, UK, Vol. 62, No. 12-B, pp. 833-852.

Majhi, J., R. Janardan, M. Smid and J. Schwerdt (1998). Multi-criteria geometric optimization problems in layered manufacturing. *Proc. Annual Symp. Computational Geometry*, pp. 19-28.

Majhi, J., R. Janardan, M. Smid and P. Gupta (1999a). On some geometric optimization problems in layered manufacturing. *Computational Geometry: Theory and Applications*, Vol. 12, No. 3-4, pp. 219-239.

Majhi, J., R. Janardan, J. Schwerdt, M. Smid and P. Gupta (1999b). Minimizing support structures and trapped area in two-dimensional layered manufacturing. *Computational Geometry: Theory and Applications*, Vol. 12, No. 3-4, pp. 241-267.

Mäkelä, I. And A. Dolenc (1993). Some efficient procedures for correcting triangulated models. *Proc. 4th Solid Freeform Fabr. Symp.*, The University of Texas at Austin, Austin, TX, pp. 126-134.

Makynen, A., J. Kostamovaara and R. Myllyla (1995). Laser-radar-based three-dimensional sensor for teaching robot paths. *Optical Eng.*, Vol. 34, No. 9, pp.2596-2602.

Mani, K., P. Kulkarni and D. Dutta (1999). Region-based adaptive slicing, *Computer-Aided Design*, Vol. 31, No. 5, pp. 317-333.

Manias E. , J. Chen, N. Fang and X. Zhang (2001). Polymeric micromechanical components with tunable stiffness. *Applied Physics Letters,* Vol. 79, No. 11, pp. 1700-1702.

Mantyla, M. (1988). *An introduction to solid modeling*, Rockville, Md.: Computer Science, 1988.

Manriquez-Frayre, A. and D.L. Bourell (1990). Selective laser sintering of pre-alloyed bronze powders. *Proc. 1st Solid Freeform Fabr. Symp.*, The University of Texas at Austin, Austin, TX, pp. 99-106.

Marinos, L. (1993). 3-D object reconstruction using 4D-Laser digitizing. In J. Rix and E.G. Schlechtendahl (Eds.), *Interfaces in Industrial Sys. for Production Eng.*, IFIP Transactions, Vol. B10, North-Holland, pp. 257-266.

Marutani, Y. and T. Nakai (1989). *Laser Research*. Vol. 17, pp. 410-418.

Mashimo, K., S. Kitabayashi and Y. Tanimura (1995). Noncontact measurement of 3-D profiles by triangulation method, *Japan Soc. Prec. Eng.*, Vol. 61, No. 2, pp. 283-286. (in Japanese).

Masood, S.H., W. Rattanawong and P. Iovenitti (2000). Part building orientations based on volumetric error in fused deposition modeling. *Int. J. Adv. Manuf. Tech.*, Vol. 16, pp. 162-.

Masood, S.H. and W. Rattanawong (2002). A generic part orientation system based on volumetric error in rapid prototyping. *Int. J. Adv. Manuf. Tech.* Vol. 19, pp. 209-216.

Mass, H. (1991). Automated photogrammetric surface reconstruction with structured light. *Proc. SPIE - Industrial Vision Metrology*, Vol. 1526, pp.70-77.

Massen, R., J. Gassler, C. Konz and H. Richter (1995). Optical digitizing with five axes: really a must?, *Computers in Industry*, Vol. 28, pp.17-22.

Masters, W.E. (1990). The ballistic particle manufacturing process. *Proc. National Conf. Rapid Prototyping*, pp. 39-48.

Materialise N.V. (1994). *CLI interface format*, Annex E to the mid-term report of BE-5930 PHIDIAS - Laser Photopolymerisation models based on medical Imaging, a Development Improving the Accuracy of Surgery, Materialise N.V., Belgium.

Matsubara, K. (1976). *Molding Method of Casting Using Photocurable Substance*. Japanese Kokai Patent Application, Sho 51[1976]-10813.

Matsumoto, M., M. Shiomi, K. Osakada and P. Abe (2001). Finite element analysis of single layer forming on metallic powder bed in rapid prototyping by selective laser processing. *Int. J. Machine Tools & Manufacture*, Vol. 42, pp. 61-67.

Mazumder, J., A. Schifferer and J. Choi (1999). Direct material deposition: design macro and Microstructure. *Material Research Innovation*, Vol. 3, pp. 118-31.

McAdams, W.H. (1954). *Heat Transmissions*. 3rd Ed., McGraw-Hill, Tokyo.

McClurkin, J.E. and D. W. Rosen (1998). Computer-aided build style decision support for stereolithography. *Rapid Prototyping J.*, Vol. 4, No. 1, pp. 4-13.

McDonald, J.P., R. Lambert and R.J. Fryer (1994). 3D measurement using stereo scene coding. *IEE Colloquium on 3D Imaging and Analysis of Depth/Range Images*, IEEE, 3/1-3/4.

McDonald, J.A., C. J. Ryall and D. I. Wimpenny, (Eds.) (2001). *Rapid Prototyping Casebook*, Professional Engineering Publ. Ltd.

McKelvey, J.M. (1962). *Polymer Processing*. John Wiley & Sons, New York, NY.

Meier, K. (1986). H.J. Zweifel. *Imaging Sci.*. Vol. 30, pp. 174.

Melvin III, L.S., Gibson and Beaman, J.J. (1991). The Electrostatic application of powder to selective laser sintering. *Proc. 2nd Solid Freeform Fabr. Symp.*, The University of Texas at Austin, Austin, TX.

Merz, R. (1994). *Shape Deposition Manufacturing*. Ph.D. Thesis, Dept of Electrical Eng., Technical University of Vienna, Vienna, Austria.

Merz, R., F.B. Prinz, K. Ramaswami, M. Terk, M. and L.E. Weiss (1994a). Shape deposition manufacturing. *Proc. 5th Solid Freeform Fabr. Symp.*, The University of Texas at Austin, Austin, TX, Vol. 5, pp. 1-8.

Merz, R., F.B. Prinz and L.E. Weiss (1994b). *Method and Apparatus for Depositing Molten Metal*. U. S. Patent 5,281,789.

Migliore, L.R. and R.W. Walker (1994). *Lasers as Tools for Manufacturing*. SPIE-Int. Soc. for Optical Eng., Bellingham, WA.

Mitchell, P.E. (Ed.) (1996). *Tool and Manufacturing Engineers Handbook, Vol. 8: Plastic Part Manuf.*, Soc. of Manuf. Engineers, Dearborn, MI.

Migliore, L.R. (1996). *Laser* Materials *Processing*. Marcel Dekker, New York, NY.

Moring, I. and H. Ailisto (1989). Active 3-D vision system for automatic model-based shape inspection. *Optics and Lasers in Eng.*, Vol. 10, pp.146-160.

Moroika, I. (1935). *Process for Manufacturing a Relief by the Aid of Photography*. US Patent 2,015,457.

Moroika, I. (1944). *Process for Plastically Reproducing Objects*. US Patent 2,350,796.

Moss, J., A. Linney, S. Grindrod and C. Moss (1989). A laser scanning system for the measurement of facial surface morphology. *Optics and Laser in Eng.*, Vol. 10, pp. 179-190.

Muller, E. (1995). Fast three-dimensional form measurement system. *Optical Eng.*, Vol. 34, No. 9, pp. 2754-2756.

Mueller, B. and D. Kochan (1999). Laminated object manufacturing for rapid tooling and pattern making in foundry industry. *Comp. in Industry.* Vol. 39, pp. 47-53.

Munz, O.J. (1956). *Photo-glyph Recording*. US Patent 2775,758.

Muthukrishnan, K. (1994). The role of diode lasers in metrology. *Proc. SPIE - Industrial Optical Sensors for Metrology and Inspection*, Vol. 2349, pp. 23-34.

Naber, H. and F. Breitinger (1996). Anwendungsmöglichkeiten der rapid prototyping technologien fur druckgrußwerkzeuge, *Aalener Gießereisymposium*, pp. 17.

NAG (1991). NAG Fortran Library Manual, Mark 15, The Numerical Algorithms Group Limited, 1991.

NAG (2002). NAG C Library, Mark 7, The Numerical Algorithms Group Limited, 2002.

Narahara, H., F. Tanka, T. Kishinami, S. Igarashi and K. Saito (1999). Reaction heat effects on linear shrinkage and deformation in stereolithography. *Rapid Prototyping J.*, Vol. 5, No. 3, pp. 120-128.

Nasri, A. and M. Sabin (2002). Taxonomy of Interpolation Constraints On Recursive Subdivision Surfaces. *The Visual Computer*, Vol. 18, No. 6, pp. 382-403.

Nau, W., R.L. Dooley and A.A. Ogal (1994*). Proc. 5th Int. Conf. Rapid Prototyping*, Dayton, OH, pp. 121.

Nee, A.Y.C., J.Y.H. Fuh and T. Miyzawa (2001). On the improvement of stereolithography process. *J. Mat. Processing Tech.*. Vol. 113, pp. 262:268.

Nelson, J.C., N.K. Vail, N.K. and J.W. Barlow (1993a). Laser sintering model for composite materials. *Proc. 4th Solid Freeform Fabr. Symp.*, The University of Texas at Austin, Austin, TX, Vol. 4, pp. 360-369.

Nelson, J.C., S. Xue, J.W. Barlow, J.J., Beaman, H.L. Marcus and D.L. Bourell (1993b). Model of t he s elective l aser s intering o f b ispheno l- A p olycarbonate. *I nd. Eng. C hem. Res.*. Vol. 32, pp. 2305-2317.

Nelson, J.C. (1993). *Selective Laser Sintering: A Definition of the Process and an Empirical Sintering Law*. Ph.D. Dissertation, The University of Texas at Austin, Austin, TX.

Nelson, J.C., N.K. Vail, J.W. Barlow, J.J. Beaman, D.L. Bourell and H.L. Marcus (1995). Selective laser sintering of polymer-coated silicon carbide powders. *Ind. Eng. Chem. Res.*. Vol. 34, pp. 1641-1651.

Nelson, T.R. and M.J. Bailey (2000). Solid object visualization of 3D ultrasound data. *Proc. SPI E Int. Soc. for Optical Eng.*, Vol. 3982, pp. 26-34.

Newman, W.S., B. B. Mathewson, Y. Zhen and S. Choi (1996). A novel selective-area gripper for layered assembly of laminated objects. *Robotics and Computer-Integrated Manuf.*, Vol. 12, No. 4, pp. 293–302.

Nguyen, H., J. Richter and P. Jacobs (1992). Diagnostic testing. In P.F. Jacobs (Ed.), *Rapid Prototyping and Manufacturing: Fundamentals of Stereolithography*. Soc. of Manuf. Engineers, Dearborn, MI.

Nickel, A. (1999). *Analysis of Thermal Stresses in Shape Deposition Manufacturing of Metal Parts*. Ph.D. Thesis, Stanford University, Stanford, CA.

Odian, G. (1981). Ring-opening polymerization. In *Principles of Polymerization*, New York, NY: Wiley-Interscience Publication, New York.

Ogale, A.A., T. Renault, R.L. Dooley, A. Bagchi, A. and C.C. Jara-Almonte (1991). 3-D photolithography for composite development: Discontinuous reinforcements. *SAMPE Quarterly*, pp. 28-38

Ogg, M. (1998). Composite - successful for tooling. *Proc. Prototyping Tech. Int. '98*. pp. 40-43.

O'Hare, M. (1988). *Innovate: How to Gain and Sustain Competitive Advantage*, Basil Blackwell, Oxford, UK.

Olabisi, O., L.M., Robenson and M.T. Shaw (1979). *Polymer-Polymet Miscibility*. Academic Press, New York, NY.

Onuh, S.O. and K.K.B. Hon (2001). Improving stereolithography part accuracy for industrial applications. *Int. J. Adv. Manuf. Tech.*, Vol. 17, pp. 61-68.

Onuh, S.O. (2001). Rapid prototyping integrated systems. *Rapid Prototyping J.*, Vol. 7, No. 4, pp. 220-223.

Osawa, S., R. Furtain, K. Kawakami, *et al*, (1995). Measurement of three dimensional objects using gray code projection. *Japan Soc. Prec. Eng.*, Vol. 61, No. 8, 1995, pp.1101-1105.

Pak, S.S. (1997). Fabrication of plastic LOM objects. *Proc. Prototyping Tech. Int.*, Vol. 1, pp. 33-35.

Pang, T.H. (1992). *G reen S trengths of S tereolithography Resins: Phenomenological Green Flexural Modulus Equations I & II*. 3D Systems Report, 3D Systems Inc., pp. 25.

Pang, T.H. and P. Jacobs (1993). Stereolithography 1993: QuickCast[TM]. *Proc. 4th Solid Freeform Fabr. Symp.*, The University of Texas at Austin, Austin, TX, Vol. 4, pp. 158-167.

Pang, T.H. (1996). Advances in stereolithography photopolymer systems. In P.F. Jacobs (Ed.), *Stereolithography and Other RP&M Technologies*. ASME Press, New York, NY.

Pang, T.H., I. Figueroa, J. Fong, A. Melisaris, R. Wang, S. Hanna, H. Nguyen, M. Guertin and M. C. Phan (1997). SL 5410: High humidity, water, and heat resistant resin for stereolithography. *Proc. 8th Solid Freeform Fabr. Symp.*, The University of Texas at Austin, Austin, TX, Vol. 8, pp. 349-363.

Pang, Y. (1993). Stereolithography epoxy resin development: Accuracy and dimensional stability. *Proc. 4th Solid Freeform Fabr. Symp.*, The University of Texas at Austin, Austin, TX, Vol. 4, pp. 11.

Pappas, S.P. (1982). *Proc. of 22nd Fall Symposium,* Soc. of Photograph. Lic. & Eng, Washington D.C., pp. 46.

Pappas, S.P. (1985). Photoinitiators for radical, cationic, and concurrent radical-cationic polymerization. In *UV Curing: Science and Technology.* Technology Marketing Corporation, Norwalk, Connecticut, pp. 11-13.

Park, J., M.J. Tari and H.T. Hahn (2000). Characterization of the laminated object manufacturing (LOM) process. *Rapid Prototyping J.*, Vol. 6, pp. 36-49.

Parker, S. (2003). *Lasers.* Thameside Press, London, England.

Patil, L., Patil, D. Dutta, A.D. Bhatt, K. Jurrens, K. Lyons, M.J. Pratt and R.D. Sriram (2002). A proposed standards-based approach for representing heterogeneous objects for layered manufacturing. *Rapid Prototyping J.*, Vol. 8, No. 3, pp. 134-146.

Perera, B.V. (1940). *Process of Making Relief Maps.* US Patent 2,189,592.

Perry, R.H. and C.H. Chilton (1973). *Chemical Engineers' Handbook*, 5th Ed., McGraw-Hill: New York, NY.

Pham, D.T., S.S. Dimov, R.S. Gault (1999). Part orientation in stereolithography. *Int. J. Adv. Manuf. Tech.* Vol. 15, No. 9, pp. 674-682.

Pham, D.T. and S. S. Dimov, *Rapid Manufacturing: The Technologies and Applications of Rapid Prototyping and Rapid Tooling*, Springer Verlag , New York, NY.

Pham, D.T., S.S. Dimov and R.S. Gault (1999). Part orientation in stereolithography. *Int. J. Advanced Manuf. Tech.*, Vol. 15, No. 9, pp. 674-682

Piegl, L. (1989a). Modifying the shape of rational B-splines. Part 1: curves. *Computer-Aided Design,* Vol. 21, No. 8, pp. 509-518.

Piegl, L. (1989b). Modifying the shape of rational B-splines. Part 2: surfaces. *Computer-Aided Design*, Vol. 21, No. 9, pp. 538-546.

Piegl, L. (1991). On NURBS: a survey. *IEEE Computer Graphics & Applications*, Vol. 11, pp. 55-71.

Piegl, L. and W. Tiller (1995). *The NURBS Book*, Springer-Verlag, Berlin.

Piegl, L. and W. Tiller (2001). Parametrization for surface fitting in reverse engineering. *Computer-Aided Design*, Vol. 33, No. 8, pp. 593-603.

Piegl, L. and W. Tiller (2002). Fitting circular arcs to measured data. *Int. J. of Shape Modeling*, Vol. 8. No. 1, pp. 1-21.

Piggott, M.R. (1994). Short fibre polymer composites, a fracture-based theory of fibre reinforcement. *J. Composite Mat..* Vol. 28, No. 7, pp. 588-606.

Pilleux, M.E., M. Allahverdi, Y. Chen, Y., Lu, M.A. Jafari and A. Safari (2002). 3-D photonic bandgap structures in the microwave regime by fused deposition of multimaterials. *Rapid Prototyping J.*, Vol. 8, No. 1, pp. 46-52.

Pinkerton, A.J. and L. Li (2003). Effects of powder geometry and composition in coaxial laser deposition of 316L steel for rapid prototyping. *Annals of CIRP*, Vol. 52, No. 1, pp. 181-184.

Pitts, D.R. and L.E. Sisson (1977). *Heat Transfer.* McGraw-Hill: New York, NY.

Pottmann, H. and S. Leopoldseder (2003). A concept for parametric surface fitting which avoids the parametrization problem. *Computer Aided Geometric Design*, Vol. 20, No. 6, pp. 343-362

Prabhu, G. and D.L. Bourell (1993). Supersolidus liquid phase selective laser sintering of prealloyed bronze powder. *Proc. 4th Solid Freeform Fabr. Symp.*, The University of Texas at Austin, Austin, TX, Vol. 4, pp. 317-324.

Pratt, M.J., A.D. Bhatt, D. Dutta, K.W. Lyons, L. Patil and R.D. Sriram (2002). Progress towards an international standard for data transfer in rapid prototyping and layered manufacturing, *Computer-Aided Design*, Vol. 34, No. 14, pp. 1111-1121.

Pratt, V. (1987). Direct least squares fitting of algebraic surfaces. *Computer Graphics*, Vol. 21, pp. 145-152.

Prinz, F.B., L.E. Weiss and D.A. Adams (1993). *Method and Apparatus for Fabrication of Three-Dimensional Articles by Spray Deposition using Masks as Support Structures*. U.S. Patent 5,203,944.

Prinz F.B. and L.E. Weiss, L.E.(1994a). *Method for Fabrication of Three-Dimensional Articles*. U. S. Patent 5,301,415.

Prinz, F.B. and L.E. Weiss (1994b). *Automated System for Forming Objects by Incremental Buildup of Layers*. U. S. Patent 5,301,863.

Prinz, F.B., L.E. Weiss and D.P. Siewiorek (1994c). *Electronic Packages and Smart Structures Formed by Thermal Spray Deposition*. U. S. Patent 5,278,442.

Prinz, F.B. (Chair) (1997). *Rapid Prototyping in Europe and Japan, Vol. II, Site Reports*. Int. Technology Research Institute, Loyola College in Baltimore, Maryland.

Qiu, D., N.A. Langrana, S.C. Danforth, A. Safari and M. Jafari (2001). Intelligent toolpath for extrusion-based LM process. *Rapid Prototyping J.*, Vol. 7, No. 1, pp. 18-23.

Qiu, D. and N.A. Langrana (2002). Void eliminating tool path for extrusion-based multi-material layered manufacturing. *Rapid Prototyping J.*, Vol. 8, No. 1, pp. 38-45.

Radstok, E. (1999). Rapid tooling. *Rapid Prototyping J.*, Vol. 5, No. 4, pp. 164-168.

Rajagopalan, M., N.M. Aziz and C.O. Huey Jr (1995). A model for interfacing geometric modeling data with rapid prototyping systems. *Advances in Engineering Software*, Vol. 23, pp. 89-96.

Ramaswami, K. (1997). *Process Planning for Shape Deposition Manufacturing*. PhD Thesis, Stanford University, Palo Alto, CA.

Rase, W.D. (2002). Physical models of GIS objects by rapidly prototyping. *Proc. Geomatics 2002*, Ottawa, Canada.

Rattanawong, W., S.H. Masood and P. Iovenitti (2001). A volumetric approach to part-build orientations in rapid prototyping. *J. Mat. Processing Tech.*, Vol. 119, Vol. 1-3, pp. 348-353.

Rayleigh, F.R.S. 1878). On the instability of jets. *Proc. London Mathematical Soc.*, Vol. 10, No. 4, pp. 4-13.

Ready, J.F. (1997). *Industrial Applications of Lasers*. 2nd Edition. Elsevier Science & Technology Books, Amsterdam.

Rees, M. (1997). Automatic additive fabrication: Realizing convoluted from and nesting in sculpture. *Potapazione and Produzione Rapida*, Italy.

Rees, M. (1999). Rapid prototyping and art. *Rapid Prototyping J.*, Vol. 5, No. 4, pp. 154-160.

Reeves, P.E. and R.C. Cobb (1997). Reducing the surface deviation of stereolithography process using in-process techniques. *Rapid Prototyping J.*, Vol. 3, No. 1, pp. 20-31.

Renap, K. and J.P. Kruth (1995). Recoating issues in stereolithography. *Rapid Prototyping J.*, Vol. 1, No. 3, pp. 4-16.

Renault, T. and A.A. Ogale (992). 3D photolithography: Mechanical properties of glass and quartz fibre composites. *Proc. ANTEC'92, SPIE.* Vol. 1, pp. 745-747.

Rennie, A.E.W., C.E. Bocking and G.R. Bennett (2001). Electroforming of rapid prototyping mandrels for electro-discharge machining electrodes. *Int. J. Adv. Manuf. Tech.* Vol. 110, pp. 186-196.

Richardson, K.E. (1991). The production of wax models by the ballistic particle manufacturing process. *Proc. 2nd Conf. Rapid Prototyping*, pp. 15-22.

Richardson, K.E. (1998). Multiple material machining: An innovative RP technology. *Rapid Prototyping,* Newsletter of the Rapid Prototyping Association of the Soc. of Manuf. Engineers. Vol. 4, No. 2, pp. 1-4.

Richter, J. and P.F. Jacobs (1992). Accuracy. In P.F. Jacobs (Ed.), *Rapid Prototyping and Manufacturing: Fundamentals of Stereolithography.* Soc. of Manuf. Engineers, Dearborn, MI.

Rioux, M. (1984). Laser range finder based on synchronized scanners. *Applied Optics*, Vol. 23, No. 21, pp. 3837-3844.

Robbins, E. (1994). *Why architects draw*, MIT Press, Cambridge, MA.

Rodrigues, S.J., R.P. Chartoff, R.P., D.A. Klosterman, M. Agarwala, M. and N. Hecht (2000). Solid freeform fabrication of functional silicon nitride ceramics by laminated object manufacturing. *Proc. 11th Solid Freeform Fabr. Symp.*, The University of Texas at Austin, Austin, TX.

Rodriguez, J.F., J.P. Thomas and J.E. Renaud (2000). Characterization of the mesostructure of fused-deposition of acrylonitrile-butadiene-styerene materials. *Rapid Prototyping J.,* Vol. 6, No. 3, pp. 175-185.

Rodriguez, J.F., J.P. Thomas, J.P. and J.E. Renaud (2002). Mechanical behavior of acrylonitrile-butadiene-styerene (ABS) fused deposition materials. Experimental Investigation. *Rapid Prototyping J.,* Vol. 7, No. 3, pp. 148-158.

Roscoe, L.E., K.L. Chalasani and T.D. Meyer (1995). Living with STL files. *Proc. 6th Int. Rapid Prototyping Conf.*, Dayton, OH., pp. 145-152.

Rosochowski, A., and A. Matuszak (2000). Rapid tooling: The state of the art. *J. Mat. Processing Tech..* Vol. 106, pp. 191-198.

Rovick, J.S. (1994). An additive fabricator for high speed production of artificial limbs. *Proc. 5th Int. Conf. Rapid Prototyping*, Dayton, OH, 1994, pp. 47-56.

Sabourin, E., S.A. Houser and J.H. Bøhn (1996). Adaptive slicing using stepwise uniform refinement. *Rapid Prototyping J.*, Vol. 2, No. 4, pp. 20-26.

Sabourin, E., S.A. Houser and J.H. Bøhn (1997). Accurate exterior, fast interior layered manufacturing. *Rapid Prototyping J.*, Vol. 3, No. 2, pp. 44-52.

Sachs, E., P. Williams, D. Brancazio, M. Cima and K. Kremmin (1990a). Three dimensional printing: Rapid prototypes and tooling directly from a CAD model. *Proc. Manuf. Int. '90*, Atlanta, GA, pp. 131-136.

Sachs, E., M. Cima and J. Cornie (1990b). Three-dimensional printing: Rapid tooling and prototypes directly from CAD model. *Annals of the CIRP*, Vol. 39, No. 1, pp. 201-204.

Sachs, E., M. Cima, P. Williams, D. Brancazio and J. Cornie (1992). Three dimensional printing: Rapid tooling and prototyping directly from a CAD model. *J. Eng. for Industry.* Vol. 114, pp. 481–488.

Sachs E, J. Haggerty, M. Cima and P. Williams (1993a). Three-dimensional printing techniques. US Patent 5, 204, 055.

Sachs, E., M. Cima, J. Cornie, D. Brancazio, J. Bredt, A. Curodeau, T. Fan, S. Kanuja, Lauder, A., J. Lee and S. Michaels (1993b). Three-dimensional printing: The physics and implications of additive manufacturing. *Annals of CIRP*, Vol. 42, No. 1, pp. 257-260.

Sachs, E.M, N.M., Partikalaki, D. Boning D, M.J. Cima MJ, T.R. Jackson and R. Resnick (1998). The distributed design and fabrication of metal parts and tooling by 3D Printing. *Proc. NSF Design and Manuf. Grantees Conf.*, Monterey, Arlington, pp. 35-36.

Safari, A. (1999). Novel piezoelectric ceramics and composites for sensor and actuator applications. *Mat. Res. Innovat.*, Vol. 2, pp. 263-269.

Sanchez-Reyes, J. (1997). A Simple Technique for NURBS Shape Modification. *IEEE Computer Graphics and Application*, Vol. 17, No. 1, pp. 52-59.

Schaer, L. (1995). Spin casting fully functional metal and plastic parts from stereolithography models. *Proc. 6th Int. Conf. Rapid Prototyping*, pp. 217-236.

Scherer, G. (1977a). Sintering of low density glasses: I, Theory. *J. Amer. Cer. Soc.*. Vol. 60, No. 5-6, pp. 236-239.

Scherer, G. (1977b). Sintering of low density glasses: II, Experimental study. *J. Amer. Cer. Soc.*. Vol. 60, No. 5-6, pp. 239-245.

Scherer, G. (1977c). Sintering of low density glasses: III, Effect of a distribution of pore size. *J. Amer. Cer. Soc.*. Vol. 60, No. 5-6, pp. 245-248.

Schlienger, E., M. Griffith, M., Oliver, J. Romero and J. Smugeresly (1998). Sacrificial materials for the fabrication of complex geometries with LENS. *Proc 9th Solid Freeform Fabr. Symp.*, The University of Texas at Austin, Austin, TX, Vol. 205–209.

Schroder, J. (1963). Apparatus for determining the thermal conductivity of solids in the temperature range from 20-100°C. *Review Scientific Instruments*. Vol. 34, No. 6, pp. 615-621.

Schwartz, A. (1993). Cubital's high throughput rapid prototyping systems: A Soider approach to modeling, *Proc. 2nd Eur. Conf. Rapid Prototyping*, University of Nottingham, Nottingham, UK, pp. 239-245.

Schwerzel, R.E., *et al.* (1994). Three-dimensional photochemical machining with lasers. *Appl. of Lasers t Ind. Chem.*, SPIE, pp. 90-97.

Segal, J.I. ad R.I. Campbell (2001). A review of research into the effects of rapid tooling on part properties. *Rapid Prototyping J.*, Vol. 7, No. 2, pp. 90-98.

Shalaby, S.W. (1979). Radiative degradation of synthetic polymers: Chemical, physical, environmental, and technological Considerations. *Macromol. Rev.* Vol. 14, pp. 419-458.

Shellabear, M. C., H.J. Langer and M. Cabrera (1992). The EOS rapid prototyping system. *Proc. 1st Euro. Conf. Rapid Prototyping*, The University of Nottingham, pp. 19-43.

Shellabear, M. (1994). Optimizing materials and process parameters for different rapid prototyping applications. *Proc. 2nd Eur. Conf. Rapid Prototyping*, University of Nottingham, Nottingham, UK, pp. 369-379.

Shi, D. and I. Gibson (2000). Improving Surface quality of selective laser sintered rapid prototype parts using robotic finishing. *J. of Eng. Manufacture*. Proc. Inst. Mech. Engrs, Part B, Vol. 214, No. 3, pp. 197-20.

Shiomi, M., A. Yoshidome, F. Abe and K. Osakada (1999). Finite element analysis of melting and solidifying processes in laser rapid prototyping of metallic powders. *Int. J. Machine Tools & Manufacture*, Vol. 39, pp. 237-252.

Shishkovsky, I. (2001a). Synthesis of functional gradient parts from RP methods. *Rapid Prototyping J.*, Vol. 7, No. 4, 207-211.

Shishkovsky, I., V. Scherbakov, and A. Petrov (2001). Laser synthesis of the functional graded filter elements from metal-polymer powder compositions. *Proc. Int. Conf. Laser Assisted Net Shape Engineering - LANE'2001*, Erlangen, Germany.

Shyamsundar, N. and R. Gadh (2002). Collaborative virtual prototyping of product assemblies over the Internet, *Computer-Aided Design*. Vol. 34, pp. 755-768.

Sih, S. and J.W. Barlow (1994). Measurement and prediction of the thermal conductivity of powders at high temperatures. *Proc. 5th Solid Freeform Fabr. Symp.*, The University of Texas at Austin, Austin, TX, Vol. 5, pp. 321-329.

Sih, S.S. and J.W. Barlow (1995). The prediction of the thermal conductivity of powders. *Proc. 6th Solid Freeform Fabr. Symp.*, The University of Texas at Austin, Austin, TX, 1995, Vol. 6, pp. 397-401.

Sih, S.S. (1996). *The Thermal Properties of Powders*. Ph.D. Dissertation, The University of Texas at Austin, Austin, TX.

Silfvast, W.T. (1996). *Laser Fundamentals.* Cambridge University Press, Cambridge, UK.

Sisias, G., R. Phillips, C.A. Dobson, M.J. Fagan and C.M. Langton (2002). Algorithms for accurate rapid prototyping replication of cancellous bone voxel maps. *Rapid Prototyping J.*, Vol. 8, No. 1, pp. 6-24.

Sitzmann, E., D. Barnes, R. Anderson, E. Dinkel, R. Srivastava, R. Haynes, G. Green and J. Kratewski (1992). Exactomer™ resins for stereolithography. *Proc. Rapid Prototyping & Manuf. Conf.*, 1992, Soc. of Manuf. Engineers, Dearborn, MI.

Siu, Y.K. and S.T. Tan (2002). Modeling the material grading and structures of heterogeneous objects for layered manufacturing. *Computer-Aided Design*, Vol. 34, pp. 705-716.

Slocum, A. (1992). *Precision Machine Design*, Prentic-Hall, Inc., New Jersey.

Smith, G.H. (1975). Belgian Patent 828,841.

Soares, O.D.D. and M. Perez-Amor (Eds.) (1987). *Applied Laser Tooling.* Martinus Nijhoff, Dordrecht, Germany.

Soo, S. C. and K.M. Yu (2001). Rapid prototyping using fractal geometry. *Proc. 12th Solid Freeform Fabr. Symp.*, The University of Texas at Austin, Austin, TX, Vol. 12, pp.424-431.

Sonmez, F.O. and H.T. Hahn (1998). Thermomechanical analysis of the laminated object manufacturing (LOM) process. *Rapid Prototyping J.*, Vol. 4, No. 1, pp. 26-36.

Spencer, J.D., R.C. Cobb and P.M. Dickens (1993). *Surface Finishing Techniques for Rapid Prototyping,* Soc. of Manuf. Engineers Technical Paper No. PE930168.

Stam, J. (1998). Exact evaluation of Catmull-Clark subdivision surfaces at arbitrary parameter values. *SIGGRAPH Computer Graphics Proc.* , ACM, pp. 395-404.

Stanley, H., R. Yancey, Q. Cao and N. Dusaussoy (1995). CT-assisted solid freeform fabrication. *Proc. 6th Int. Conf. Rapid Prototyping*, Dayton, OH pp. 267-274.

Steen, W.M. (1997). *Laser Materials Processing.* 2nd Edition. Springer Verlag, New York, NY.

Stevens, M.P. *Polymer Chemistry.* 2nd Edition, Oxford University Press: New York, 1990.

Steinbichler, H. (1994). *Method and Apparatus for Ascertaining the Absolute Coordinates of an Object.* U.S. Patent, No. 5,289,264.

Stoddart, R.D. (1997). Drop-on-Demand yields precision tooling patterns. Proc. *Prototyping Tech. Int. '97*, 1997, pp. 224-227.

Stout, K.J., W.P. Dong, L. Blunt, E. Mainsal and P.J. Sullivan (1994). *Three Dimensional Surface Topography: Measurement, Interpretation and Applications, A Survey and Bibliography*, Penton Press, London.

Stratasys Inc. (1991). Fast, precise, safe prototypes with FDM. *Proc. 2nd Solid Freeform Fabr. Symp.*, The University of Texas at Austin, Austin, TX, Vol. 2, pp. 115-122.

Stratasys, Inc. (1997). *QuickSlice and FDM Manuals*, Stratasys Inc.

Styger, L. (1994). Firming designs. *IEE Review.* Vol. 41, No. 1, pp. :38-39.

Subramanian, P.K., G. Zong, N.K. Vail and J.W. Barlow (1993). Selective laser sintering of Al_2O_3. *Proc. 4th Solid Freeform Fabr. Symp.*, The University of Texas at Austin, Austin, TX, Vol. 4, pp. 350-359.

Subramanian, P.K., N.K. Vail, J.W. Barlow and H.L. Marcus (1995). Selective laser sintering of alumina with polymer binders. *Rapid Prototyping J.*, Vol. 1, No. 2, pp. 24-35.

Suh, Y.S. and M.J. Wozny (1994). Adaptive slicing of solid freeform processes. *Proc. 5th Solid Freeform Fabr. Symp.*, The University of Texas at Austin, Austin, TX, Vol. 5, pp. 404-411.

Suh, Y.S. and M.J. Wozny (1995). Integration of a solid freeform fabrication process into a feature-based design CAD system environment. *Proc. 6th Solid Freeform Fabr. Symp.*, The University of Texas at Austin, Austin, TX, Vol. 6, pp. 334-341.

Sui, G., W. Zhang, W. and M.C. Leu (2000). Study on water deposit in rapid freeze prototyping. *Proc. 11th Solid Freeform Fabr. Symp.*, The University of Texas at Austin, Austin, TX, Vol. 11, pp. 342-349.

Sun, M.M., J.C. Nelson, J.J. Beaman and J.W. Barlow (1991). *Proc. 2nd Solid Freeform Fabr. Symp.*, The University of Texas at Austin, Austin, TX, Vol. 2, pp. 46.

Sun, M.M. (1992). *Physical Modeling of the Selective Laser Sintering Process*. Ph.D. Dissertation, The University of Texas at Austin, Austin, TX.

Sun, C. and X. Zhang (2002). Experimental and numerical investigations on microstereolithography of ceramics. *J. Applied Physics*, Vol. 92, No. 15, pp. 4796-4802 .

Suzuki, H., S. Takeuchi and T. Kanai (1999). Subdivision surface fitting to a range of points. *Proc. 7th Pacific Conf. Graphics and Applications (Pacific Graphics'99)*, IEEE Computer Society, pp. 158-167.

Swaelens, B., J. Pauwels and W. Vancraen (1995). Support generation for rapid prototyping. *Proc. 6th Int. Conf. Rapid Prototyping*, Dayton, Ohio, pp. 115-121.

Swaelens, B., J. Pauwels and W. Vancraen (1997). *Method for supporting an object made by means of stereolithography or another rapid prototype production method*, US Patent Number 5,595,703, January 21.

Svelto, O. (1998). *Principles of Lasers*. Kluwer Academic Publishers, Boston, MA.

Swainson, W.K. (1977). *Method, Medium and Apparatus for Producing Three-dimensional Figure Product*. US Patent 4,041,476.

Swann, S. (1996). Integration of MRI and stereolithography to build medical models: A case study. *Rapid Prototyping J.*, Vol.2, No. 4, pp. 41-46.

Swinkels, F.B. and M.F. Ashby (1981). *Acta Metall*. Vol. 29, pp. 259.

Tadmore, Z. and C.G. Gogos (1979). *Principles of Polymer Processing*. Wiley-Interscience: New York, NY.

Takagi, T. and N. Nakajima (1993). Photoforming applied to fine machining. *Proc. IEEE Micro Electro Mech. Sys.*, Fort Lauderdale, Florida, pp. 173-178.

Takamasu, K., T. Uekawa, K. Kawakaimi and S. Ozono (1993). Profile measurement using multi-gray scale pattern projection. *Proc. SPIE - Measurement Technology and Intelligent Instruments*, Vol. 2101, pp. 1093-1099.

Takashi, M. (1999). An overview of offset curves and surfaces. *Computer Aided Design*, Vol. 31 No. 4, pp. 165-73.

Tangelder, J. W. H. and J.S.M. Vergeest (1994). Robust NC path generation for rapid shape prototyping. *J. Des. Manuf.*, Vol. 4, pp. 281-292.

Tangelder, J. W. H., J.S.M. Vergeest and M.H. Overmars (1998). Interference-free NC machining using spatial planning and Minkowski operations. *Computer-Aided Design*, Vol. 30, No. 4, pp. 277-286.

Tata, K., G. Fadel, A. Bagchi and N. Aziz. Efficient slicing for layered manufacturing. *Rapid Prototyping J.*, Vol. 4, No. 4, pp. 151-167.

Tay, E.H., M.A. Manna and L.X. Liu (2002). A CASD/CASM method for prosthetic socket fabrication using the FDM technology. *Rapid Prototyping*, Vol. 8, No. 4, pp. 258-262.

Taylor, J.B. and D.R. Cormier, (2001). A process for solvent welded rapid prototyping tooling. *Robotics Computer-Integrated Manuf.*, Vol. 17, pp. 151-157.

Terzides, C., B. Schunck and E. G. Vakalo (1993). Towards reverse engineering: reconstruction of objects using deformable models, *Proc. 4th Eur. Int. Conf. on the Application of Artificial Intelligence, Robotics and Image Processing to Architecture, Building Engineering, Civil Engineering, Urban Design and Urban Planning*, Elsevier, Amsterdam, pp. 425-434.

Theodoracatos, V. (1994). Occlusion-free monocular three-dimensional vision system. *Optical Eng.*, Vol. 33, No. 10, pp. 3476-3483.

Thoma, D., C. Charbon, C., G. Lewis and R. Nemec (1996). Directed light fabrication of iron-based materials. In *Advanced Laser Processing of Materials - Fundamentals and Applications*. Boston, MA, pp. 341-246.

Thompson, D.C. and R.H. Crawford (1995). Optimizing part quality with orientation. *Proc. 6th Solid Freeform Fabr. Symp.*, The University of Texas at Austin, Austin, TX, pp. *362-368*.

Tobin, J.R., B. Badrinarayan, J.W. Barlow, J.J. Beaman and D.L. Bourell (1993). Indirect metal composite part manufacture using the SLS process. *Proc. 4th^h Solid Freeform Fabr. Symp.*, The University of Texas at Austin, Austin, TX, Vol. 4, pp. 303.

Tobolsky, A.V. (1960). *Properties and Structure of Polymers*. John Wiley & Sons:, New York, NY.

Tolochko, N.K., T. Laoui, Y.V. Khlopkov, S.E. Mozzharov, V.I. Titov and M.B. Ignatiev (2000). Absorptance of powder materials suitable for laser sintering. *Rapid Prototyping J.,,* Vol. 6, No. 3, pp. 155-161.

Too, M.H., K.F. Leong, C.K. Chua, Z.H. Du, S.F. Yang, C.M. Cheah and S.L. Ho (2002). Investigation of 3D non-random porous structures by fused deposition modeling. *Int. J. Adv. Manuf. Tech.* Vol. 19, pp. 217-223.

Tseng, A.A. and M. Tanaka (2001). Advanced deposition techniques for freeform fabrication of metal and ceramic parts. *Rapid Prototyping J.,* Vol. 7, No. 1, pp. 6-17.

Tyberg, F. and F.H. Bohn (1998). Local adaptive slicing. *Rapid Prototyping J.*, Vol. 4, No. 3, pp.118-127.

Ullett, J.S., R.P. Chartoff, A.J. Lightman, J.P. Murphy and J. Li (1994). Reducing warpage in stereolithography through novel draw styles. *Proc. 5th Int. Conf. Rapid Prototyping*, Dayton, OH, 109-125.

Ullett, J.S., B.-T. Tia, J.W. Schultz and R.P. Chartoff (1999). Thermal-expansion and fracture toughness properties of parts made from liquid crystal stereolithography resins. *Materials & Design*. Vol. 20, No. 2, pp. 3:91-397.

Ullett, J.S., B.-T. Tia, J.W. Schultz and R.P. Chartoff (2000). Novel liquid crystal resins for stereolithography - Processing parameters and mechanical analysis. *Rapid Prototyping J.,* Vol. 6, No. 1, pp. 8-17.

Vail, N.K. and J.W. Barlow (1990). Microencapsulation of finely divided ceramic powders. *Proc. 1st Solid Freeform Fabr. Symp.*, The University of Texas at Austin, Austin, TX, Vol. 1, pp. 8-15.

Vail, N.K. and J.W. Barlow (1991). Ceramic Structures by selective laser sintering of microencapsulated, finely divided ceramic materials. *Proc. 2nd Solid Freeform Fabr. Symp.*, The University of Texas at Austin, Austin, TX, Vol. 2, pp. 195-204.

Vail, N.K. and J.W. Barlow (1992). Effect of polymer coatings on sintering of ceramic parts. *Proc. 3rd Solid Freeform Fabr. Symp.*, The University of Texas at Austin, Austin, TX, Vol. 3, pp. 124-130.

Vail, N.K., J.W. Barlow, J.W. and H.L. Marcus (1993). Silicon carbide preforms for metal infiltration by selective laser sintering. *Proc. 4th Solid Freeform Fabr. Symp.*, The University of Texas at Austin, Austin, TX, Vol. 4, pp. 204-214.

Vail, N.K. (1994). *Preparation and Characterization of Microencapsulated, Finely Divided Ceramic Powders for Selective Laser Sintering.* Ph.D. Dissertation, The University of Texas at Austin, Austin, TX.

Vail, N.K., B. Balasubramanian, J.W. Barlow and H.L. Marcus (1996). A thermal model of polymer degradation during selective laser sintering of polymer coated ceramic powders. *Rapid Prototyping J.*, Vol. 2, No. 3, pp. 24-40.

Van der Schueren, B. and J.P. Kruth (1995). Powder deposition in selective metal powder sintering. *Rapid Prototyping J.*, Vol. 1, No. 3, pp. 23-31.

Vancraen, W., B. Swaelens and J. Pauwels (1994). Contour interfacing in rapid prototyping - tools that make it work. *Proc. 3rd European Conf. Rapid Prototyping and Manufacturing*, The University of Nottingham, U.K., pp. 25-33.

Varadan, V.K., X. Jiang and V.V. Varadan, V.V. (2001). *Microstereolithography and Other Fabrication Techniques for 3D MEMS.* John Wiley & Sons, New York, NY.

Vàrady, T., R. R. Martin and J. Cox (1997). Reverse engineering of geometric models - an introduction. *Computer-Aided Design*, Vol. 29, No. 4, pp. 255-268.

Vàrady, T., P. Benkó and G. Kos (1998). Reverse engineering regular objects: simple segmentation and surface fitting procedures. *Int. J. of Shape Modeling*, Vol. 4, pp. 127-141.

Venuvinod, P.K. (1993). Automated analysis of 3D polyhedral assemblies: Assembly directions and sequences. *Trans. NAMRI/SME*, Vol. 21, pp. 399-406. Republished in the *J. of* Manuf. Sys., Vol. 12, No. 3, 1993, pp. 246-252.

Venuvinod, P.K, and C.F. Yuen (1994). Efficient automated geometric feature recognition through feature coding. *Annals of CIRP*, Vol. 43, No. 1, pp. 413-416.

Venuvinod, P.K. and S.Y. Wong (1995). A graph-based expert system approach to geometric feature recognition. *J. Intelligent* Manuf., Vol. 6, pp. 155-162.

Venuvinod, P. K. (1999). Intelligent production machines: Benefiting from synergy amongst modeling, sensing and learning. In Clarence de Silva (Ed.), *Intelligent Machines: Myths and Realities.* CRC Press LLC, Chapter 7, pp. 207-244.

Venuvinod, P.K. and H. Sun (2000). Corporate cultures in the eras of productivity, quality, and innovation: A perspective from Hong Kong. *Proc. Asian Academy Seminar*, Hyderabad, India.

Vuyyuru, P., C. Kirschman, G.M. Fadel, A. Bagchi and C. Jara-Almonte (1994). A NURBS based approach for rapid prototyping realization. *Proc. 5th Int. Conf. Rapid Prototyping*, Dayton, OH. pp. 229-240.

Wai, H.C. (2001). RP in art and conceptual design. *Rapid Prototyping J..* Vol. 7, No. 4, pp. 217-219.

Wang, H. (1995). Long-range optical triangulation utilizing collimated probe beam. *Optics and Laser in Eng.*, Vol. 23, pp. 41-52

Wang, W., J.G. Conley and H.W. Stoll (1999). Rapid tooling for sand casting using laminated object manufacturing process. *Rapid Prototyping J.*, Vol. 5, No. 3, pp. 134-140.

Wang, X.C., T. Laoui, J. Bonse, J.P. Kruth, B. Lauwers and L. Froyen (2002). Direct selective laser sintering of hard metal powders: Experimental study and simulation. *Int. J. Adv. Manuf. Tech.* Vol. 19, pp. 351-357.

Wani, M. and B. G. Batchelor (1994). Edge-Region-Based Segmentation of Range Images. *IEEE Trans. Pattern Analysis and Machine Intelligence*, Vol. 16, No. 3, pp. 314-319.

Wasser, T., A. Dhar and C. Jayal and Pistor (1999). Implementation and evaluation of novel buildstyles in fused deposition modeling (FDM). *Proc. 10th Solid Freeform Fabr. Symp.*, The University of Texas at Austin, Austin, TX, pp. 95-102.

Webb, D. and V. Gerdes (1994). Computer-aided support structure design for stereolithography models. *Proc. 5th Int. Conf. Rapid Prototyping*, Dayton, OH, pp. 221-228.

Weiss, L.E., F.B. Prinz and D.P. Siewiorek (1991). A framework for thermal spray deposition: The MD* system. *Proc. 2nd Solid Freeform Fabr. Symp.*, The University of Texas at Austin, Austin, TX, Vol. 2, pp. 178-186.

Weiss, L.E., F.B. Prinz and L.E. Gursoz (1992). *Method and Apparatus for Fabrication of Three-Dimensional Articles by Thermal Spray Deposition.* U. S. Patent 5,126,529.

Weiss, L., F.B. Prinz, G. Neplotnik, P. Padmanabhan, L. Schultz and L. R. Merz (1996). Shape deposition manufacturing of wearable computers. *Proc. 7th Solid Freeform Fabr. Symp.*, The University of Texas at Austin, Austin, TX, Vol. 7, pp. 34-35.

Weiss, L., R. Merz, F.B. Prinz, G. Neplotnik, P. Padmanabhan, L. Schultz and K. Ramaswami (1997). Shape deposition manufacturing of heterogeneous structures. *SME J. Manuf. Sys.*, 1997, Vol. 16, No. 4, pp. 239-248.

Weiss, L.E. (1997). Process o verviews. I n F riedrich B . Prinz, C hair, *R apid P rototyping i n Europe and Japan, Vol. I, Analytical Chapters.* Baltimore: Int. Technology Research Institute, Loyola College in Maryland.

Wilkinson, J. H. and C. Reinsch (1971). *Linear Algiba*, Vol. II, Springer-Verlag, USA.

Wlliams, J.D. and C.R. Deckard (1998). Advances in modeling the effects of selected parameters in the SLS process. *Rapid Prototyping J.*, Vol. 4, No. 2, pp. 90-100.

West, A.P., S.P. Sambu and D.W. Rosen (2001). A process planning method for improving build performance in stereolithography. *Computer-Aided Design.* Vol. 33, pp. 65-79.

Wick, C., J. Benedict and R. Veilleux (1994). *Tool and Manufacturing Engineers Handbook, Vol.2: Forming*, Soc. of Manuf. Engineers, Dearborn, MI.

Wiedemann, B. and H.-A. Jantzen (1999). Strategies and applications for rapid product and process development in Daimler-Benz AG. *Comp. in Industry*, Vol. 39, pp. 11-25.

Williams, R.E. and V.L. Melton (1998). Abrasive flow finishing of stereolithography prototypes. *Rapid Prototyping J.*, Vol. 4, No. 2, pp.56-67.

Wise, S. (1996). Net shape nickel ceramic c omposite tooling from RP models. *Proc. SME Rapid Prototyping and Manuf. Conf.*, 1996.

Wise, S. (2000). Nickel ceramic tooling from RP&M models. In P.D. Hilton and P.F. Jacobs (Eds.), *Rapid Tooling: Technologies and Applications.* Marcel Dekker, Inc., New York, NY, 2000, pp. 121-144.

Wodziak, J., G. Fadel and C. Kirschman (1994). A genetic algorithm for optimizing multiple part placement to reduce build time. *Proc. 5th Int. Conf. Rapid Prototyping*, Dayton, OH.

Wohlers, T. (1994). 3D digitizing systems. *Computer Graphics World*, Vol. 17, No. 4, April.

Wohlers, T. (1995). 3D digitizing for reverse engineering. *Computer Graphics World*, Vol. 18, No. 3. March.

Wohlers, T. (2000). *Wohlers Report 2000 - Rapid Prototyping & Tooling State of the Industry.* Wohlers Associates Inc. Fort Collins, Colorado.

Wohlers, T. (2002). *Wohlers Report 2002 - Rapid Prototyping & Tooling State of the Industry.* Wohlers Associates Inc. Fort Collins, Colorado.

Wohlers, T. and T. Grimm (2001). Here lies...obituaries of RP manufacturers. Column titled "Perspectives" in March issue of *Time-Compression Technologies North America Magazine*, Communication Technologies, Inc., West Doylestown, PA.

Wohlers, T. and T. Grimm (2002). A year filled with promising R&D. Monthly column titled "Perspectives" in *Time-Compression Technologies North America Magazine*, Communication Technologies, Inc., West Doylestown, PA.

Wood, L. (1993). *Rapid Automated Prototyping: An Introduction*, Industrial Press, New York, NY.

Wu, S. (1981). *Polymer Interface and Adhesion*, Marcell Dekker, New York, NY.

Wu, G., N.A. Langrana, R. Sadanji and S. Danforth (2002). Solid freeform fabrication of metal components using fused deposition of metals. *Materials & Design*. Vol. 23, pp. 97-105.

Xu, F., Y.S. Wong, H.T. Loh, J.Y.H. Fuh and T. Miyazawa (1997). Optimal orientation with variable slicing in stereolithography. *Rapid Prototyping J.*, Vol. 3, No. 3, pp. 76-88.

Xu, F., H.T. Loh and Y.S. Wong (1999). Considerations and selection of optimal orientation for different rapid prototyping systems. *Rapid Prototyping J.*, Vol. 5, No. 2, pp. 54-60.

Xue, S.S. and J.W. Barlow (1990). Thermal properties of powders. *Proc. 1st Solid Freeform Fabr. Symp.*, The University of Texas at Austin, Austin, TX, Vol. 1, pp. 179-185.

Xue, S.S., J.W. Barlow (1991). Models for the prediction of the thermal conditions of powders. *Proc. 2nd Solid Freeform Fabr. Symp.*, The University of Texas at Austin, Austin, TX, Vol. 2, pp. 62-69.

Yagui, S. and D. Kunni (1957). Studies in the effective thermal conductivities in packed beds. *Amer. Inst. Chem. Eng. J.*, Vol. 3, No. 3, pp. 373-381.

Yancey, R.N., D.S. Eliasen, S.T. Neel, J.H. Stanley and R. Dzugan (1994). Reverse Engineering using computed tomography, Proc. *5th Int. Conf. Rapid Prototyping*, The University of Dayton, OH, pp. 141-149.

Yancey, R.N., D.S. Eliasen, S.T. Neel, R. Dzugan and P. Jacobs (1995). Integration of reverse engineering, solidification modeling, and rapid prototyping techniques for the production of net-shape cast tooling, *Proc. 6th Int. Conf. Rapid Prototyping*, The University of Dayton, OH, pp. 275-281.

Yang, Y, H.T. Loh, J.Y.H. Fuh and Y.G. Wang (2002a). Equidistant path generation for improving scanning efficiency in layered manufacturing. *Rapid Prototyping J.*, Vol. 8, No. 1, pp. 30-37.

Yang, Z.Y., Y.H. Chen and W.S. Sze (2002b). Layer-based machining: recent development and support structure design. *Proc. I. Mech. E., Part B: J. Eng. Manuf.*, Vol. 216, Vol. 7, pp. 979-991.

Yang, Z.Y., Y.H. Chen and W.S. Sze (2002c). Layer-based machining: recent development and support structure design. *Proc. Inst. Mech. Engrs*, Vol. Vol. 216, Part B, pp. 979-991.

Yao, A.W.L. and Y.C. Tseng (2002). A robust process optimization for a powder type rapid prototyper. *Rapid Prototyping J.*, Vol. 8, No. 3, pp. 180-189.

Yardimci, M.A. and S. Güçeri (1996). Conceptual framework for the thermal process modeling of fused deposition. *Rapid Prototyping J.*, Vol. 2, No. 2, pp. 26-31.

Yarlagadda, P.K.D.V., J.P. Ilyas and P. Christodoulou (2001). Development of rapid tooling for sheet metal drawing using nickel electroforming and stereolithography processes. *J. Mat. Processing Tech.*. Vol. 111, No. 1-3, pp. 286-294.

Yuen, C.F. and P.K. Venuvinod (1999). Geometric feature recognition: Coping with the complexity and infinite variety of features. *Int. J. Computer Integrated Manufacture*, Vol. 12, No.5, pp. 439-452.

Yuen, C.F., S.Y. Wong and P.K. Venuvinod (2003). Development of a generic computer-aided process planning support system. *J. Mat. Processing Tech.*, Volume 139, No. 1-3, pp. 394-401.

Zak, G. (1999). *Rapid Layered Manufacturing of Short Fibre-reinforced Parts*. Ph.D. Thesis, Dept of Mech. and Ind. Eng., University of Toronto, Toronto, Canada.

Zak, G., M. Haberer, C.B. Park and B. Benhabib (2000). Mechanical properties of short-fibre layered composites: Prediction and experiment. *Rapid Prototyping J.*, Vol. 6, No. 2, pp. 107-118.

Zang, E.E. (1940). *Vitavue Relief Model Technique*. US Patent 3,137,080.

Zehner, P. and E.U. Schlünder (1970). Warmeleifahigkeit von schüttungen bei mässigen temperatturen. *Chemie-Ing. Techn.* Vol. 42, No. 17, pp. 933-941.

Zeid, Z. (1991). *CAD/CAM Theory and Practice*, McGraw-Hill, Inc.

Zhang, W., M.C. Leu, Z. Ji and Y. Yan (1999). Rapid freezing prototyping with water. *Materials & Design*. Vol. 20, pp. 139-145.

Zhang, X., X. N. Jiang and C. Sun (1999). Micro-stereolithography of polymeric and ceramic microstructures. *Sensors and Actuators A: Physical*, Vol. 77, No. 2, pp. 149-156.

Zhang, G., Y.T. Tsou and A.L. Rosenberger (2000). Reconstruction of the homunculus skull using a combined scanning and stereolithography process. *Rapid Prototyping J.*, Vol. 6, No. 4, pp. 267-275.

Zhang, Y., X. He, J. Han, S. Du and J. Zhang (2001). Al2O3 ceramics preparation by LOM (laminated object manufacturing). *Int. J. Adv. Manuf. Tech.* Vol. 17, pp. 531-534.

Zhou, J.G. and Z. He (1999). A new rapid tooling technique and its special binder study. *Rapid Prototyping J.*, Vol. 5, No. 2, pp. 82-88.

Zhu, W. M. and K.M. Yu (1998). Non-manifold representation for rapid prototyping multi-material. *Proc. 7th Eur. Conf. Rapid Prototyping & Manuf.*, Aachen, Germany, pp. 35-47.

Zhu, W. M. and K.M. Yu (2000). Modeling multi-material assembly for rapid manufacture. *Proc. 9th Eur. Conf. Rapid Prototyping and Manuf.*, Athens, Greece, pp. .97-112.

Zhu, W. (2002) *Rapid Manufacturing of Multi-material Objects*. Ph.D. Thesis, The Hong Kong Polytechnic University, Hong Kong.

Zhuang, B.H., J.H. Zhang, C.Z. Jiang, Z. Li and W.W. Zhang (1994). Precision laser triangulation range sensor with double detectors for measurement on CMMs. *Proc. SPIE - Industrial Optical Sensors for Metrology and Inspection*, Vol. 2349, pp. 44-52.

Ziabicki, A., I. Jarecki and A. Wasiak (1998). Dynamic modeling of melt spinning. *Computer Theoret. Polymer Science*. Vol. 8, pp. 143-157.

Ziemian, C.W. and P.M. Crawn III (2001). Computer aided decision support for fused deposition modeling. *Rapid Prototyping J.*, Vol. 7, No. 3, pp. 138-147.

Zorin, D. and P. Schröder (2000). *Subdivision for Modeling and Animation*, SIGGRAPH 2000 Course Notes, ACM.